HORMONAL BIOCHEMISTRY

HORMONAL BIOCHEMISTRY

By

Dr. Lata Bhattacharya

Professor
School of Studies in Zoology & Biotechnology
Vikram University
Ujjain (M.P.)
(India)

DISCOVERY PUBLISHING HOUSE PVT. LTD.
NEW DELHI-110 002

Published by:

DISCOVERY PUBLISHING HOUSE
4383/4B, Ansari Road, Darya Ganj
New Delhi-110 002 (India)
Phone: +91-11-23279245; 23253475; 43596065
Mobile: +91 9811179893 / +91 9871656464
E-mail: discoverybooksindia@gmail.com
orderdphbooks@gmail.com
namitwasan9@gmail.com
web: www.discoverypublishinggroup.com

First Published: 2010
Reprinted: 2025

ISBN: 978-81-8356-646-9

Hormonal Biochemistry

Printed at:
Infinity Imaging Systems
Delhi

Preface

Biochemistry today has made spectacular progress in unraveling the mysteries of animate nature. This progress has allowed us to gain deeper insight into the principles of vital activity and has to a very significant extent stimulated the development of applied disciplines, especially medicine. The present title *"Hormonal Biochemistry"* is intended for those who wish to understand living organisms, especially man. Biochemistry is essential for this purpose, but it would be almost impossible for a student to survey on his own the massive body of existing knowledge, constantly augmented by a remarkable torrent of brilliat discoveries. The purpose of the book, then is to organize our knowledge into something that can be comprehended in a relatively short time and still convey a reasonable complete picture of the chemical structure and function of man. Readability without sacrifice of coverage has been a prime goal. An important device in gaining that goal is to keep attention constantly focused on function, with repeated use of rationalization to show that the chemical facts are not isolated, but part of a whole.

Biochemistry has two major goals as a fundamental science, it treats the vital functions from the standpoint of physical chemistry, and as an applied discipline, it points out practical applications for the wealth of scientific knowledge it has acquired. The dual purpose has been, as far as possible, take into account in the writing of this book. We have also tried to summarize our pedagogical experience in teaching biochemistry to students specializing in medicine and pharmacy. The material of this book has been organized according to the principle of functionally in order to trace the close relationship between the functions of the living organism and the structures of its constituents molecules as well as the chemical and physico-chemical process in which they are involved.

The aim of this book is to present a core of biochemical knowledge that is desirable for undergraduate and postgraduate students and also those involved in the field of medical, microbiology, biotechnology and pharmaceutical. Every attempt has been made to keep abrest of the advances in the subject and at the same time to include the fundamentals.

To make the work more comprehensive and informative, the author has consulted many authoritative books, research journals, abstracts, monographs etc. He is grateful to all those great scholars whose work are cited or substantially reproduced.

There can be no claim to originality except in the manner of treatment and much of the information has been obtained from the books and scientific journals available in the different libraries.

The author expresses his thanks to his friends and colleagues whose continue inspirations have initiated him to bring out this book.

The author expresses his gratitude to Mr. Wasan and Staff of M/s Discovery Publishing House Pvt. Ltd., for their whole hearted co-operation in the publication of this book.

In the mean time, the author will remain sincerely responsible for any shortcomings of the book and be grateful to the readers for their suggestions and constructive criticism for the continuous betterment of the book. He takes this opportunity to appeal to the readers to send their suggestions straightway to his publisher.

Author

CONTENTS

The Neurohormones

CHARACTERISTICS OF NEUROHORMONES

Neurohormones, in the broad sense of the word, comprise all the physiologically active products of the nerve cells. Irrespective of the quantity of its secretion, each nerve cell secretes actively in this sense, at least at its nerve ending, synapses, etc. The enzymatic synthesis and decomposition of specific materials such as acetylcholine and noradrenaline at both sensory and motor nerve endings and synapses is the source of changes in the electric potential. These changes form the basis of nervous activity. In the older literature these substances produced by normal nerve cells are included in the term neurohormones. In the last two decades, however, the term has been given a new and narrower meaning, to exclude the aforementioned ultramicroscopic secretions, for which the term neurohumoral factors is used. The neurohumoral factors are not true hormones for the purposes of this book, as they produce their effect at their place of origin and thus belong to the category of protohormones. The neurohormones *sensu stricto,* on the other hand, are a special group of true hormones combining the characters of both tissue hormones, being produced in tissues with functions other than hormonal, and gland hormones, originating in specialized secretory (neurosecretory) cells. From the true neurosecretory cells there can further be distinguished those cells of nervous origin which have become completely specialized to endocrine neuro-secretory activity and have lost their nervous function during this phylogenetic adaptation, e.g. the cells of the medullary part of the adrenal bodies producing noradrenaline. There is, however, a clear phylogenetic link between these three types of cells which is also evident from their development in ontogeny. It follows from phylogenetic considerations that the same active substance may occur in one case as a neurohumoral factor in the endings of normal nerve cells, and in another as a neurohormone in a neurosecretory cell.

The true neurosecretory cell may be defined as a special type of nerve cell characterized, in addition to its nervous function, by the production of a secretion with specific properties. This secretion originates in the cytoplasm of the neurosecretory neurones in the form of granules of variable size, characterized by their reaction to particular staining methods and by their shining white appearance under dark-ground illumination. Another important feature of the neurosecretory granules observed in most of the known neurosecretory cells is their movement along the axons of the neurosecretory cells, usually to places where they accumulate, i.e. the paired neurohaemal organs, e.g. the corpora cardiaca of insects, the neurohypophysis of vertebrates or the sinus glands of crustaceans. From the site of accumulation they pass into the circulatory system at specific periods. This movement of neurosecretion has been shown experimentally in the neurosecretory cells of various animals, including insects. As mentioned previously the cutting of one of the nervi corporum cardiacarum results in the accumulation of the neurosecretory materials in the cut end of the proximal part of the nerves, whereas it gradually disappears in the distal part including the corpus cardiacum and corpus allatum. The movement of granules has also been observed in living cells in tissue cultures. A speed of movement of about 3 mm in 24 hours has been observed.

Johnson (1963) later showed another way in which a neurosecretion can reach its target tissue. On the basis of observations in aphids, he concluded that the neurosecretory substances originating from the brain neurosecretory cells pass through the corpora cardiaca and are distributed to various internal organs inside the nervous pathways leading from these structures. This conclusion is based on convincing histological evidence obtained in several aphid species, in which the granules could be observed along the branches of all three nerves leaving the corpus cardiacum, right to their endings in the muscles and other organs. Johnson attempted to generalize his findings and suggested that the reason why similar observations in other insects bore negative results might be that 'in insects other than aphids, either insufficient material is normally present in the axons for it to be readily detected by histochemical means, or it does not stain differentially with the known staining procedures'.

Serious evidence seems, however, to contradict this conclusion. Firstly, there is irrefutable experimental evidence of the role of the haemolymph in the action of some of the brain hormones. Secondly, organs such as the corpora allata, supposed by some authors to be regulated by the import of neuro-secretory material, show the same cyclical activity after transplantation, indicating that activation hormone is present in the haemolymph. Thirdly, there is even less evidence of the direct utilization of neurosecretory material in tissue than of its passage into the haemolymph. Regarding the last, the evidence obtained by Ganguly and Banerjee (1961) in the pyrrhocorid bug *Macrocercea grandis* should be mentioned.

It seems much more likely, as suggested by Highnam (1965, personal communication), that the condition found in aphids is an exception among insects, due perhaps to far-reaching changes in the amount of body fluid associated with sucking plant sap, which could otherwise unfavourably influence the hormonal balance. It might rather be deduced that the development of neurohaemal organs such as the corpora cardiaca is a later acquisition during evolution, so that the state in aphids could simply have originated as a return to a more primitive stage, as is the case for a number of other aphid structures. This does not, however, necessarily mean that direct transport through nerve endings does not also exist, in many other groups of insects, in addition to humoral transport, since it is unlikely that neurohaemal organs like the corpora cardiaca in aphids should be without any function.

As suggested for the first time by Ernst Scharrer (1952), the sum total of the neurosecretory cells in the organism, together with their specific neurohormones, can be regarded as a special, third correlation system of the animal organism mediating relationships between the nervous and endocrine systems. Its function consists in mediation between slow, repeated, prolonged environmental stimuli of low intensity, e.g. changes in photoperiod, and long-term reactions by the transformation of similar, weak nerve impulses to slow endocrine responses resulting in physiological adaptation of long duration, e.g. gonadal activity. Attention has recently been concentrated mainly on various transport mechanisms and release mechanisms at the ultra-structural level indicating the need for a new classification of modes of communication between neurons and non-neuronal effectors.

The Carrier Substance

No clear data are yet available concerning the chemical composition of the neurohormonal granules. It has, however, been adequately substantiated that the granules are not pure neurohormone; they are composed of stainable material, the carrier substance, on which the corresponding neurohormone is bound either chemically or physically, e.g. adsorbed. The neurohormones appear first to be freed from the csrrier substsnce when passing into the blood (haemolymph), perhaps freed by the enzymolytic activity of the haemocytes. Several types of neurosecretory cells can be distinguished by the staining properties of the carrier substance. No data are available on the connection between these differences in the carrier substance and differences between the neurohormones in the different neurosecretory cells. In contrast with its original concept, the carrier substance is now supposed to be insoluble in lipoid solvents and is probably polypeptide or protein in character. The molecules of the neurohormones, relatively small in size, seem to be attached to the large protein molecules of the carrier substance. The neurohormones themselves seem to have rather different chemical properties. Some of them have been said to have lipoid character, e.g. a substance with the effect of AH was claimed to be identical with cholesterol. This has since been questioned by Gersch (1961) by his results on the identity of AH with the neurohormone D_1. Other neurohormones have been shown to be substances soluble in water *and* alcohol with acid or alkaline properties. Apart from less feasible hypotheses, the opinion that at least some insect and other invertebrate neurohormones, like the main vertebrate hormones, are of a polypeptide nature now seems to be thoroughly substantiated. Indirect evidence of its veracity was obtained in the eyestalks of crustaceans. It *has* been suggested by Novak and Gutmann (1962) that the carrier substance is closely related with the so-called gliosecretion of *Periplaneta* and other arthropods. This independent phylogenetic origin of the carrier substance and the neurohormones suggests that the carrier substance probably represents the original secretory activity of the primary ectoderm cells.

A detailed cytochemical study of the neurosecretory granules and other neuroplasmic inclusions in *Periplaneta americana* was carried out by Pipa (1961, 1962). In this, gliosecretion, as well as other structures, were discussed. In another paper the ultrastructure of the gliosecretory granules was investigated (Pipa *et al.,* 1962). The differences found between these and the neurosecretory material seem to be irrelevant from the point of view of the conclusions reached by Novak and Gutmann (1962), as suggested by Novak (1964). For further papers dealing with the ultrastructure of the neurosecretory system.

The Occurrence of Neurohormones

Histologically typical neurosecretory cells have been found in all the main metazoan Phyla. In insects, they have been found in all groups where they have been seriously looked for. This also applies in crustaceans and other arthropods. There is now no doubt that at least three or four different neurohormones occur in all arthropods.

The General Significance of Neurosecretion in the Organism

Neurohormones act either directly on the effector organs, as in myotropic or chromatophorotropic effects, or indirectly by the secretion of secondary hormone, or again by influencing the production of a tertiary hormone regulated by the secondary, dependent hormone. This last applies to the control of the adrenal hormone in vertebrates by the ACTH of the adenohypophysis which is itself under the control of the hypothalamus. Several examples of environmental influences mediated in such a way have been described in insects, e.g. the influence of the distension of the abdomen on AH production in *Rhodnius*, and the corresponding influence of photoperiod in diapausing insects, etc. Doubtless the various cyclic phenomena in the production of myotropic and chromatophorotropic effects observed in various insects by Gersch *et al.* (1960) are also of this type.

A general concept of the photo-neuro-endocrine systems and the question of their evolution is discussed by E. Scharrer (1964). He emphasizes the role of light as a synchronizer of circadian and annual rhythms in many important functions of living organisms, such as reproduction, moulting, migration, metabolic adjustment, etc. He distinguishes the following steps in the evolution of complex neuro-neurosecretory-endocrine control, as known in vertebrates: (*i*) the transparent animal in which the organs are exposed to changes in illumination; (*ii*) animals with an opaque integument excluding light except through a transparent window overlying the central nervous system; (*iii*) animals in which the retinal and central neurones transmit stimuli directly to endocrines affecting the light-dependent target organs; (*iv*) animals in which the products of the neurosecretory cells intervene between the nervous and endocrine systems. He emphasizes, however, that photoperiodicity is only one of the phenomena handled by neuroendocrine mechanisms.

Several papers deal with *the general significance and origin of the neurosecretory cells.* Bern (1963) views the 'neurosecretory neurone' as a doubly specialized cell having the features necessary both for the conduction of impulses and for the production of secretion. He discusses the ultrastructural characteristics of neurosecretory cells, with special reference to the formation of elementary neurosecretory granules by the Golgi apparatus and to the fate of neurosecretory granules in nerve endings (the so-called synaptic vesicles of the neurosecretory cells). He states that since the membranes of the neurosecretory axon bulbs are consistently smooth-surfaced, there is no evidence of pores or canals which might serve as channels for the discharge of neurosecretory granules formed, and no definite evidence of the extraneuronal occurrence of neurosecretory granules. On the other hand, unlike Johnson (see above) he emphasizes the hormogenic function (in the sense of a true endocrine organ) of the neurosecretory cells as one of their unique features. He also discusses the factors possibly determining the development of specialized 'hormone-producing neurones amid their ordinary neighbours'.

The Effects of Neurohormones

Ignoring the neurohumoral factors concerned with the nature of nervous activity, the following categories of effects caused by neurohormones can be distinguished: (1) Myotropic effects (influencing the frequency and amplitude of the spontaneous rhythmic activity of muscle, e.g. in the heart, Malpighian tubules, oviducts, gut, etc.). (2) Chromatophorotropic effects (influencing pigment movements, *i.e.* dispersion and concentration, in epidermal cells and chromatophores; also the development of various pigments in the epidermis). (3) Daily activity pattern. (4) Influencing water balance. (5) The activation of enzymes. (6) The activation of the secretion of other endocrine glands. (7) The induction of diapause (the humoral effect of the suboesophageal ganglion in the female moth pupa on the diapause of the embryos of the next generation. (8)

Influencing the development of the gonads and related effects, e.g. the suboesophageal castration cells in females of *Leucophaea maderae*. There remains little doubt that several of these activities are due to one and the same hormone. Thus, for example, the myotropic neurohormones C and D of Gersch have been shown to produce corresponding effects on pigment movements in *Carausius* and *Corethra*. It seems probable that one of these hormones (D_1 is identical with the AH of the pars intercerebralis. But as there is still little clear evidence regarding the other activities it seems preferable to discuss the various activities separately, irrespective of whether they are due to different hormones or to the same hormone.

As distinct from the metamorphosis hormones, insect neurohormones came from various groups of neurosecretory cells (nsc) detected by histological methods, but the functions of the relevant neurohormones are not known. On the other hand, the effects of extracts of various parts of the nervous system are known, although the nsc which produce them have not been identified. There are very few exceptions to this rule, the chief ones being the activation hormone produced by nsc of the pars intercerebralis and the diapause hormone produced by the paired groups of cells in the suboesophageal ganglion. It thus seems suitable for the aims of this book that the distribution and various features of the neurosecretory cells and the corresponding structures in connection with their nerve endings should be discussed separately from individual functions whose source is still unknown.

DISTRIBUTION OF NEUROSECRETORY CELLS

The Cerebral Ganglion

The main paired group of nsc is localized within the pars intercerebralis. It usually contains several types of nsc, as described by Johansson. In addition to the most conspicuous, which are generally held to be the source of AH, there are a number of smaller, less conspicuous cells to which various other neurohormones or functions have been ascribed. The structure, staining properties and ultrastructure of various nsc were studied in detail by Panov and his co-workers in different insect groups. Gersch has carried out an immunological study of the neurosecretory system of *Periplaneta americana*. He found high specificity against the neuro-secretory material and the antibodies selectively labelled the secretory neurones in the brain and the secretory products of its nsc in the retrocerebral complex.

A detailed topography of the neurosecretory system in the head of pyrrhocorid bug *Macrocercea grandis* (Gray) was elaborated by Ganguly and Banerjee (1961). In addition to the neurosecretory cells already known, three new groups were described and their homology with the X-organ, sinus gland and epineurium of crustaceans was claimed. The term *Wigglesworth organ* was suggested for the supposed X-organ homologue and the name *Hanstrom organ* for the hypothetical sinus gland homologue. Both structures were later rediscovered in the brain of the adult silkworm *(Bombyx man)* by Ganguly and Basu (19623). A detailed cytochemical analysis of the various types of neurosecretory cells was carried out in the last mentioned and several further papers (Ganguly and Deb, 1960; Ganguly and Basu, 1962b).

The formation of the neurosecretory system and its association with diapause in the last larval instar were studied in the moth *Ostrinia (= Pyrausta) nubilalis*. No specific differences were found between the amounts of neurosecretory material in diapausing and non-diapausing brains in any of the groups of cells. A comparative cytological investigation of the brain neurosecretory cells was made by Panov and Kind (1963) in 20 different species of Lepidoptera. Six groups of these cells were found in each hemisphere of the suboesophageal ganglion and far-reaching agreement in their structure was found in all the species studied.

A series of very detailed investigations on the brain nsc in the water-beetle *Dytiscus marginalis* and their daily and seasonal cycles of activity were published by Abraham (1964a, b, 1966a, b).

Various authors have studied cyclical changes in these cells in different insects during larval instars and metamorphosis, at both the optic and the electron microscope level. Accumulation of neurosecretory granules, which differed in different types of neurosecretory cells, was observed during diapause. The author concludes that some of these cells have a specific role in diapause development and that diapause is not a simple consequence of the absence of AH (called 'growth and development hormone' by the author). Here, however, a distinction must be made between accumulation of granules in the neurosecretory cells and actual secretory activity, including the axoplasmic movement of the materials produced and their passage into the haemolymph. As shown by Highnam (1965), the actual rate of the endocrine activity of a neuro-secretory cell is affected by the following three factors, which may vary independently: (*i*) the rate of synthesis of material in the cell; (*ii*) the rate of transport of the material along the axon; (*iii*) the rate of release of the transported material into the circulatory system. Their interaction is therefore complex.

Four specific large neurosecretory cells have been described by Girardie and Girardie (1970, 1972) from the protocerebrum of *Locusta migratoria,* located in the anteroventral area of the protocerebrum, beneath the nerve of the median ocellus. Their axons join the nervi corporum cardiacarum I. They are positive to paraldehyde fuchsin, alcyan blue and Victoria blue. Their electrocoagulation results in an increase of volume of the haemolymph and with a distension of the abdomen. The implantation of a set of these cells compensates these changes. The innervation of the lateral ocelli and their connection with the retrocerebral neuroendocrine system by means of the nervi corporum cardiacarum II was studied by Brousse-Gaury (1971) in the same laboratory.

The Suboesophageal Ganglion

Several paired groups of neurosecretory cells have also been described in the suboesophageal ganglion in various insect orders. Among the nsc in the suboesophageal ganglion of female cockroaches *(Leucophaea maderae),* B. Scharrer (1955a, b) found four type A-cells with a profusion of neurosecretory granules staining purple with paraldehyde fuchsin, two ventromedial type B-cells with red-staining granules occurring only on the surface layer of the cytoplasm, and two smaller type C-cells, also with red-staining granules. Fukuda and Takeuchi (1967a, b) described another type of cell in this ganglion, which supposedly produced the diapause hormone a pair of conspicuous nsc staining red with azocarmine and situated posteriorly to the centre of the ganglion. So far nothing is known of the way in which the neurosecretory sub-stance of these cells reaches the target organs, *i.e.* the ovaries. A distinct difference in their appearance was found in pupae conditioned to lay non-diapause eggs and those laying diapause eggs. In the former they are filled with neurosecretory material, which evidently remains unreleased, while in specimens producing diapause eggs the amount of neurosecretory material is very small, so that it is probably released immediately into the haemolymph. The place where this occurs has so far not been determined.

The Ventral Nerve Cord

Data on the fate and function of the neurosecretory cells in the ventral ganglia are still very scarce. The secretory products from all three types of neurosecretory cells found within the thoracic and abdominal ganglia of the roach *Blaberus craniifer* were observed inside axons in the connectives between ganglia. Ligation of the connectives between ganglia showed that the neurosecretion moved in both anterior and posterior directions from the ganglia in which it was produced. A detailed study of neurosecretory cells in the ventral ganglia of *Schistocerca gregaria* Forsk. was carried out by Delphin (1963). He described the location of three types of A-cells (A_1 A_2, A_3), two types of B-cells (B_1 B_2), and one type each of C- and D-cells. The axonal transport of neurosecretory material was studied by ligation experiments, in nerve sections and by autoradiographic investigations using ^{35}S-DL-cystine. The granules were shown to pass backwards and forwards along the connectives and along at least some of the nerves originating from the ganglia. Differences in the secretory activity of various cells were observed. Extirpation and reimplantation of the last abdominal ganglion suggested that a hormonal factor which helps to control egg maturation is produced. Egg-laying was notably stimulated by implanting the ganglion from a mature female into a newly moulted one. The author also claims that the A_1 neurosecretory cells of the abdominal ganglia secrete an antidiuretic factor which is released under conditions in which water conservation might be necessary for the insect.

Raabe (1965,1966) and some of her co-workers (Besse, 1967; Chalaye, 1967), give a detailed description of the nsc in the ventral nerve cord ganglia. Raabe distinguished four categories of neurosecretory-granule-containing cells in the individual ganglia and paid special attention to the C nsc. These differ from the type A, B_1 and B_2 nsc by being negative to Gomori's chrome-haematoxylin-phloxine and staining conspicuously only with azan. The distribution of the various types of nsc in the ventral nerve cord of the phasmids *Clitumnus extradentatus* and *Carausius (= Dixippus) morosus* was described in detail. Here there are 3 pairs of A-cells in the suboesophageal ganglion, 2 pairs in the thoracic ganglia and I pair in the first four abdominal ganglia. From the changes observed in their granule content, it was concluded that they have a bearing on the regulation of water metabolism. B_1 cells were found only in the suboesophageal ganglion (I pair). Three pairs of small B_2-cells were found in each of the abdominal ganglia. Their activity is assumed to be related to sexual function in females.

The azan-staining C-cells are the most abundant nsc in the abdominal ganglia. They are present in all the ganglia of the ventral nerve cord and also in the tritocerebrum. Two lateral groups of 6 to 8 small cells were found in each of the thoracic ganglia and 20 cells in the central area of the abdominal ganglia. In addition, some small anterior and posterior cells and a pair of larger posterior cells were reported in the suboesophageal ganglion of *Clitumnus.* In this association, special paired neurohaemal organs filled with particles of azan-staining neurosecretory material were found in the paired lateral branches of each ganglion of the ventral nerve cord.

The Visceral (Sympathetic) Nervous System

As is generally known, the insect visceral nervous system starts with the frontal ganglion (in front of the brain) and proceeds posteriorly via the recurrent nerve, to constitute the hypocerebral ganglion, to which the paired corpora cardiaca are joined, being of nervous origin (except their glandular parts, which are of different origin), usually by means of short nerves. From here, one to three pairs of nerves lead posteriorly to one or two pairs of further small ganglia (visceral and sometimes ingluvial ganglia). The median visceral nerve also originates from the hypocerebral ganglion, continues into the prothoracic ganglion and then connects the individual thoracic and abdominal ganglia in turn.

The data obtained in different groups of insects differ as to the occurrence of the nsc in the individual visceral ganglia. This may be due first of all to the staining methods used, and also to the cycles of their activity, in which there may be periods when the cell perikarya do not contain a significant amount of neurosecretion. Clarke and Langley (1960), using paraldehyde fuchsin and chrome-haematoxylin-phloxine, found no intrinsic nsc in the frontal or any other ganglion of the visceral (sympathetic) system in locusts. Van Loof (1960) found nsc in the frontal ganglion

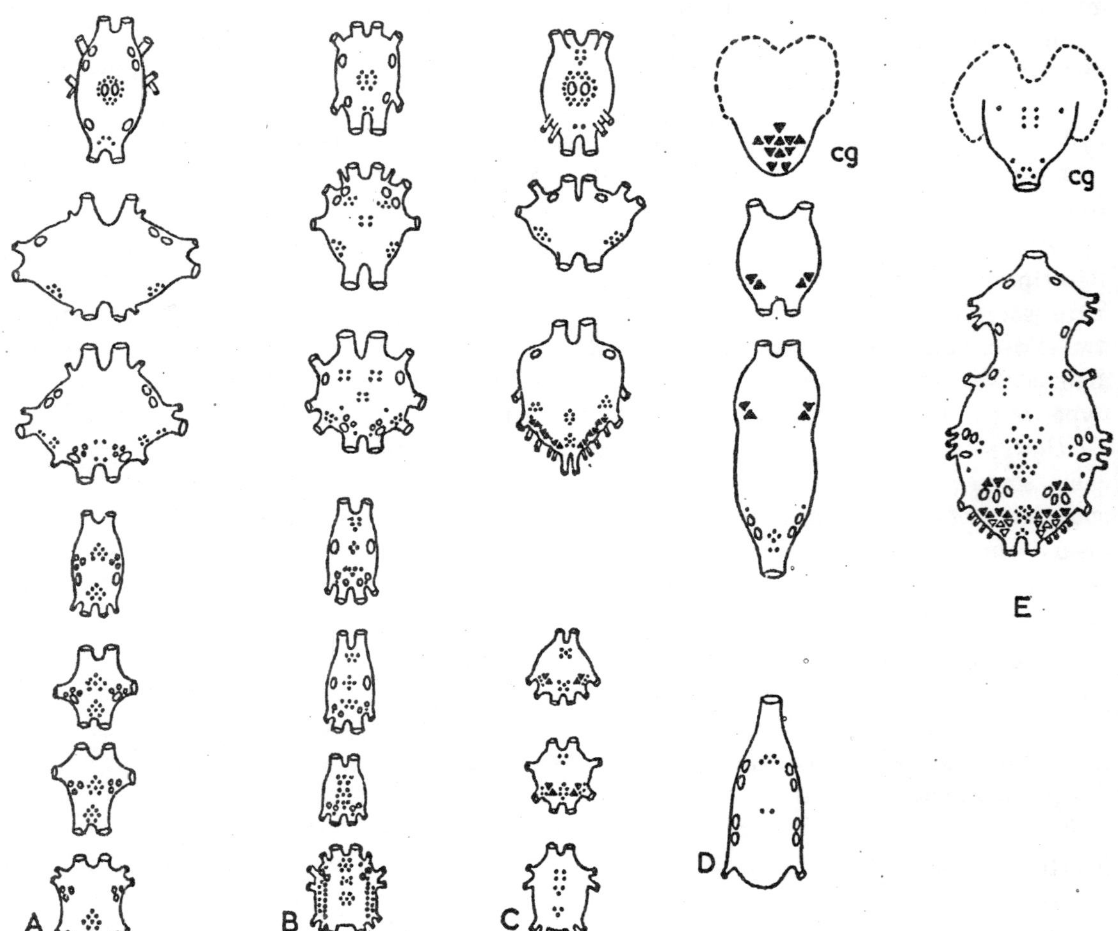

Fig. 1.1. Distribution of different types of neurosecretory cells in ganglia of the ventral cord in various insects. (A) *Clitumnus extradentatus,* (B) *Periplaneta americana,* (C) *Locusta migratoria,* (D) *Galleria mellonella,* (E) *Rhodnius prolixus.* cg-cerebral ganglion, (From Raabe, 1971).

of *Bombyx mori*. On the other hand, Strong (1966) reported neurosecretory granules in brain nsc axons leading to the hypocerebral ganglion and continuing through it as the inner and outer oesophageal nerve supplying the wall of the foregut and the ingluvial ganglia.

The Metameric (Perisympathetic) Neurohaemal Organs

Unlike crustaceans, no neurohaemal organs except the corpora cardiaca were known in insects until 1965, when Raabe described a pair of specific thickenings in the paired transverse branches of the median visceral nerve in the thoracic and each of the abdominal metameres of the stick-insects *Carausius morosus* and *Clitumnus extradentatus*. The reason for this delay was probably their inconspicuousness and the fact that they do not stain with specific neurosecretory stains such as chromehaematoxylin-phloxine and paraldehyde fuchsin. Since then, however, they have been found by Raabe and her many co-workers in most insect orders and actually in all in which a serious search for them was made (Raabe *et al.*, 1971).

Morphology

In the simplest, most primitive case, the metameric neurohaemal organs (mno) are a pair of symmetrical ovoid thickenings, one in each lateral transverse branch of the median visceral nerve, in front of the ganglion, in each thoracic and abdominal segment. This is the situation in Phasmida *(Carausius, Clitumnus,* etc., see Raabe, 1965, 1966, I97ib; Raabe and Monjo, 1970), and in some Orthoptera, e.g. *Acheta*. The transverse nerves innervate the spiracles and some of the other pleural muscles. In other cases the thickenings may move closer together and unite at the spot where the branches leave the median nerve, as in cockroaches. Later, the originally paired structures may amalgamate to form a single ovoid swelling of the median nerve, near its gangliar attachment and behind its branching point, as in *Locusta* (Chalay, 1967), or, in a different form, *Aeschna* (Charlet, 1972).

In 'higher' orders we find more elaborate structures of the same type, such as the bilobular organ of Lepidoptera, e.g. *Galleria* (Provansal, 1972) or the spherical median organ of some Coleoptera (Grillot, 1968, 1970). In Hymenoptera, the visceral nerve tends to remain within the central nervous system (the connectives as well as the ganglia). Here the mno arise from the connectives or from the wall of the ganglion (as in *Diprion,* see Provansal, 1971), or they may appear as longitudinal thickenings along the anterior surface of the main lateral nerves ('les nerfs segmentaires' of Provansal, 1968). In this case the median visceral nerve seems to branch within each ganglion, so that this part is invisible externally.

In nerves dissected in physiological saline, the neurohaemal organs can be distinguished by their bluish white, opalescent colouration, which is similar to that of the corpora cardiaca. This is no doubt due not to the actual neurosecretory substance, but to the pterine accompanying it.

Histology

Histologically, any mno consists of characteristic neurohaemal tissue. In the centre it contains ramified neurosecretory fibres surrounded by glial cells. The whole structure is enclosed in a thin surface membrane, often with deep imaginations, which is doubtless a secretory product of the glial cells. It sometimes includes individual nerve cells and neurosecretory cells, as in some of the organs of Hymenoptera, Heteroptera and Coleoptera (Raabe, 1971a), although in most other orders these cells are absent (Brady and Maddrell, 1967; Finlayson and Osborne, 1970).

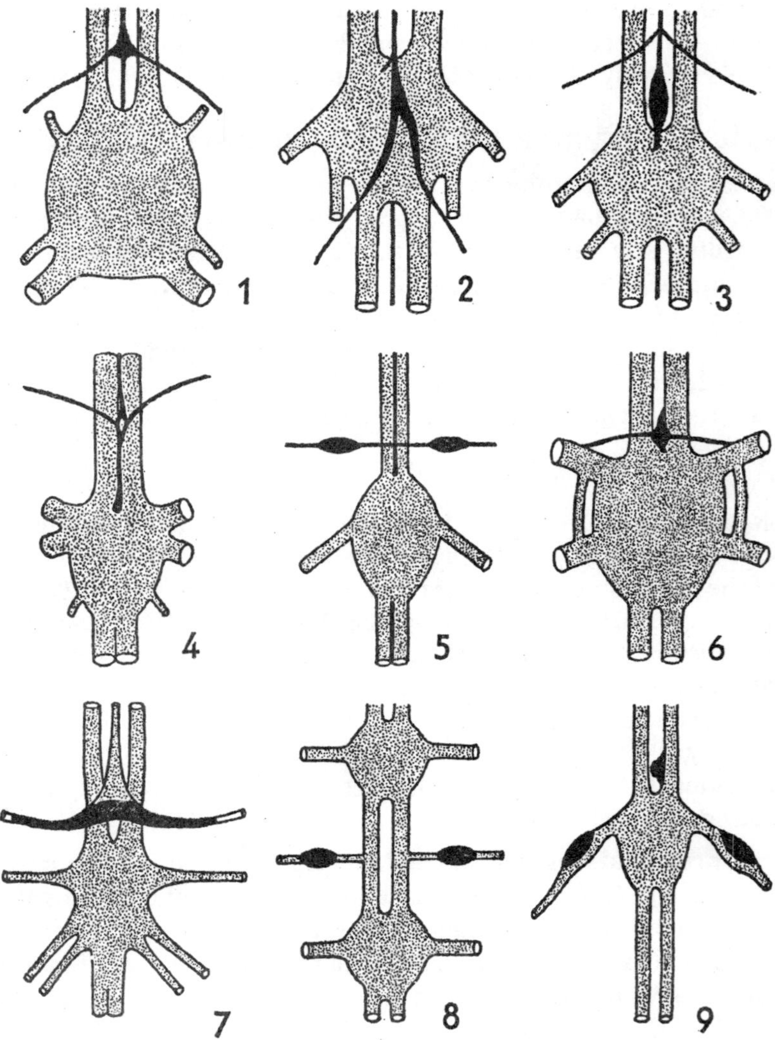

Fi;. 1.2. Metameric neurohaemal organs of various groups of insects. I, *Leucophaea maderae;* 2. *Periplaneta americana; 3. Locusta migratoria; 4. Aeschna* sp.; 5. *Clitumnus extradentatus;* 6. *Chrysoscarabus* sp.; 7. *Galleria mellonella;* 8. *Diprion pini;* 9. *Vespa crabro.* (Adapted from various papers by Raabe *et al.*).

The neurosecretion is transported to the organs along the neurosecretory axons and in most cases the greater part accumulates in the surface layer of the organ in the form of floccules or droplets. It is completely Gomori and paraldehyde-fuchsin negative and stains more readily with azocarmine (Dupont-Raabe, 1956; Raabe 1965; cf. Raabe and Monjo, 1970). It also stains, though less readily, with phloxine. The fact, that it stains deeply with bromophenol blue, Glenner and Lillie's dimethylaminobenzaldehyde-diazo-S acid and fast green FCF indicates that it probably contains proteins with pyrrole or indole groups. The last two stains mentioned do not act on the phloxinophilic surface floccules, however. As emphasized by Rasbe and Monjo (1971), it should be noted that the neurosecretory substance in the metameric organs, like that in type C neurosecretory cells, is not visualized after fixation with Bouin's fluid, but is clearly discernible after fixation with plain formalin or with Helly's fluid.

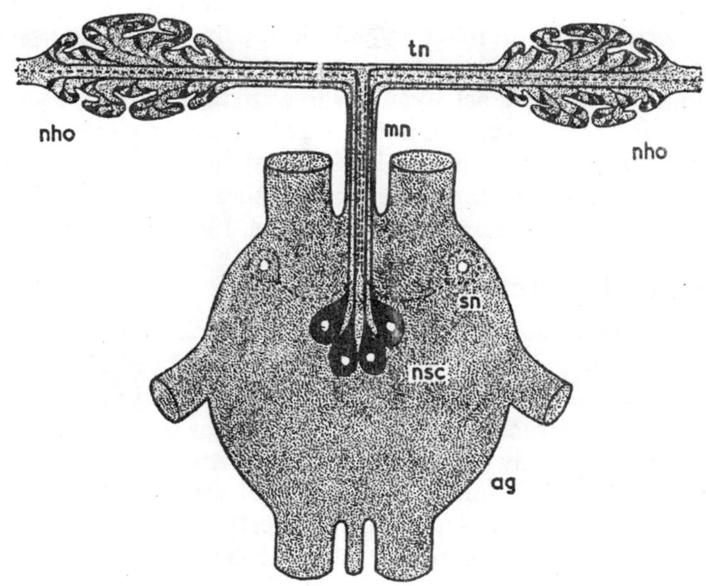

Fig. 1.3. Scheme of a pair of metameric neurohaemal organs *(nho)* with the corresponding abdominal ganglion *(ag)* in Phasmidae. *mn* - median visceral nerve, *tn* - transversal visceral nerve, *sn* -sympathetic neurons, *nsc* - neurosecretory cells. (From Raabe, 1971.)

Raabe (1965) claimed that the products of the neurosecretory cells of the ventral ganglia were stored in the mno and passed into the haemolymph. This no doubt refers to type C neurosecretory cells, which are rather abundant there. The general validity of this statement is, however, made questionable by the findings of Finlayson and Osborne (1968, 1970), who showed in *Carausius* that several neurosecretory perikarya occurred along the link nerve, *i.e.* a thin nerve branch connecting the transverse nerve, together with its neurohaemal organ, with the main segmental nerve of the particular ganglion. These perikarya correspond to the number of axons forming the transverse nerve. The definitive solution of this problem is a main task in further research.

Ultrastructure

A detailed electron microscopy study of mno in Phasmids *(Carausius* and *Clitumnus)* by Raabe and Ramade (1967) fully confirmed their neurohaemal character. They contain numerous neurosecretory fibres and their relatively thick surface membrane has large numbers of deep invaginations constituting a genuine internal stroma. The passage of the neurosecretory substance into the haemolymph probably takes place in their many sinuses, which have a very thin surface membrane. Several types of glia cells, vacuoles and lipid and glycogen granules, etc., were found. The ultrastructural appearance of the two types of mno in Hymenoptera *(Vespa* and *Vespula)* is similar (Provansal, Baudry and Raabe, 1970; Raabe, Baudry and Provansal, 1970). The lateral longitudinal organs attached to the main segmental nerves in this group differ from the median neurohaemal organs, as they have no surface glial membrane and practically no glia cells. They are packed with neurosecretory nerve fibre endings and have large numbers of sinuses. A specific neurosecretory cell containing small granules was also observed.

Physiology

Little, if anything, is known of the physiological significance and function of the mno in the insect organism. The chief reason is the impossibility of completely removing such a complicated, ramified structure from the living insect. Some data were obtained by injecting extracts into insects during an inactive period, and by incubating explanted tissues, e.g. heart, Malpighian tubules and rectum, in such extracts (Raabe *et al.*, 1971). It was demonstrated by these methods that the extracts act upon cardiac rhythm (Raabe, Cazal, Chalay and Besse, 1966) and upon tanning of the cuticle. On the other hand, physiological colour changes and the trehalose content were unaffected (Chalay, 1969). Changes in the diurnal activity of the type C neurosecretory cells supposed to deposit their secretion here, suggest that there is some interference with the activity of the diurnal rhythm (Raabe, 1966).

Electrical Activity

'Spontaneous' electrical activity, in the form of action potentials, was demonstrated in *Carausius morosus* mno by Finlayson and Osborne (1970). The action potentials resembled those observed in other neurosecretory cells, in that their spike duration was longer than for ordinary neurones. They were produced by the neurohaemal organs irrespective of the presence or absence of any perikarya in the preparation and thus appear to arise in the organ itself. They may have a bearing on the release of secretion. This would be in agreement with the earlier observation (Highnam, 1961) that electrical stimulation results in emptying of the neurosecretory granules from the neurosecretory cells and corpora cardiaca.

Some conclusions as to how these organs function can be drawn from the following observations by Raabe (1966-1971). If the median nerve is sectioned in front of a ganglion, reduction of the azan-positive material in the neurohaemal organs of the corresponding ganglion is observed, together with intensive accumulation of the neurosecretion at the proximal end of the sectioned nerve. The author concluded that the substances contained in the neurohaemal organs originate in the neurosecretory cells of the ganglia. Whether this is also true of all other C-cells still remains to be determined. It definitely does not apply to the type A- and B-cells in these species, since their neurosecretion is Gomori- and PAF-positive and so such material was found in the neurohaemal organs. On the other hand, Musko and Novak (1973) reported a certain amount of Gomori- and PAF-positive granules in the neurohaemal organs in *Gryllotalpa*.

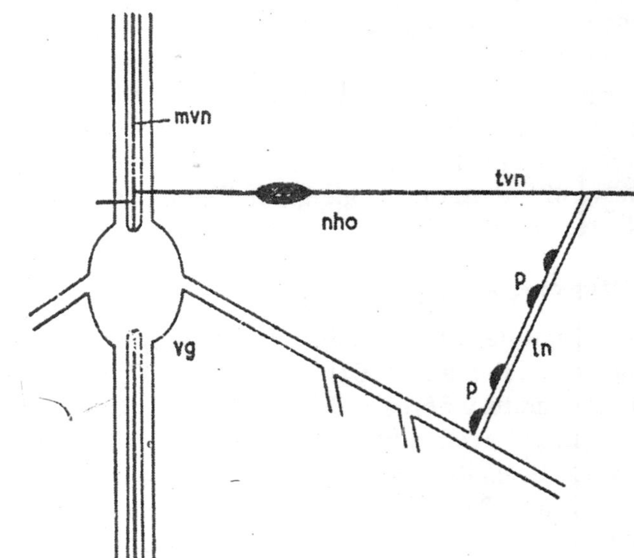

Fig. 1.4. Scheme of innervation of a metameric neurohaemal organ (nho). ***vg*** **- ganglion of ventral cord,** ***mvn*** **-medina visceral nerve,** ***tvn*** **- transversal nerve,** ***In*** **- link nerve,** ***p*** **- perikaryons of link nerve. (From Osborne and Finlayson, 1970.)**

Smalley (1970) studied the histology and ultrastructure of the nsc of the abdominal ganglia of *Periplaneta americana* whose axons enter the paired metameric neurohaemal organ, which has united here to form a single structure in the median line. Three groups of neurosecretory cells in the ganglion posterior to the organ were found to be concerned: the midline median nerve cells situated between the connectives entering the ganglion, the nsc group behind the ventral neuropile and a third group below the centre of the dorsal surface of the ganglion. The cells were negative in specific staining reactions, but took up ^{3}H from injected tritiated dopamine more readily than the other cells of the ganglion. Electron-transparent vesicles were observed in their endings in the neurohaemal organs.

A secondary neurohaemal organ,, in the form of a thickening of the paired nervus allatus II, observed by Novak (1959), was studied by Belyaeva (1966) and Theodorescu and Novak (1973) in the same species and by Awasthi (1968, 1971) in *Gryllodes sigillatus*. Each of the paired dilations is spindle-shaped and contains, around the central swollen nerve fibres, large numbers of glia cells covered by a sheath of the same nature as the neural lamella and, in fact, is continuous with it by means of the sheath of the nervus allatus II. The organ is filled mainly with a Gomori-positive neurosecretion with different sized granules, coming from the brain via the nervi corporum cardiacarum, the corpus cardiacum, the nervus allatus I, the corpus allatum and the nervus allatus II. No neurosecretion was observed in the last (and thinnest) segment of the nerve (II D) between the thickening and the suboesophageal ganglion. Unlike the metameric neurohaemal organs, the thickenings of the nervus allatus II have a regular structure.

In their study of the metameric neurohaemal organs in *Gryllotalpa gryllotalpa,* Musko and Novak (1973) detected a pair of large neurosecretory cells staining red with azan, but Gomori- and PAF-negative, at the posterior end of each ganglion of the ventral nerve cord, in the median line between the two connectives. The axon of each of these cells entered the transverse branch of the median nerve on its body side. They formed part of a small group of neurones (10 to 20) surrounded by small cells of a glial character and enclosed in a thin sheath staining blue with azan (like the neural lamella). The whole structure gave the impression of a small incorporated ganglion connected to the median visceral nerve. On the basis of their findings, the authors formulated the hypothesis that this is a relic of the original state of the median visceral system, which they suppose originally had a small ganglion in each body segment. Following this concept, they expressed the view that the frontal and hypocerebral ganglia are simply specialized forms of similar ganglia of the cerebral and suboesophageal ganglia.

The Ganglionic Neurohaemal Organs

A special pair of neurohaemal organs have recently been found in each of the ganglia of the ventral nerve cord (Musko and Novak, 1973). They are placed each on a thin paired nerve rising from the laterodorsal region of the ganglion tightly to its surface. These organs are thicker, more spacious and with less deep constrictions than the perisympathetic neurohaemal organs. They seem to drain neurosecretion from nsc of the lateral parts of the ganglia. Unlike the perisympathetic organs, they contain granules of paraldehyde-fuchsine (Gomori-positive) neurosecretion beside of the azan-positive (phloxinophylic) one. The term ganglionic neurohaemal organs has been suggested for them.

SURVEY OF NEUROHORMONES

Myotropic Effects

The starting point for experiments on the myotropic effects of neurohormones are Loewi's classic experiments. In insects such effects may be demonstrated on organs with a spontaneous rhythmic activity such as the heart, Malpighian tubules, oviducts and gut. One of the first authors to show in insects the hormone dependence of the pulsations in isolated Malpighian tubules, oviducts and gut was Koller (1948). He showed that extracts of brains and corpora cardiaca accelerate the contractions of isolated Malpighian tubules in insect-Ringer solution, whereas the same extracts reduce the peristaltic movements of the gut under similar conditions. In contrast, stimulation of gut pulsations are caused by extracts of the gut and corpora allata. Enders (1956) has found the opposite effect of extracts from brains, corpora cardiaca and corpora allata on the peristalsis of the oviducts in the stick-insect *Carausius morosus.* Brain extracts had a positive effect, whereas the effect of extracts of corpora cardiaca and corpora allata was negative.

Experiments with Corethra. A detailed account of research into myotropic effects in the transparent larvae of *Corethra (Chaoborus crystallinus)* has been given by Gersch (1952,1955,1960a, b) and his co-workers (Fuller, 1960). This species has the great advantage for this kind of work that the effect of various extracts on the gut can be made visible by the use of various vital stains and can be easily observed *in vivo* owing to the transparency of the larva's body. A characteristic feature of the species is the strong antiperistaltic movements which follow each food acceptance. The function of these movements is to bring the digestive enzymes into the pharynx, where the process of digestion in this insectivorous species commences. Mechanical or thermal irritation of the brain and of single ganglia in the ventral nerve cord has demonstrated the dependence of these antiperistaltic movements on specific ganglia. There were three indications in favour of a humoral character for this effect: (1) The effect of the irritation only appeared after a definite 'latent' period; it subsequently increased and continued to affect the movement for about one hour. (2) The observed effect was in no way reduced when the adjacent ganglion on either side had been destroyed by burning so that the connectives were interrupted. (3) No nervous branches entering the gut walls could be found arising from these ganglia, the only nerve supply to the intestine being the stomatogastric nervous system arising from the brain. A similar effect was obtained by irritating the cerebral ganglion, with the only difference that in this case, the rise in the rate of peristaltic movement was preceded by a wave of increased activity of short duration. This was no doubt due to a nervous impulse from the stomatogastric system. The proof of a humoral action of the ganglia via the haemolymph was obtained by experiments with extracts, not only from the various ganglia of the nervous system of the same species *(Chaoborus),* but also from ganglia of the cockroach *Blatta orientalie* and the stick-insect *Carausius morosus.* These extracts were equally active on the isolated gut in insect-Ringer. The relevant active substances were obtained from all parts of the nervous system, and were specific neither for species nor order. Later research has shown that they affect the heart-beat not only in other arthropods but even in other animal Phyla such as molluscs and vertebrates.

A Humoral Factor Obtained by Homogenization of the Corpora Cardiaca

This substance stimulates the pericardial cells to produce a heart-accelerating principle in *Periplaneta americana,* was described by Davey (1961, 1962, 1963a, b) in great detail. Pericardial cells treated with the factor *in vitro* show histological signs of increased secretory activity and produce a substance increasing the heart rate, which was identified as indolalkylamine. In living animals, release of this factor from the corpora can be induced by feeding. As suggested by Gersch (1964) there are no grounds for assuming that this form of interaction of the pericardial cells is general in insects, and there is good evidence of the direct effect on the heart rate of the neurohormones isolated, amongst others, from the corpora cardiaca. As emphasized by Evans (1962) and Gersch (1964), it would be very interesting to determine the relationship of Davey's 'corpus cardiacum hormone' to the neuro-hormones C_1 and D_1 and possibly to the other active principles previously shown to occur in the corpora cardiaca as well as to the secretion of the pericardial cells.

The Isolation of Neurohormones

The aforementioned effects were investigated by Gersch's co-worker Unger (1956), who used isolated hearts of *Blatta orientalis* and *Periplaneta americana* as test material. Insect-Ringer extracts from various parts of the nervous system were used. The heart of insects is known to be innervated partly by the stomatogastric system and partly by thin nerve branches from the corresponding abdominal ganglia. The pulsation may therefore be influenced directly by either of the nervous paths (cf. Florey, 1952). However, it was shown in the insect-Ringer preparations that the heart-beat could be stimulated by means of the same extracts as had been found to influence gut peristalsis. In contrast, extracts of the corpora allata subdued the heart-beat. This effect also showed a very wide non-specificity; the cockroach heart preparations were influenced not only by extracts from cockroaches but also by extracts from stick-insects, dragonfly larvae and blowflies *(Calliphord)*. The detailed analysis of such extracts and their separation by paper chromatography has revealed that no less than three physiologically different substances can be isolated. One of these substances, found in all parts of the ventral nerve cord but missing in the brain and the retrocerebral glands, was shown to be identical with acetylcholine. It stimulated the heart-beat and gut peristalsis. Another substance occurring in the brain and the corpora cardiaca as well as in the ventral nerve cord was called neurohormone D by Gersch and Unger (1951). This equally stimulated the frequency of the heart beat but at the same time increased its amplitude. At high concentrations it stopped the heart-beat during diastole. Its characteristic fluorescence was later shown to be due to a mixture of pterines which are very hard to separate. A third substance, partially antagonistic to neurohormone D and denoted neurohormone C by the same authors, was found in all parts of the nervous system and in the corpora cardiaca like neurohormone D, but was also found in the corpora allata. It too increased the frequency of the heart-beat, but decreased its amplitude, and at higher concentrations, stopped the heart during systole. In this respect the effect of neuro-hormone C is antagonistic to the effect of neurohormone D. All these three substances were also isolated from the haemolymph.

Further research along these lines enabled Gersch and his collaborators (Gersch, Fischer, Unger, Koch, 1960; cf. Gersch, 1960) to separate the original neurohormone D into two different substances D_1 and D_2 and the original C into hormones C_1 and C_2. These were extracted from 3200 lyophilized central nervous systems of *Periplaneta americana* with a succession of petroleum ether, ethyl ether, ethyl acetate, ethanol, water, acetic acid and pyridine. The largest amount of the physiologically active component was obtained from the water fraction, and was found to be identical with the ethyl alcohol fraction. Paper chromatography of the water-soluble fraction with butanol-pyridine-water (i : 2.5 : 2.5) clearly showed two substances: one with the activity of neurohormone C (at an RF of 0.17) and the other with the activity of neurohormone D (at an RF of 0.33). They were eluted with water and shown by bioassay to correspond with the previously described hormones. The substance with the physiological action of neurohormone D was recrystallized several times to yield about 50 µg. of fine spheroid crystals with a fairly constant form. This substance was designated as neurohormone D_1 and some of its physical properties were determined. The absorption peaks in the infrared were found at 1140,1400,1480,1640, 3400 and 3600 cm^{-1}. In this relatively pure form it was slightly soluble in water, hardly soluble in ethyl alcohol, and insoluble in organic solvents. When peptides are added, however, it readily dissolves in water. The second substance recrystallized in needle form, showing a tendency to fuse, and corresponded with neurohormone C in physiological activity and was designated as neurohormone C_1.

Two further substances with similar though distinctly different features were obtained from the ethyl acetate fraction by paper chromatography with butanol-ethanol-acetic acid-water. One of them, with an RF of 0-17, was designated as neurohormone C_2 and agreed with C_1 in its positive effects on the frequency of the heart-beat without inducing systolic contraction. The other substance D_2 with an RF of 0.25, corresponded (approximately) with D_1 in its crystal form, and showed a similar stimulation of the heart-beat. Unlike C_1 and D_1, neither C_2 nor D_2 produced any effect on colour change. The possibility must be taken into account that they may be either decomposition products or intermediate compounds in the synthesis of C_1 and D_1. None of the four substances could be identified with either adrenaline, noradrenaline or histamine (Unger, 1956). On the other hand, a series of substances of synthetic or vertebrate origin have proved to be active in isolated gut or heart experiments. For example acetylcholine, doryl, physostigmine and histamine have been found to stimulate antiperistaltic movements, the most active being acetylcholine which is still active at a concentration of IO^{-7}. In contrast, the effect of adrenaline and pilocarpine is distinctly negative, damping the movement (cf. Gersch, 1960). Extracts of the nerve ring and the ventral ganglion of *Ascaris* affect the heart-beat of the snail *Helix pomatia.* Similarly, extracts from the snail *Aplysia* affect the heart-beat of *Chaoborus.* This shows the marked non-specificity of neurohormones.

Gersch (1962) made a detailed study of the possible identity of some of these neurohormones with the AH produced by the medial neuro-secretory cells of the brain, using last instar larvae of *Calliphora erythrocephala* and *Lymantria dispar.* A series of experiments involving the injection of brain extracts, injection of the isolated neurohormones C_1, D_1, C_2 and D_2, as well as implantation of immature ring glands suggested that neurohormone D_1 was identical with AH. It was possible to induce pupation in brainless *Lymantria dispar* larvae by injecting the purified neurohormone D_1. Similar results were obtained in *Calliphora.* The neurohormone C_1, however, showed no such activity either in *Lymantria* or, with two or three exceptions, in *Calliphora* (cf. P- 49).

Recently, however, further research using electrophoretic separation of brain extracts on polyacrylamide gels (Sephadex G - 100) has shown that AH is different from neurohormone D, despite the fact that they are both of a peptide nature. Neurohormone D was found to be an oligopeptide with a molecular weight of about 2000, with specific biochemical properties (thermolability, acid lability, quick decomposition by trypsin, etc.). SH groups are an important component of its molecule, as they are essential for its action. It thus to some extent resembles the vasopressin of vertebrates (Richter, 1973). In addition to the above properties, it was shown to reduce significantly the membrane potential of the prothoracic glands, while activation hormone I had no effect, and activation hormone II raised the membrane potential (Gersch and Birkenbeil, 1973). It was active in vertebrates as well as in spiders and crustaceans, affecting water flux and Na transport across the wall of the urinary bladder of the toad in the same way as vasopressin (Richter, 1973).

Six factors having an accelerator action on the isolated heart of *Periplaneta* were extracted in ethyl acetate, ethyl acetate-ethanol mixtures, and ethanol from the corpora cardiaca-allata of the cockroach by Ralph (1962). Another accelerator, soluble in saline solutions, was found in all neural structures examined. Factors, possibly five in number, that caused heart deceleration were found by the same solvent extraction method in the brain, and the suboesophageal and ventral cord ganglia. Little, if anything, is known about the actual function of the myotropic neurohormones in the insect organism. It can be assumed, however, that an increase or decrease in heart rhythm and peristaltic movements is of adaptive value for insects under certain specific conditions. The mechanisms of any such correlations still remain to be determined. Another function could be the induction of specific processes, such as ecdysis, by the ingestion of air or water, fledging movements, etc., or of oviposition, with the necessary contraction of the oviduct and abdominal body wall. The latter can be deduced from papers like those by Enders (1955), Davey (1965, 1967), de Besse (1965). Thomas and Mesnier (1973) analysed this mechanism in stick-insects (*Carausius morosus* and *Clitumnus extradentatus*). Two types of neuro-secretory factors were shown to be concerned with egg-laying in these species. One is a stimulant substance produced by A-nsc in the cerebral ganglion and the suboesophageal, thoracic and abdominal ganglia, while the other (of an inhibitory character) has been localized solely in the brain. These two factors seem to combine to induce the circadian rhythm of egg-laying in these insects.

Rhythms of Activity

A series of papers has shown a causal connection between cyclical, seasonal and diurnal activity and the activity of the neurosecretory cells. Klug (1958) has shown that there is a correlation between winter and summer diapause and the volume of the nuclei of the corpora allata. Also the neurosecretory cells of the pars intercerebralis were found to be full of secretion at noon, but were almost empty at 4 o'clock in the morning.

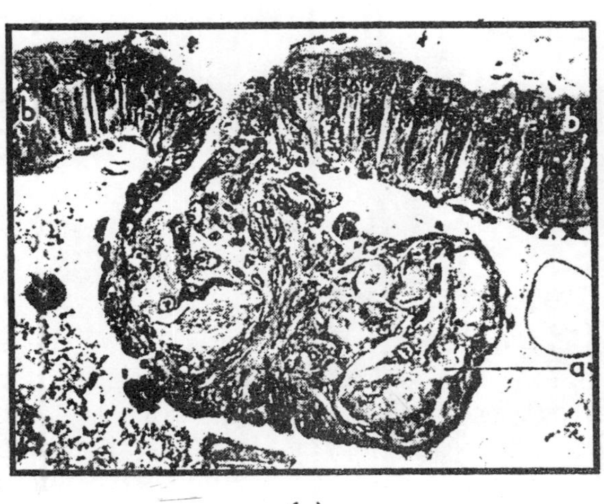

(a)

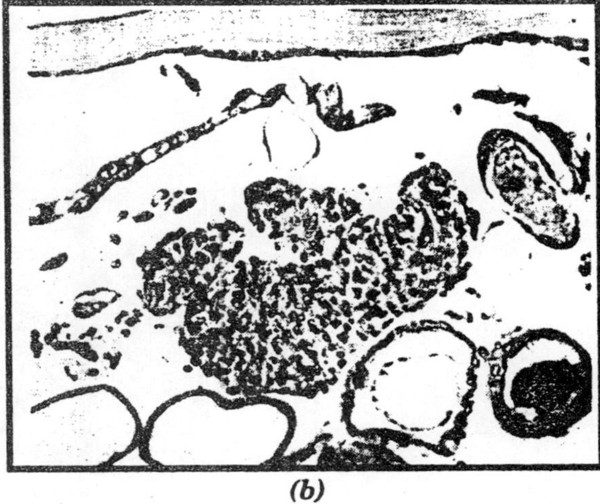

(b)

PLATE 22. (*a*) Hypertrophy of the gut wall of *Carausius morosus* following allatectomy in an early larval instar. *a*—cancer-like growth of the gut wall; *b*—gut epithelium. (From Pflugfelder, 1958.) (*b*) Hypertrophy of lympho-cytes following allatectomy in *Carausius morosus*. (From Pflugfelder, 1938.)

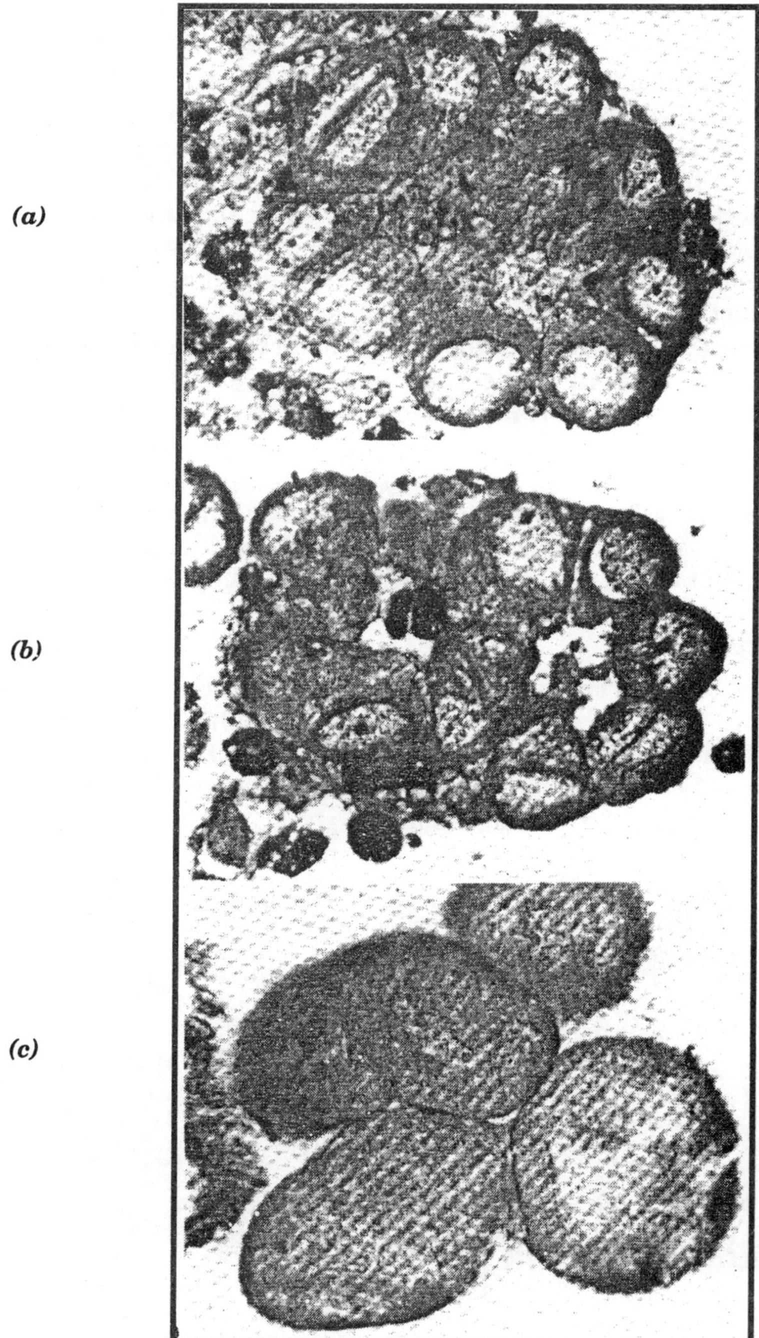

PLATE 23. Corpus allatum in *Mammestra brassicae* pupae and adult. (*a*) beginning of pupal instar, with degenerating surface membrane and phagocytes on surface of gland only, (*b*) 4th day of instar, with disappearance of surface membrane and digestion of intercellular matter by individual phagocytes which have penetrated the gland, (*c*) 4 cells of an adult c. allatum completely devoid of intercellular matter, same magnification × 800. (From Novak et. al., 1973.)

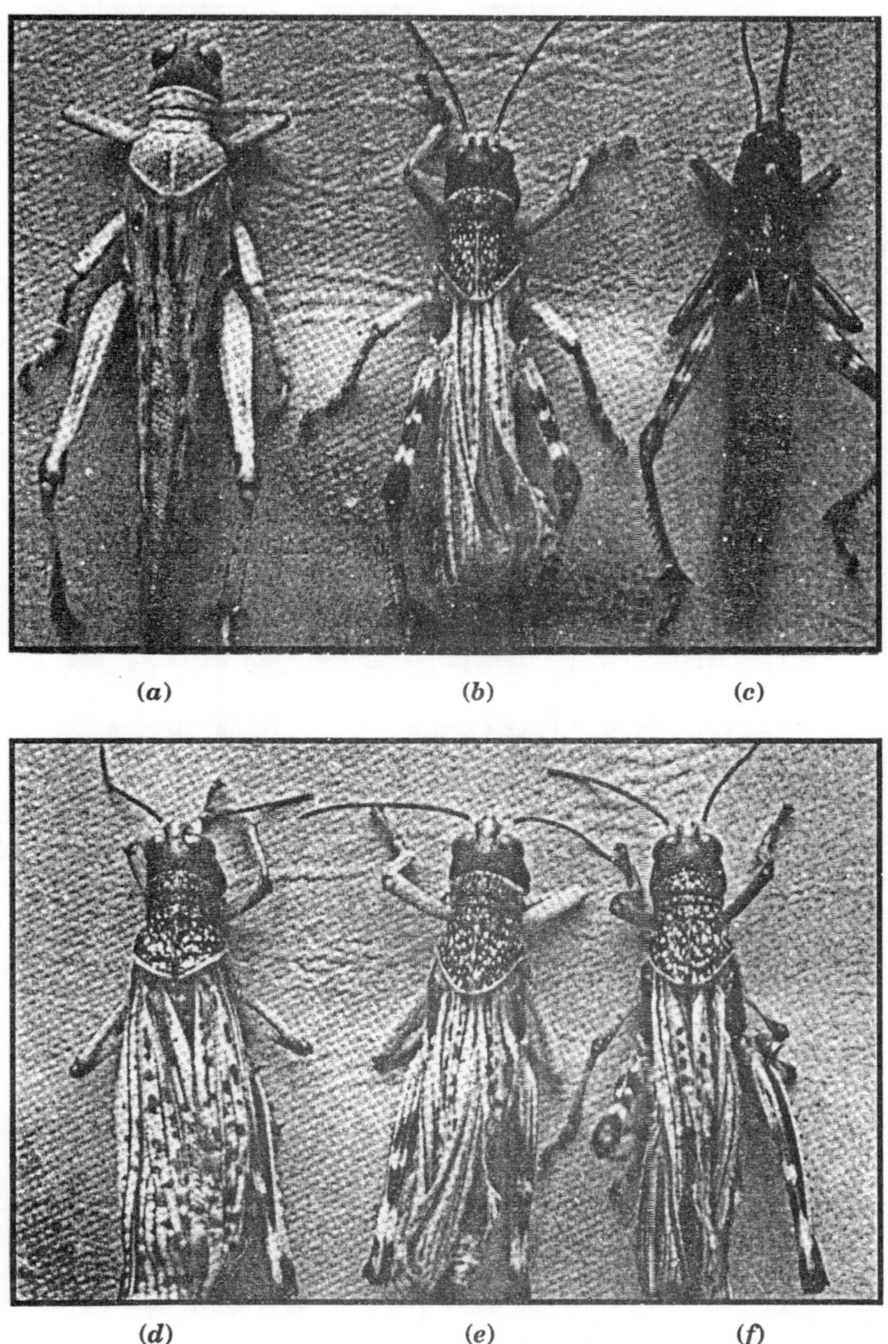

PLATE 24. Intermediates between normal adult (*a*) and Vth (last) instar nymph (*c*) of *Schistocerca gragaria.* (*b, d, e, f*) intermediates of varying degrees, produced by topical application of juvenoid. Note the larval shape and coloration of the pronotum in the intermediates. (Orig. Slama.)

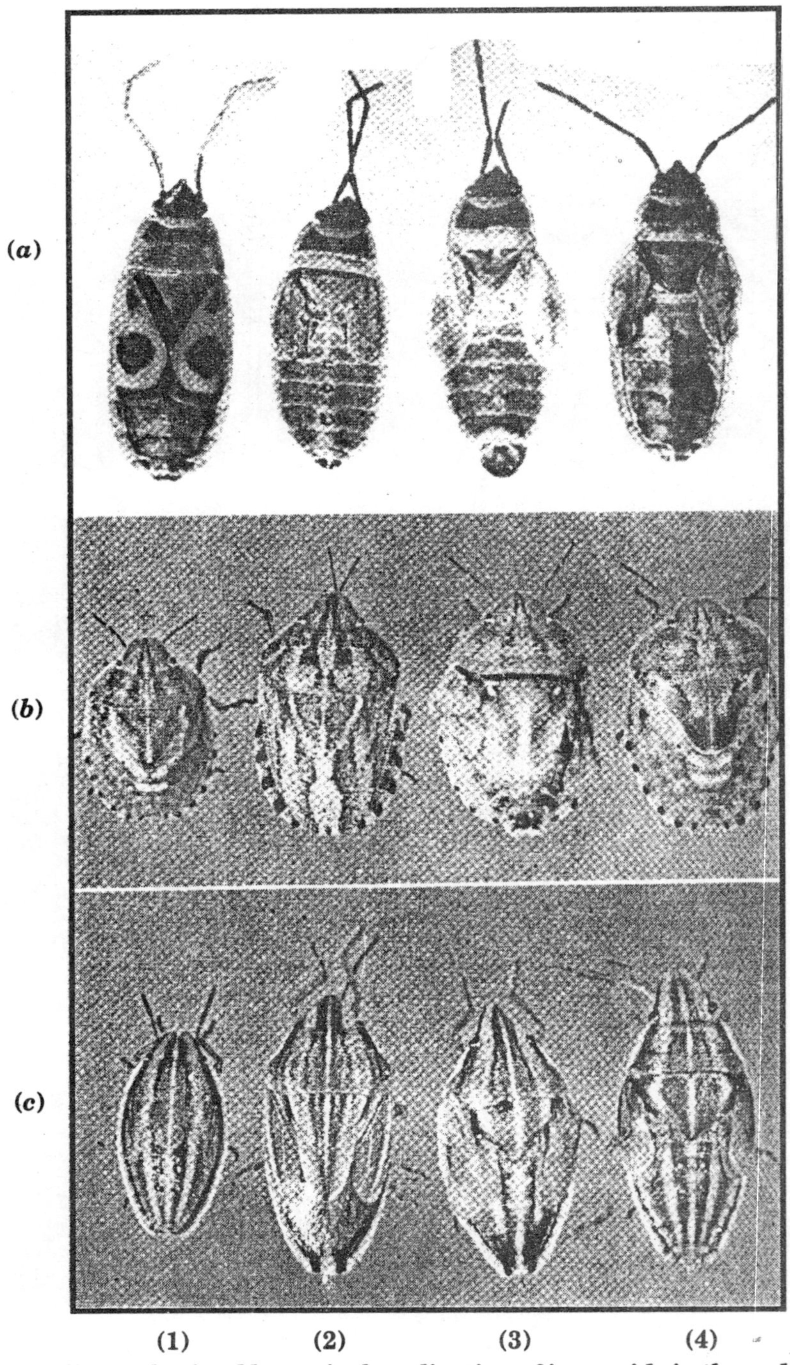

PLATE 25. Intermediates obtained by topical application of juvenoids in three different species of bugs. (*a*) *Pyrrhocoris apterus* (1-normal female adult, 2-normal Vth instar nymph, 3, 4-intermediates); (*b*) *Eurygaster integriceps* (1-normal Vth instar nymph, 2-normal adult, 3, 4-intermediates); (*c*) *Aelia lineata* (I-normal last instar nymph, 2-normal adult, 3 4-intermediates). (Orig. Slama.)

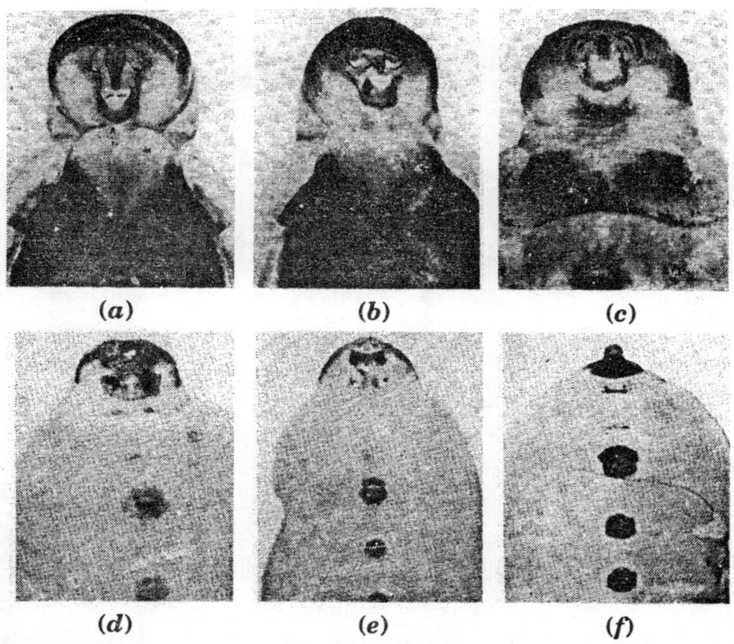

PLATE 26. Abdomina of intermediate forms (*b-e*) between last instar larva (*a*) and adult (*f*) of *Pyrrhocoris* apterus (males) following the implantation of a corpus allatum at different stages in the course of the last larval instar.

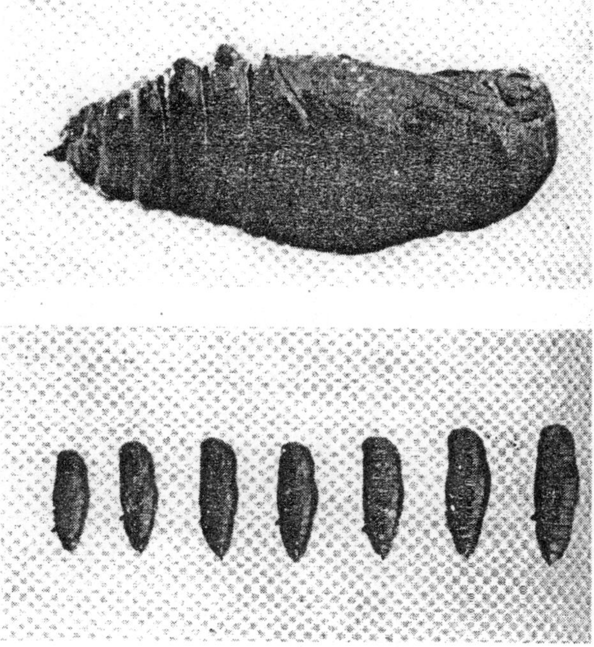

PLATE 27. Developmental anomalies of a hysterothetelic nature. The remains of larval pseudopodia in *Hypantria cumea* pupae.

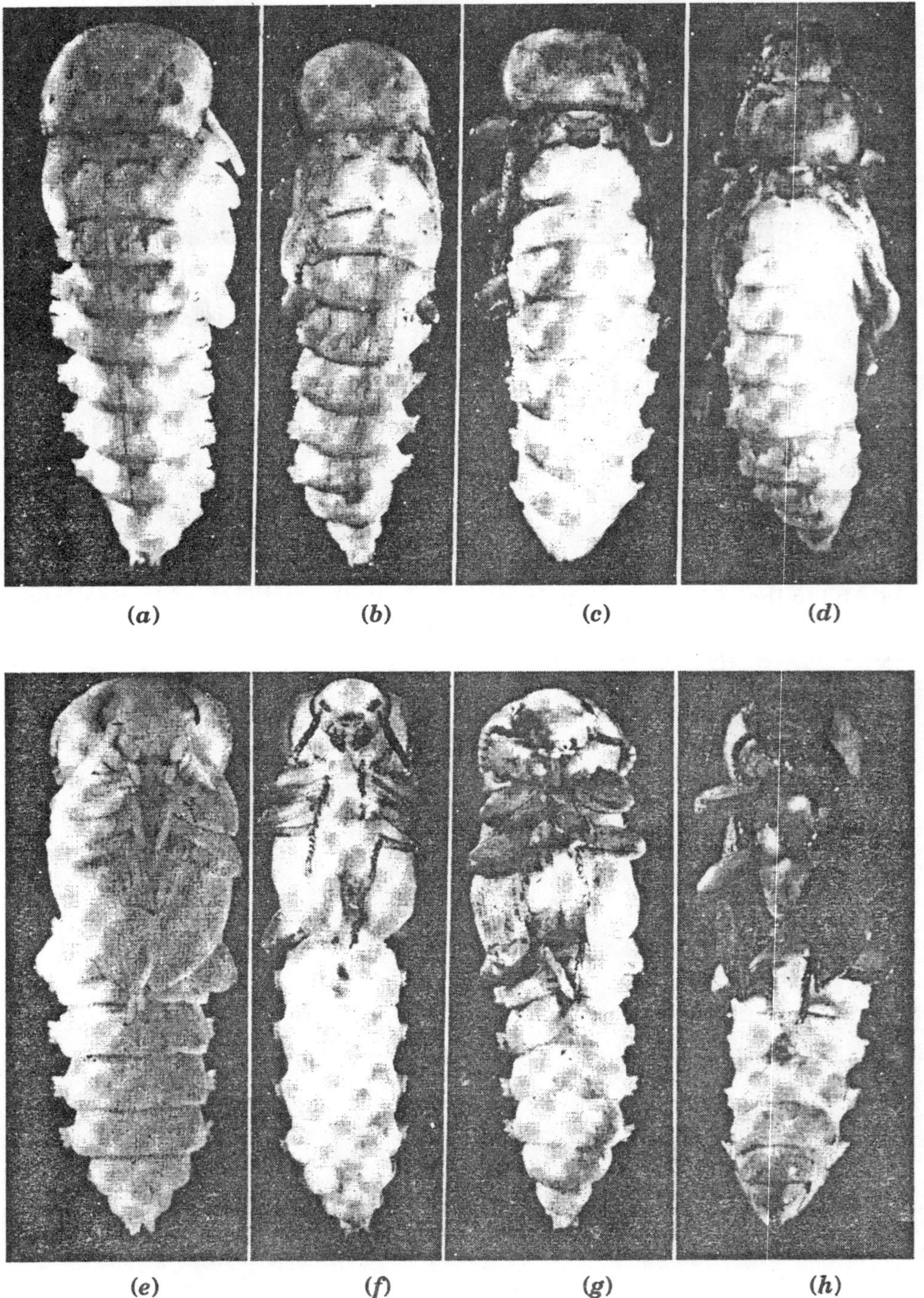

PLATE 28. *Tenebrio molitor* pupae affected by topical juvenoid treatment. (a-d) dorsal views, (e-h) ventral views. (From Socha and Sehnal, 1972.)

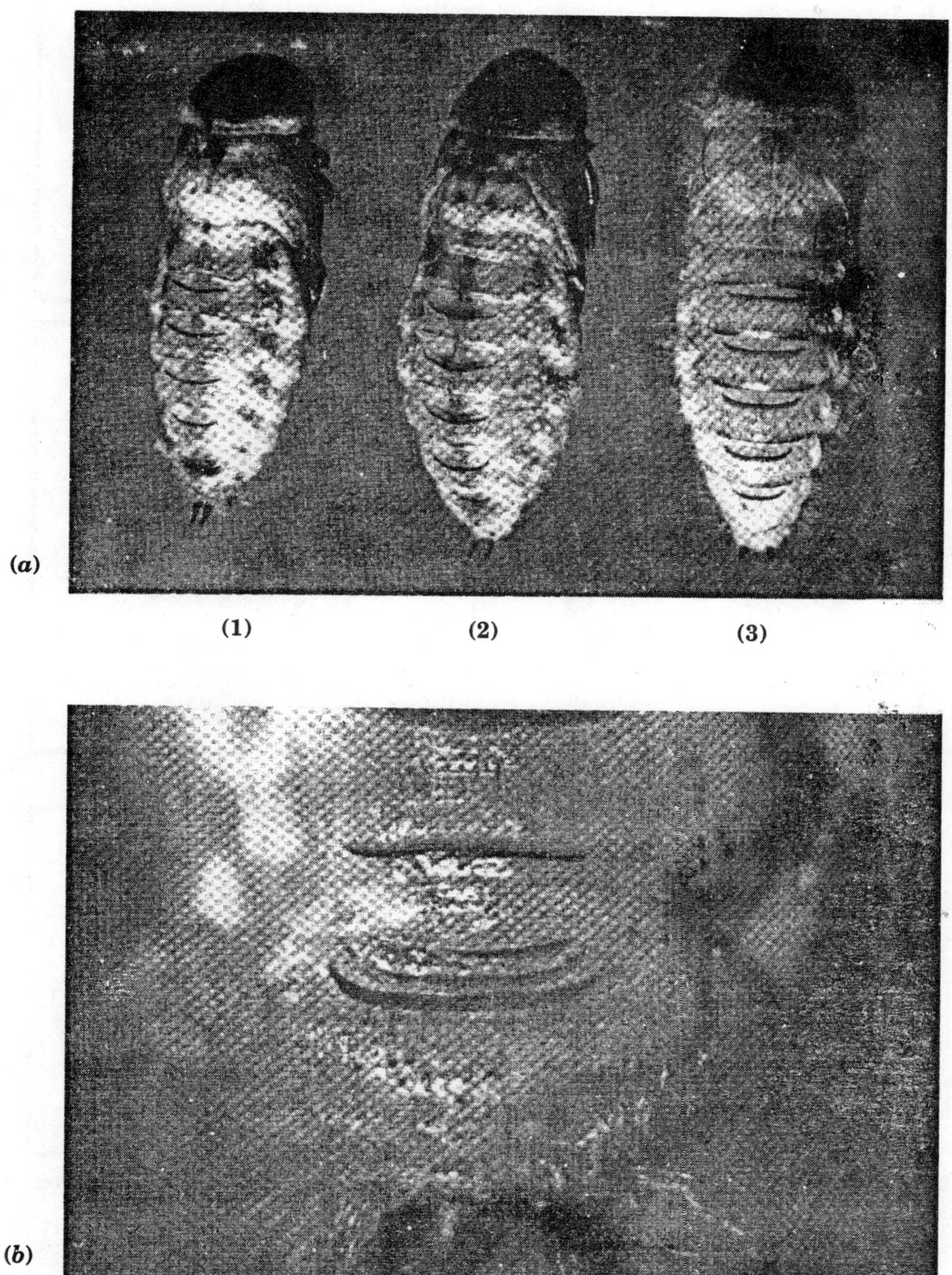

Plate 29. ***Dermestes vulpinus*** **pupae affected by juvenoid treatment. (*a*) I-normal pupa 2-second pupa, 3-third pupa. (*b*) three last abdominal terga of third pupa. Note the three chitinou ridges on the last tergites, which are specific for the double pharate pupa. (Orig. Slama.)**

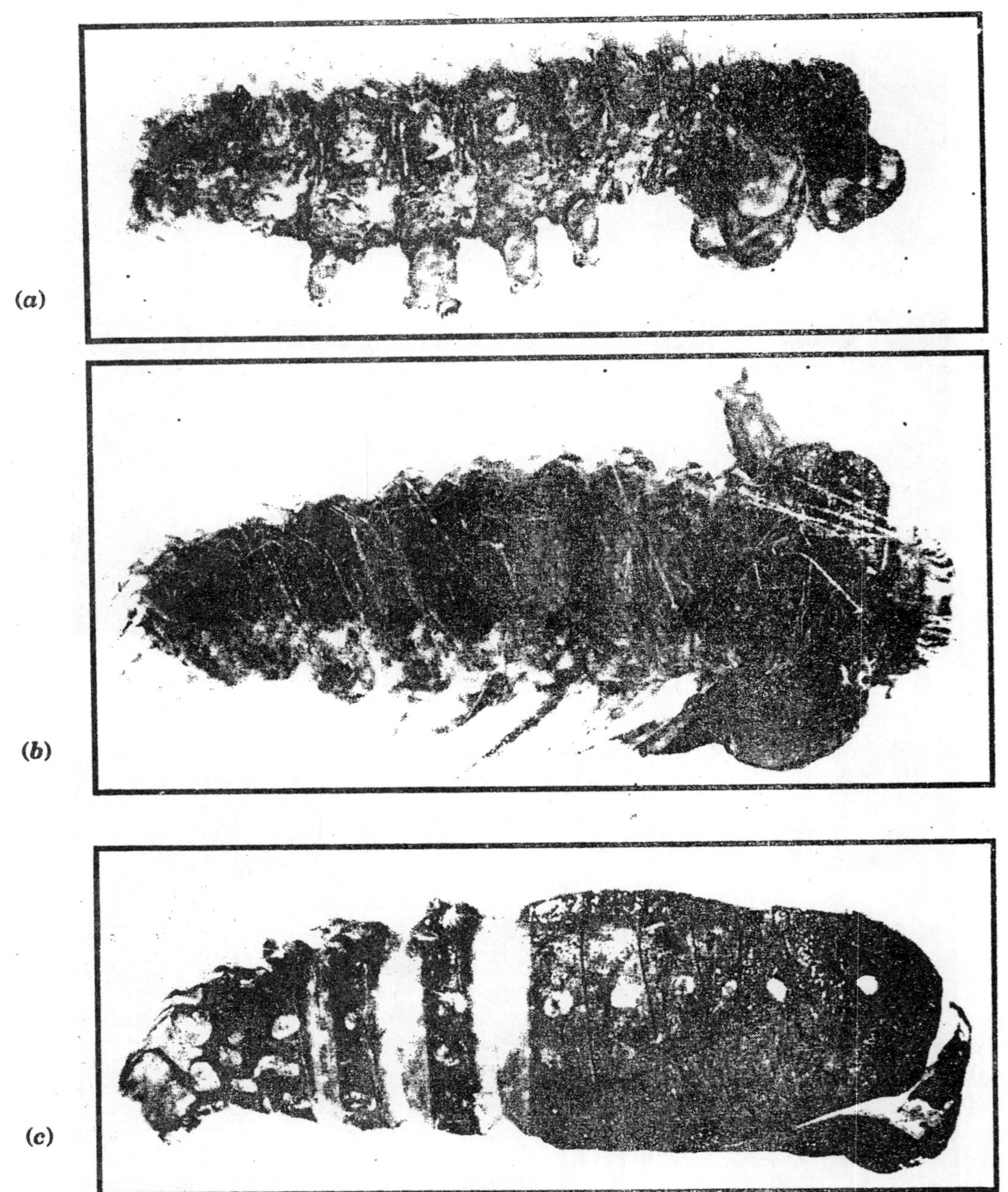

PLATE 30. Three *Lymantria dispar* larvo-pupal intermediates produced by juvenoid treatment. (a) most strongly affected, (c) least strongly affected. (Orig. Slama.)

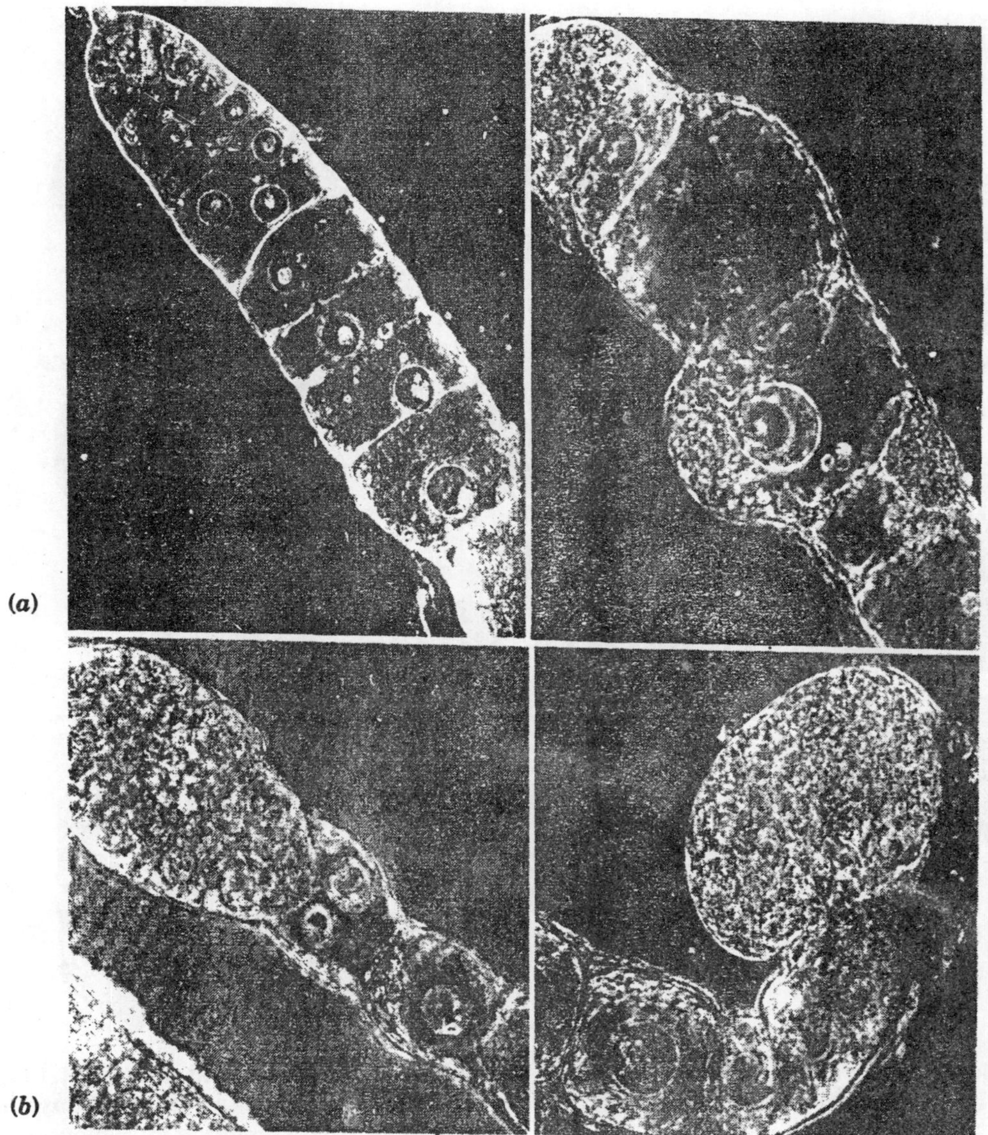

Plate 31. Ovarioles of *Thermobia domestica* after juvenoid treatment. (*a*) control female (plain acetone 42 days previously), (*b*) ovarioles of experimental specimens, 42 days after treatment; note the zones of proliferating pre-follicular tissue along the previtellar oocytes, (*c, d*) the same, showing varying degrees of hypertrophy of the germarium. Living tissue, phase contrast. × 320. (Orig. Rohdendorf.)

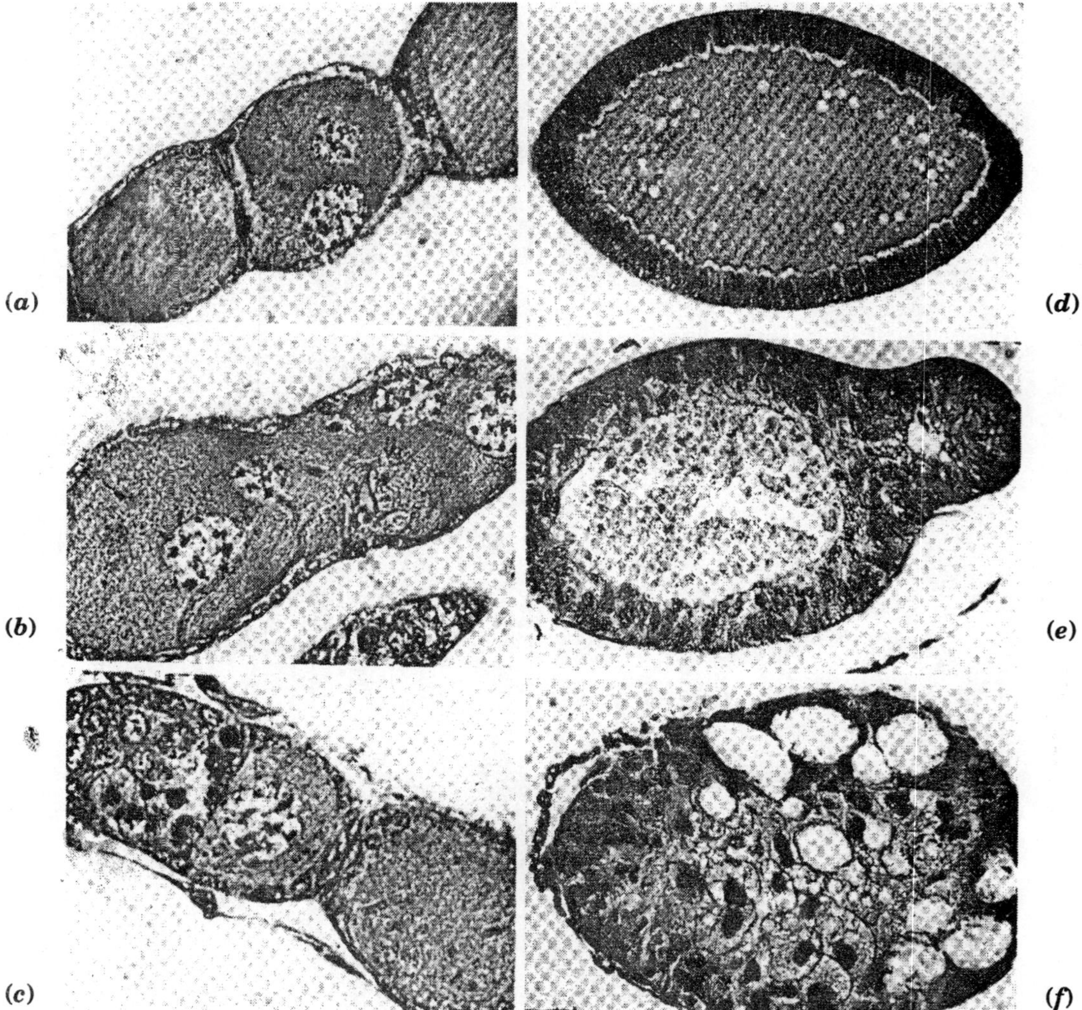

PLATE 32. Sections of ovarioles of *Dixippus morosus* specimens after juvenoid treatment. (*a*) part of vitellarium of intermediate produced by applying JHa to last instar nymph (6 days after moulting, 8 days after application); one follicle with two nuclei, involvement of follicular zone. (*b*) more severely affected specimen, with obliteration of interfollicular epithelim. (*c*) section of germarium and part of previtellarium of 20-day last instar nymphs treated on 15th day after ecdysis; both normal and injured oogonia. (*d*) distal follicle in late previtellogenic stage in 8-day nymph treated immediately after ecdysis. (*e*) severely affected distal follicle of nymph 14 days after ecdysis (treated between 2nd and 12th day after ecdysis.) (*f*) terminal phase of resorption of distal follicle in I-day-old abnormal adult obtained from supernumerary nymph. (From Socha, 1973.)

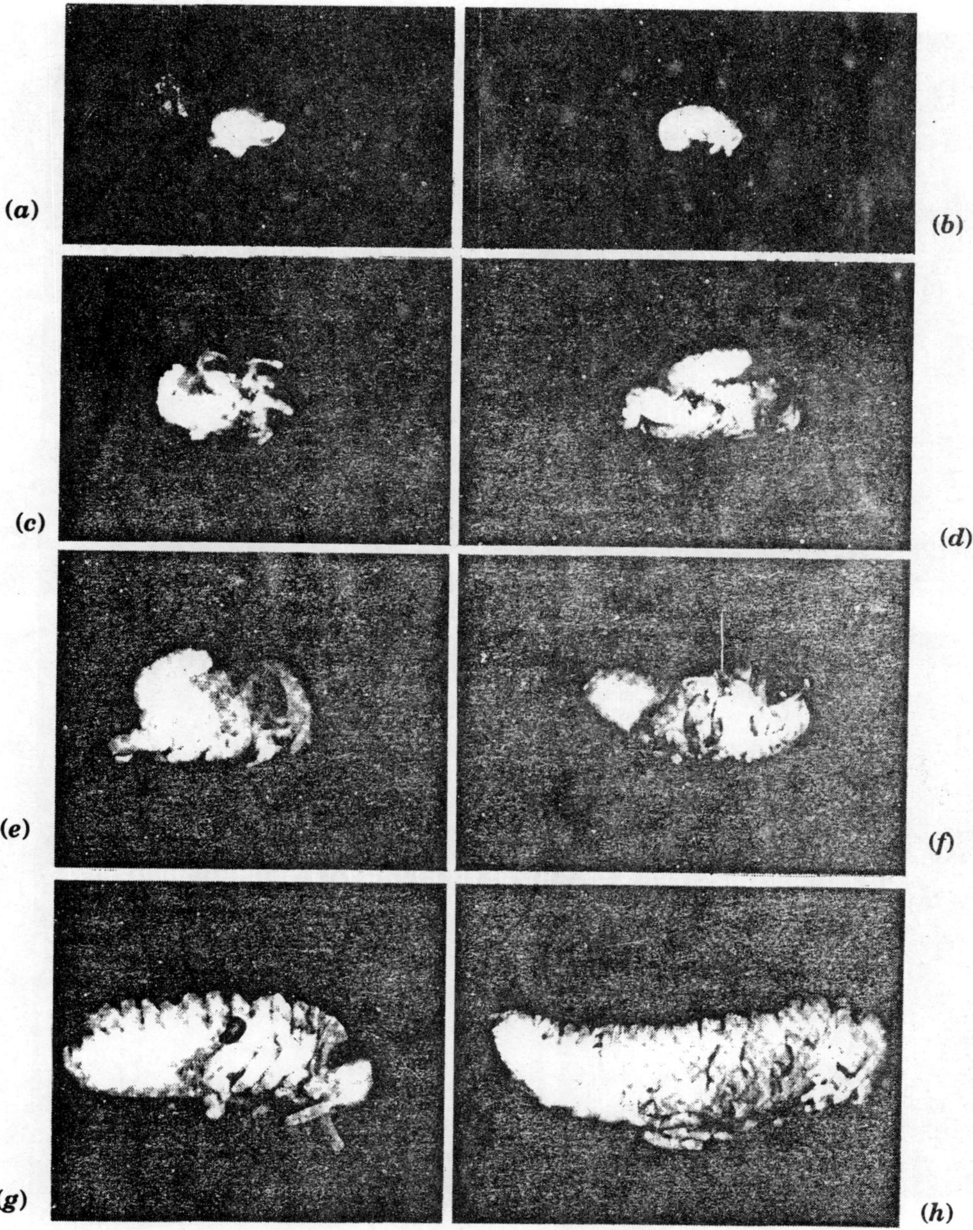

PLATE 33. Juvenilized Schistocerca gregaria embryos disected from eggs after treatment with farnesyl methyl ether and fixed at time when controls were hatched. (*a*) apodous stage with discernible eye pigment, (*b*) protopod stage, (*c*) about half way through blastokinesis, (*d*) second half of blastokinesis (note marked sclerotization and pigmentation as distinct from normal specimens of same age, see Fig. 56), (*e*) miniature embryo at end of blastokinesis; (*f*) the same stage, slightly later, embryo larger and of almost normal size, with pigmented pleuropodium, but with stunted appendages. (From Novak, 1969.)

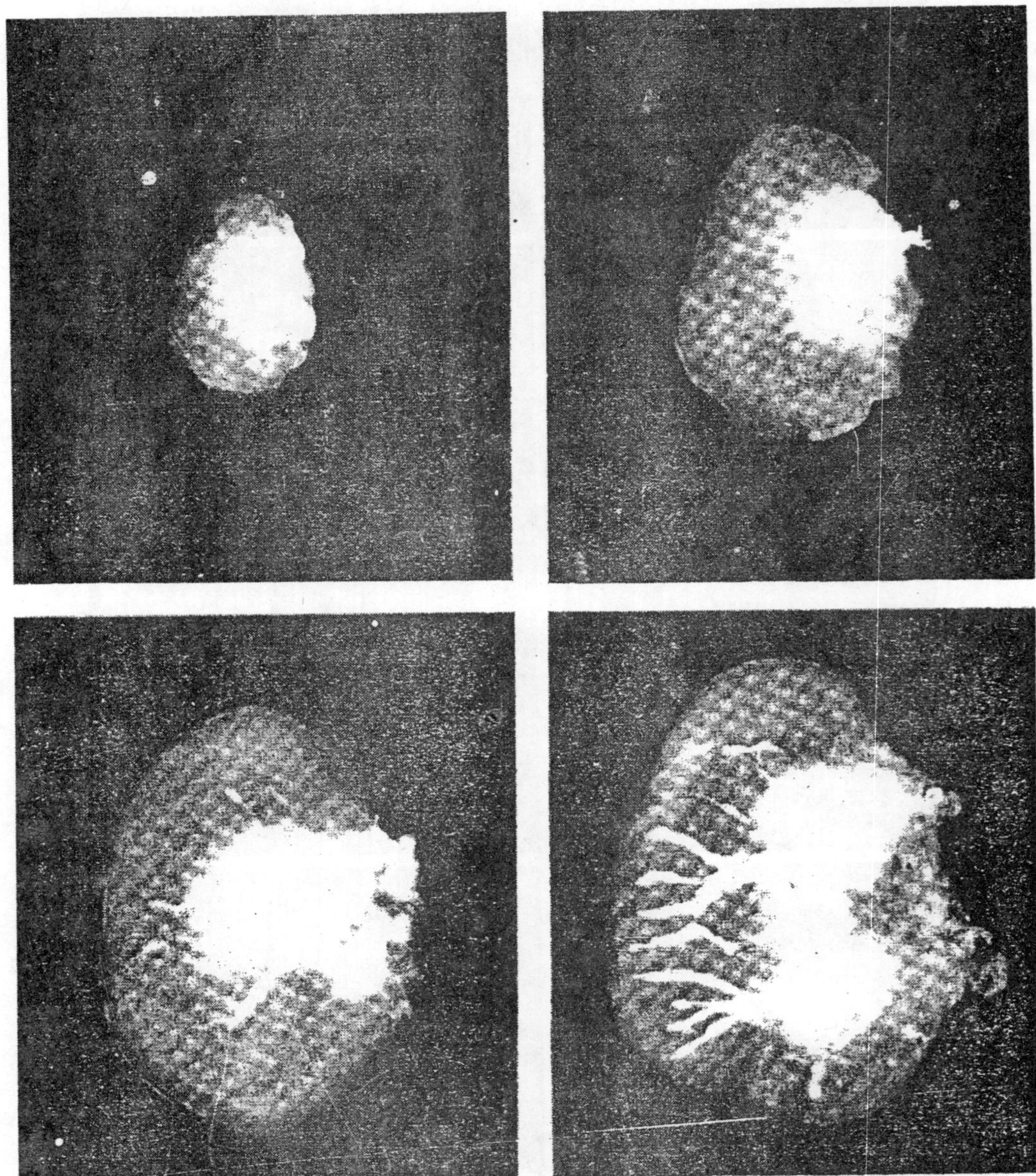

PLATE 34. Imaginal discs at four successive stages of development of the last larval instar of *Galleria mellonella*.

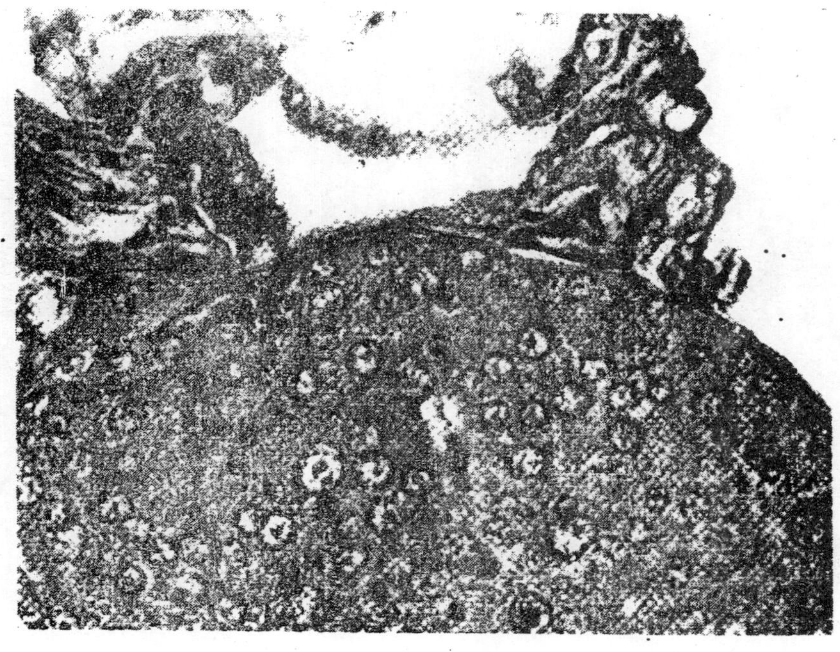

(*a*)

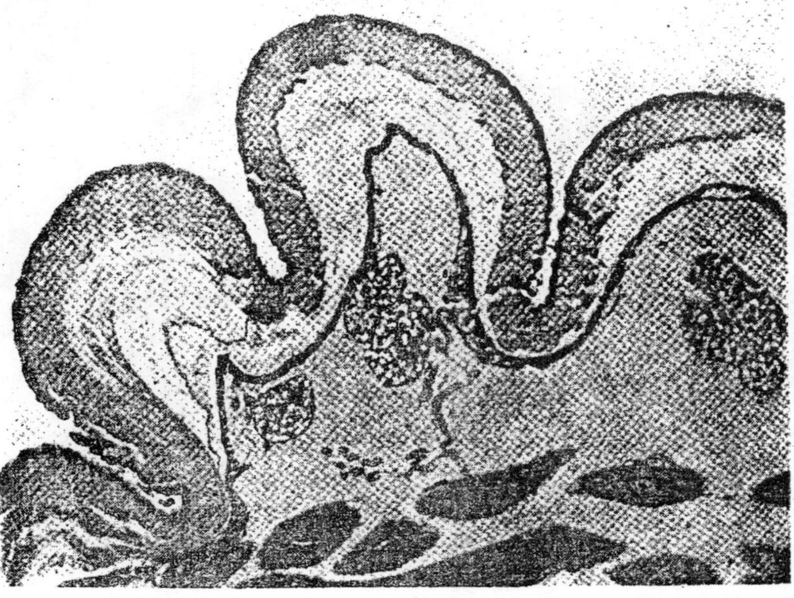

(*b*)

PLATE 35. **(*a*) Part of a median section through the corpus allatum of an adult female oncopeltus fasciatus, stained with Dobell's ammonium molybdenate-haematoxylin method. × 1000. (*b*) Sagittal section through the epidermis of a diapausing pronymph of *Cephalcia abietis*.**

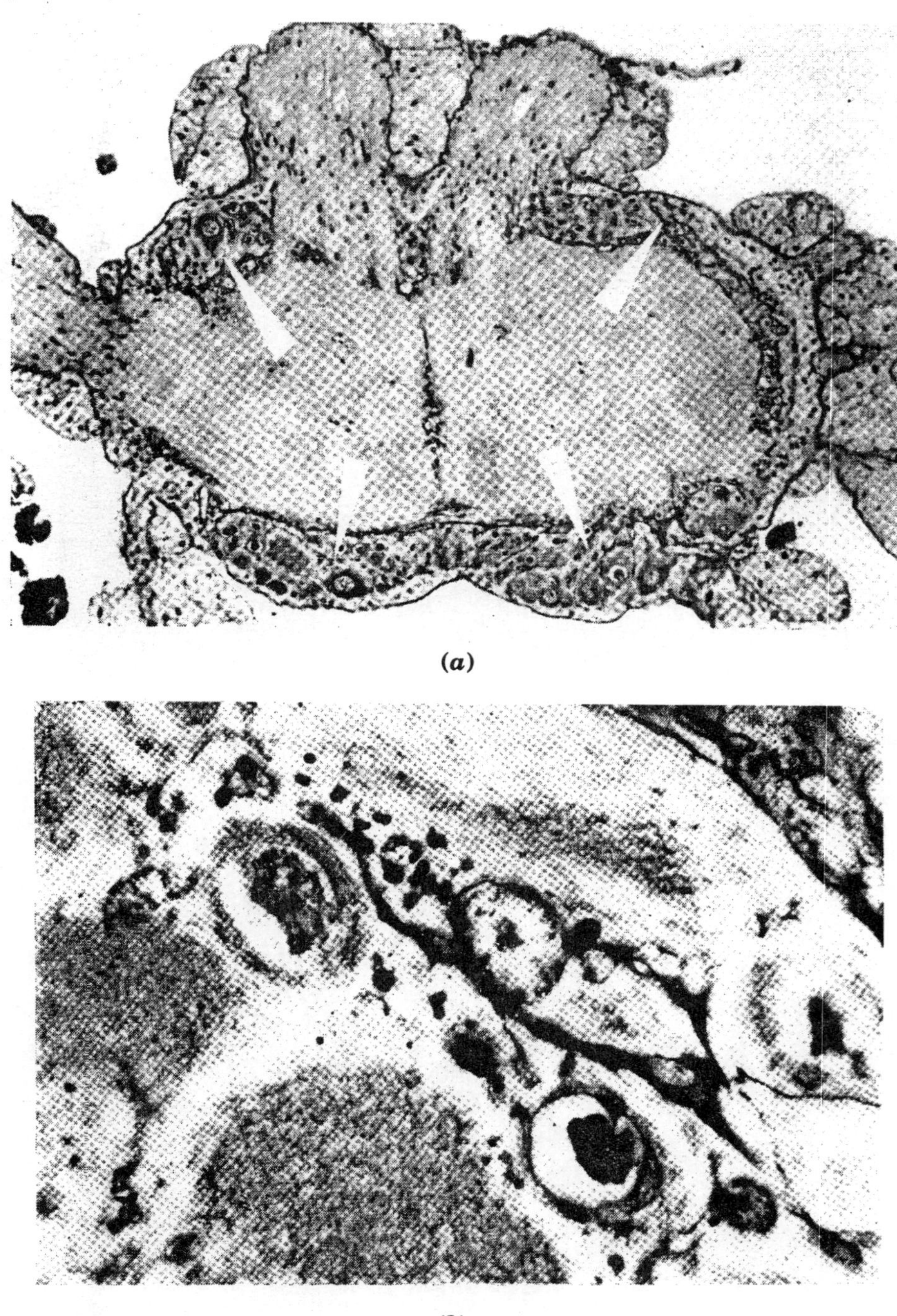

(*a*)

(*b*)

PLATE 36. Gliosecretion in *Periplaneta americana,* Xth instar nymph. (*a*) Frontal section of the mesothoracic ganglion. Arrows indicate the main accumulations of the gliosecretion. 120 : I. (*b*) Detail from the above, 1500 : I. Interneuronal gliosecretion around gliocyte nuclei and trabeculocytes with trabeculae.

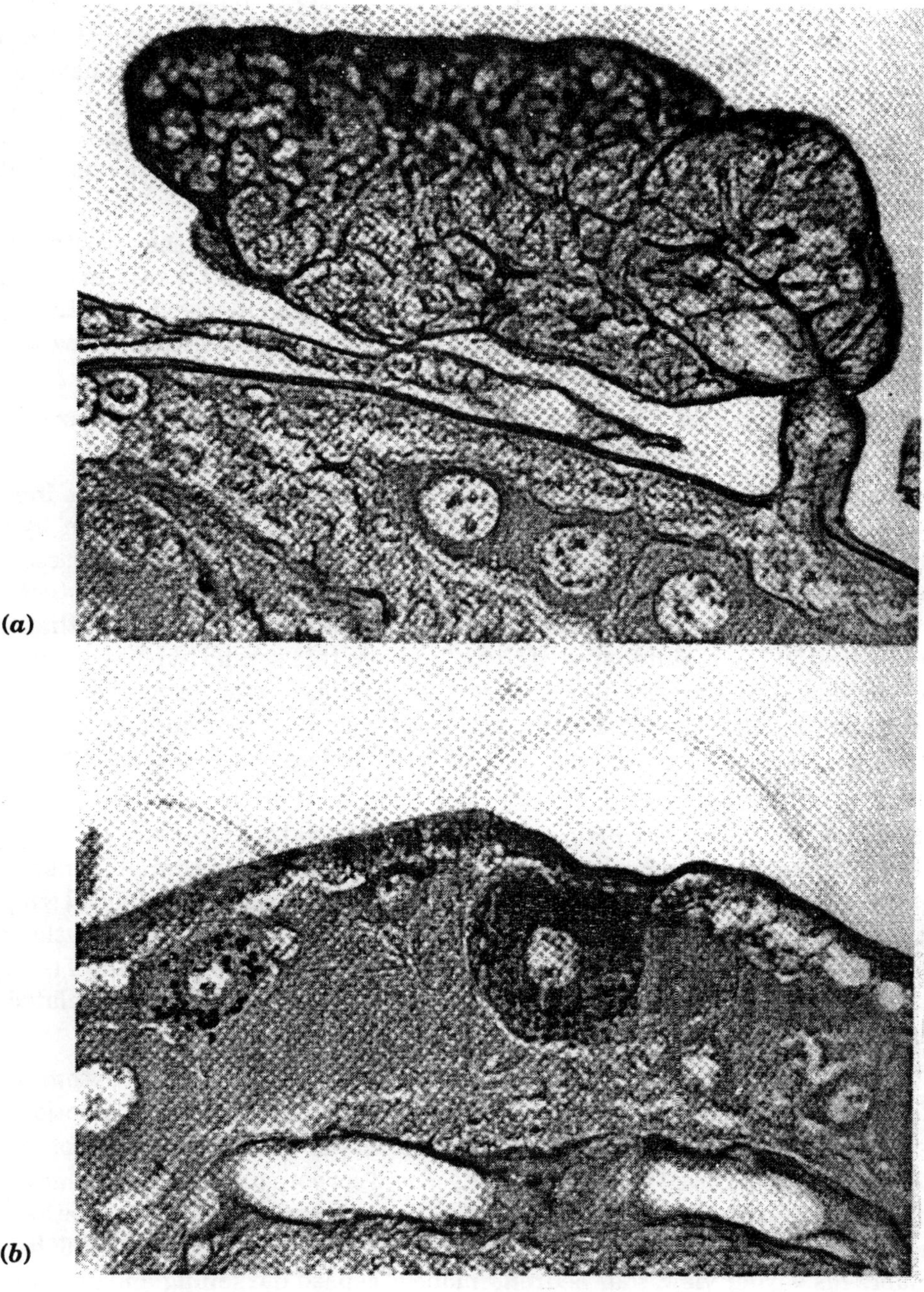

PLATE 37. (*a*) Ganglionic metameric organ of the IInd abdominal ganglion of Gryllotalpa gryllotalpa L. with lateral part of the ganglion. Medium section. Paraldehydefuchsine after Bouin fixation. × 675. (From Musko, Novak, 1975.) (*b*) Lateral part of the IInd abdominal ganglion of *Gryllotalpa gryllotalpa* L. with two different kinds of neurosecretory cells. The same preparation and magnification, from the same paper as (*a*).

Oezbas and Hodgson (1958) have shown that an extract of the corpora cardiaca decreases the 'spontaneous' electrical activity in isolated ganglia of the ventral nerve cord in *Periplaneta americana,* whereas extracts of the brain, abdominal ganglia and corpora allata have no effect. Cutting the nerve between the brain and corpora cardiaca made extracts of the latter less effective. When the corpora cardiaca extracts were injected into living animals, movements became un-coordinated and stereotyped locomotor behaviour appeared several hours after the treatment and lasted for between one and three days.

Harker (1955, 1956, 1958, 1959, 1961) studied the endocrine control of the activity rhythm of the same species, using parabiosis and transplantation experiments. Her results showed that the suboesophageal ganglion was the source of the active substance involved. Injecting extracts of this ganglion into decapitated insects produced a period of increased activity, the duration of which depended on the time of day when the injection was carried out.

The role of neurosecretion and the interaction of other parts of the endocrine and nervous system in controlling circadian rhythms of activity

These were studied in detail by Harker (1960a, b, 1961) in *Periplaneta americana.* Immediate regulation is effected by cyclical release in a circadian rhythm of secretion by paired neurosecretory cells situated on the ventral and slightly lateral surface of the suboesophageal ganglion. The neurosecretion which induces this phase of activity is produced daily in correlation with the onset of darkness. However, this dependence is not autonomous. It seems that another substance is involved in the maintenance of their function; this second substance comes from the corpora cardiaca and enters the suboesophageal ganglion via the corpus allatum-suboesophageal ganglion nerve, in the form of methylene-blue-positive granules. Once induced, the rhythm of neurosecretory activity continues for a number of days, even if the nervous connections of the ganglion are broken.

Midgut tumours were produced artificially by the same author (Harker, 1958, 1960b) in *Periplaneta americana,* by experimentally breaking the above rhythms. Transplantable metastasing tumours were induced by the repeated implantation of suboesophageal ganglia out of phase with the animal's own ganglion and by breaking the neurosecretory cycle through specific environmental light conditions. It has been suggested that the production of tumours by recurrent nerve severance, as shown by B. Scharrer (1945) (cf. p. 306), is also related to the timing of sub-oesophageal ganglion secretory activity (Harker, 1960b).

Mothes (1960) carried out a detailed study of diurnal activity and the rhythm of colour change in *Carausius morosus,* using the *in vitro* colour change test already mentioned. The quantities of neurohormones C_1 and D_1 in Ringer extracts of the brain, the suboesophageal ganglion, the ventral nerve cord, the corpora cardiaca and corpora allata, and the haemolymph were calculated. These experiments produced the following results: The concentration of both hormones in the haemo-lymph was very high at night and low by day (cf. Gersch, 1960b). Gersch (1960b) considers his earlier view, that neurohormone C caused darkening and neurohormone D lightening of the body, to have been disproved. Rather the period of decreased activity and light cuticle is when the quantity of the neurohormones in the haemolymph is low (*i.e.* by day), while the night is the period of high activity and dark cuticle.

Brain extracts prepared from individuals at the end of the night (early in the morning) showed low activity, whereas the activity of extracts increased during the day.

Extracts of the suboesophageal ganglion and the ventral nerve cord showed the same activity (high at night and low by day) as the haemolymph, whereas extracts of the corpora cardiaca and corpora allata showed no diurnal change in activity.

Several weeks in reversed conditions of day and night caused the animals to change their rhythm of activity, the activity of extracts also became reversed; *i.e.* low in the cerebral ganglion while high in the haemolymph and ventral ganglia by day and *vice versa* at night. The activity of extracts of the corpora cardiaca and corpora allata again remained unchanged.

Harker's experiments were discussed in a review by Truman (1971), who concluded that the results had not been confirmed by other authors. The site of the cockroach 'clock' is now thought to be in the optic lobes of the cerebral ganglion in these insects, as indicated by the experiments of Nishiitsutsuji-Uwo and Pittendrigh (1968). Extirpation of the lobes results in loss of rhythmicity, not due to surgical trauma, in the experimental insects. Blackening the eyes or severing the optic nerve has a similar effect, even if the ocelli are intact. On the basis of later studies (Brady, 1960), nervous control with synaptic transmission is presumed in this case.

The Eclosion Hormone

Another type of circadian rhythm is the one influencing the incidence of eclosion (*i.e.* adult emergence) and moulting. Here again, the control mechanism (the 'clock') seems to be localized in the cerebral ganglion. In this case, phase setting is accomplished by nervous perception of light by the brain, which, in turn, activates release of the eclosion hormone by neurosecretory cells. The hormone is produced by the brain and reaches the circulation by means of the corpora cardiaca, where it accumulates and is emptied into the haemolymph on the day of eclosion. As well as freeing the adult body from the pupal exuviae, the eclosion hormone is concerned with the switch from pupal to imaginal behaviour, even if the pupal cuticle is removed experimentally at an earlier stage. The eclosion hormone has been found in the silk moth *Antheraea pernyi* and in the tobacco hornworm *Manduca sexta* (Truman, 1971a, b, 1972a).

Circadian hormonal timing of *larval* ecdyses in the *same* species is somewhat different. Ecdyses occur at specific times of the day which are instar and species specific and depend upon the photoperiod and temperature in the relevant determination period. Determination occurs by influencing the timing of production of the hormones which initiate moulting. It seems to be AH which is responsible for production of the other two metamorphosis hormones and hence for the timing of the moulting process (Truman, 1972b).

Neurohormones and Hardening of the Cuticle

Immediately after ecdysis, the insect cuticle is mostly soft and white. It was shown in a number of insects that the process of hardening and darkening the cuticle is also under neurohumoral control from the brain through a neurohormone known as *bursicon.* On the basis of earlier findings by Fraenkel (1935) and Cottrell (1962), bursicon was found to act in adult cyclorrhaphous flies immediately after emergence (Fraenkel and Hsiao, 1963, 1965). It is claimed that bursicon is produced both by the median nsc of the cerebral ganglion and by the compound thoracic ganglion (from which it has been extracted).

It has been concluded that it is a polypeptide with a molecular weight of *c.* 40,000. It has also been shown in *Periplaneta americana,* in *Tenebrio molitor, Galleria mellonella, Oncopeltus fasciatus* and *Schistocerca gregaria.*

The hormone is found in the haemolymph several minutes after ecdysis, reaches its maximum 30 to 60 min. later and disappears completely in a matter of hours. Similar bursicon timing was observed in cockroaches (Minks, 1965, 1967), where it seems to occur in most of the nerve ganglia, but is supposed to be released in the largest amounts from the last abdominal ganglion.

Fogal and Fraenkel (1969) claim that it also influences melanization and continued cuticle formation in the adult fleshfly *Sarcophaga bullata*. Some questions regarding the action of bursicon still remain open, *e.g.* its relationship to AH and MH action, the hormonal dependencies of cuticle deposition, which start long before bursicon appears, in the preceding instar. The dependence of a single process like cuticle deposition on two quite different hormones (MH, a steroid, and bursicon, a peptide) is likewise not easy to understand. There are further cases in which no similar factor is evidently necessary, *e.g.* the ecdysis of isolated pupal abdomens in the experiments of Williams (1947, 1948), and other authors.

Several papers have been published on the effect of MH on pupariation in cyclorrhaphous flies. It was suggested that MH is not the only factor responsible for puparium formation, but that a brain neurohormone different from AH also participates. This hypothesis is based on the observation that if blowfly larvae are ligated after the critical period for MH, the anterior part of the body pupariates several hours sooner than the posterior part. If the posterior part is injected with haemolymph from a pupariating larva, or with active brain extract (in insect-Ringer), however, the posterior end can be induced to moult sooner than the anterior end Zdarek and Fraenkel, 1969). The authors conclude that a specific hormone, which accelerates puparium formation and the inhibition of tanning and potentiates the action of MH, is produced in the pars intercerebralis. This hormone is assumed to be different from both AH and bursicon.

It has been suggested that other processes of insect ontogenesis might also be under neurosecretory control, such as 'plasticization' (softening) of the cuticle in blood-sucking insects like *Rhodnius prolixus,* which facilitates distension of the abdomen and thereby the intake of larger amounts of blood. Neurosecretory axons of the abdominal nerve supplying the epidermis were found in this connection (Wigglesworth, 1970). Active absorption of the fluid filling the trachea before eclosion from the egg and in connection with larval ecdysis has also been considered as being under hormonal control, in which case bursicon might again be responsible (Wigglesworth, 1970).

Changes in the rate of heart-beat during postembryonic development were studied in detail by Roussel (1971) in *Locusta migratoria.* It was found to diminish progressively with the approach of metamorphosis and its course was similar in each instar: it increased at the beginning of the instar, reached the maximum about one-third of the way through the intermoult period and then decreased, attaining a minimum at the end of the second third and then rising again slightly towards ecdysis. It was faster in adult insects and was slightly higher in males than in females. It rose with the temperature up to a given limit (42°C) and was reduced by CO_2 and starvation. It was also affected by the corpora allata and, to a lesser degree, by the prothoracic glands. The implantation of nsc of the pars intercerebralis and of the cc had virtually no effect.

Neurohormones and the Colour Change

Two types of colour change occur in animals in general. In one change of colour (or in insects mostly only the change in intensity of colour) results either from the concentration or the dispersion of pigment granules contained in special pigment cells (or in all epidermal cells as in *Carausius morosus),* or from the concentration or expansion of melanophores as in the tracheal air sacs of *Corethra* larvae *(Chaoborus crystallinus).* This type of colour change is a short-term process which is usually cyclic, mainly in a 24-hour period. It results in a rapid adaptation by the insects to the colour or light intensity of the environment. This type of adaptation is called physiological colour change. By contrast, the term morphological colour change refers to long-term or irrevocable changes in the colour of the integument. These are produced in the course of ontogenetic development, often as a response by the organism to specific conditions of temperature, humidity, light or other conditions. Such changes have an equally adaptive character. There are, of course, examples of colour change which are intermediate between these two types. In any case, both physiological and morphological colour changes are regulated by hormones.

All types of colour change are of significance from the aspect of natural selection. The change may consist in the adoption of protective (cryptic) colouration in response to circadian or other environmental changes, *e.g.* changes caused by differences in humidity (Joly, 1968), temperature, or it may be related to thermoregulation, or to display associated with mating and parental behaviour (see Highnam, 1969). The rhythmicity and intensity of the change is usually closely correlated with the relevant environmental changes.

Physiological colour change

Colour change of the stick-insect epidermis

Relatively few cases of physiological colour change occur in insects, in contrast to crustaceans where it is a very common feature. Only two cases have so far been described in some details. One of these is the 24-hour change in the intensity of colouration in *Carausius morosus.* This colour change is caused by the movement of pigment granules inside the epidermal cells. Four different pigments are found in *Carausius:* a green and a yellow pigment which are distributed uniformly in the cytoplasm of the epidermal cells, an orange pigment dispersing and concentrating horizontally on a plane with the nucleus, and a dark brown pigment moving in a vertical direction.

By day the dark brown pigment is found at the internal end of each epidermal cell while the orange pigment is concentrated around the nucleus. This explains the light colouration of the insect in day-time. At night, the dark brown pigment moves towards the outer end of the cell and the orange granules spread out evenly, horizontally across the cell. As a result the integument of the insect darkens.

This movement of pigments is related to the light conditions. If stick-insects are illuminated at night they become lighter and, by contrast, stick-insects kept in darkness during the day become darker in colour. These changes, however, only become apparent after a while. If the insects are kept in constant darkness the rhythm of colour change takes several weeks to disappear completely. About the same length of time is necessary to reverse the rhythm by keeping the insects in the dark by day and in the light by night. The reactions of the insects to humidity are comparable with their reactions to changes in illumination. In a high relative humidity the colouration becomes darker, and becomes lighter in a low humidity. The colour change is clearly adaptive. Stick-insects, sitting on the twigs of a shrub and with a remarkable cryptic body form and behaviour, are very much less noticeable to their enemies if they are lighter by day and darker at night.

The possibility of a humoral mechanism for the colour change in *Carausius* was first suggested by Giersberg (1928) and later by Atzler (1930). A detailed study of the problem was carried out by Janda (1934, 1936), who showed histologically that there was no connection between the nerve endings and the epidermal cells. In this way he proved that a direct nervous influence on colour change was impossible. In addition, he carried out a series of ligature experiments. When a nymph was ligatured at the middle of the body so as to prevent haemolymph passing from the anterior to the posterior part of the body, the diurnal colour change continued only in the anterior part of the body while the posterior part remained constantly light in colour. The ligature caused no apparent harm to the insect even after many days. When the ligature was loosened, the rhythm of colour change reappeared. After extirpation of the brain the whole body became an even light colour. Janda concluded from this that the colour change is initiated by a substance which is released from the brain and transported by the haemolymph. This was confirmed by experiments in which a piece of integument darkened when transplanted from a light-adapted to a dark-adapted insect and, conversely, a piece from a dark-adapted to a light-adapted specimen became lighter in colour.

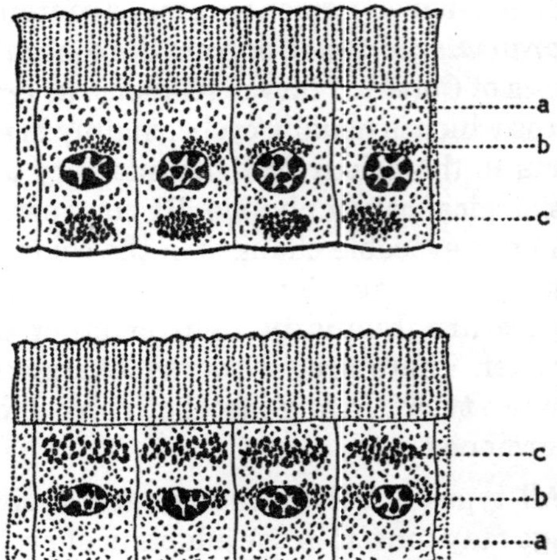

Fig. 1.5. *Carausius morosus*—pigment movement in the epidermal cells. *Above*—light adaptation; *below*—dark adaptation, (*a*) green and yellow pigment, motionless, *(b)* orange-red pigment, moving in a horizontal level, (*c*) brown pigment, moving vertically. (After Giersberg, 1928.)

The conclusions of Janda were confirmed and extended by the work of Dupont-Raabe (1951a, b, cf. 1959). She showed that specimens made light in colour by extirpation of the brain become very dark if implanted with the brains of dark-adapted individuals. The same effect could be caused by injecting extracts prepared from such brains. The high activity of the extracts was evident from the discovery that even one-twentieth of the amount obtained from a single brain was sufficient to induce darkening. The effect of an extract of the suboesophageal ganglion was equally intensive. A similar, but weaker effect was caused by extracts of the corpora cardiaca and of the ganglia of the ventral nerve cord. Removal of various parts of the brain has demonstrated the interesting fact that the active principle is produced neither by the neurosecretory cells of the pars intercerebralis nor by those of the lateral groups of the protocerebrum, but by specific neurosecretory cells in the tritocerebrum. The existence of these cells was shown by a special histological technique. Raabe (1959) named this substance neurohormone C and concluded that the active substance could not be considered identical with AH .This conclusion, however, conflicts with results obtained by Gersch (1961).

Another active substance with a similar but weaker effect was obtained, also by Raabe (1949-58), from the corpora cardiaca and was designated substance (neurohormone) A. Its identity with the A-substance obtained from the eye stalks of various Crustacea by Carlisle and his co-workers (1955) has been suggested (Raabe, 1959), but this is still inconclusive (Carlisle and

Knowles, 1959). It does, however, seem to be identical with an active substance obtained from the corpora cardiaca of various insects by Hanstrom (1936, 1940) and M. Thomsen (1943). This latter substance activates the chromatophores of crustaceans. The relationship of these active substances to the C and D neurohormones of Gersch is not yet clear, because of the inadequacy of the technical data; although according to Raabe (1959) it seems possible to distinguish between at least some of them.

In a series of papers, Gersch and his collaborators have been able to show that the myotropic substances C and D mentioned above are identical with the substances affecting colour change in *Carausius.*

A very sensitive *in vitro* test was developed, which made possible a detailed analysis of chromatophorotropic activity. Small pieces *of Carausius* integument in insect-Ringer proved to respond readily to neurohormone extracts in far lower concentrations than these which produced a response in whole animals. With this technique it could be shown that neurohormone C_1 causes an intense darkening of the integument whereas D_1 produces less intense darkening at low concentrations and a lightening at higher concentrations (Gersch and Mothes, 1956).

The term 'normal solution' was employed to refer to the quantity of both neurohormones obtained from 10 specimens (*i.e.* from the cerebral ganglion including the corpora cardiaca and corpora allata, from the suboesophageal ganglion, the whole ventral nerve cord and as much haemolymph as possible from each specimen) dissolved in 1 ml of Ringer solution. The term 'biological unit' was used to refer to the smallest quantity of either hormone which, dissolved in 1 ml Ringer solution, still produced a darkening of the integument. About 1,000 biological units of the two hormones together were found in one specimen of *Periplaneta americana* tested with *Carausius* integument. As 50 μg of crystallized neurohormone D_1 were recovered from an extract of 3,200 cockroaches, each specimen appears to contain 0-015 μg.

A colour change similar in type to the one in stick-insects also occurs in Mantids. It was shown that light stimuli reach the nervous system solely through the eye. If the eye stalks are cut, the animals do not respond to light changes. Hormonal regulation has not yet been fully elucidated in this case, however (Wigglesworth, 1965).

The *in vitro* test was used to show that neurohormone D_1 is located in the pars intercerebralis (although similar quantities occur also in other parts of the nervous system) whereas C_1 is found in the deuterocerebral and tritocerebral parts of the brain. This would suggest that C_1 and D_1 are identical with the C and A, respectively of Raabe. The lack of an effect of extracts from the pars intercerebralis in Raabe's experiments is possibly explained by the fact that at high concentrations its effect is to lighten the integument (Raabe used darkening as a measure of activity). As already mentioned, neurohormones C_2 and D_2 were found to have no effect on colour change.

Colour change in Corethra Larvae

Another example of a physiological colour change which has been thoroughly investigated occurs in the larvae of Corethra *(Chaoborus cristallinus).* In these transparent aquatic larvae the colour change is limited to the surface of the air-sacs. There are two pairs of kidney-shaped sacs (in the thorax and in the seventh abdominal segment) of tracheal origin and with a hydrostatic function. All four sacs are covered with cells containing a black pigment and which are capable of considerable distension and contraction. Against a dark background (e.g. the bottom of the pool in which the *larva lives)* the chromatophores are distended so that they cover the entire surface of the air-sacs. When a larva is transferred to a light background the chromatophores quickly contract, becoming rounded to form only small black spots on the surface of the bladders. Thus the colour change is caused by the movement of whole cells, the melanophores. Research during the last decade has shown that, in this case also, a humoral control is involved (Kopenetz, 1949; Hadorn, 1949; cf. Raabe, 1959).

Methodical research on the mechanism of the colour change in *Chaoborus* was carried out by Gersch (1956; cf. 1960a, b) and his school. A series of experiments, involving the stimulation or removal of individual ganglia in the nerve cord as well as the injection of extracts from various parts of the nervous system, have shown an interaction of at least three neurohormones with partially antagonistic effects. Sub-sequent separation of extracts with paper chromatography and electrophoresis has shown that the principal substances concerned are identical with neurohormones C_1 and D_1 and acetylcholine. The effect of these substances on the melanophores corresponds with their effect on the movement of pigment in *Carausius.* Neurohormone C_1 causes expansion of the melanophores and therefore a darkening of the air-sacs. The effect of acetylcholine is very similar, whereas neurohormone D produces a lightening of the bladder colour.

Although physiological colour change is a common phenomenon in Malacostraca, in some Cephalopoda and in vertebrates, it has only been observed in a few additional species of insects. In none of these has the physiology yet been studied. Thus it has been found in some grasshoppers (Acrididae) including *Kosciuskola tristis* (Key and Day, 1954), in the family Oedipodidae, in some mantids and in many other phasmids besides *Carausius.* There is little doubt that neurohormones are again involved in the colour change of these species.

Principles affecting Colour Change

Principles affecting colour change from both insects and crustaceans were isolated by the above-mentioned methods and were tested on crustacean and insect chromatophores, and muscle contraction rhythms in a detailed comparative study by Gersch, Unger, Fischer and Kapitza (1964). Paper chromatography and paper electrophoresis of various parts of the neurosecretory system of both crustaceans *(Leander adspersus* and *Crangon vulgaris)* and insects *(Periplaneta americana* and *Carausius morosus)* showed the same four different substances in each. group: two neurohormones, C_1 and D_1, which already have been found to be of a peptide nature, acetylcholine and serotonin. The principles obtained from insects were in complete agreement with those from crustaceans, both as regards their physico-chemical properties and their effects in biological tests on chromatophores in crustaceans, on pigment movement in the epidermis of *Carausius* and on the heart rate in *Periplaneta.* On this basis, suggestions for the chemical identification of various colour change factors described in crustaceans by earlier authors were made.

Morphological Colour Change

Examples of morphological colour change are much more numerous in insects than physiological colour change. In addition to the clearly adaptive true colour change occurring in some Orthoptera and Lepidoptera, changes in the size and form of the pigment (melanin) pattern which occur in many insect groups such as grasshoppers, bugs, moths, beetles and wasps may also be included. It can be supposed that hormones are involved in these changes also, even though not all have yet been studied from this angle.

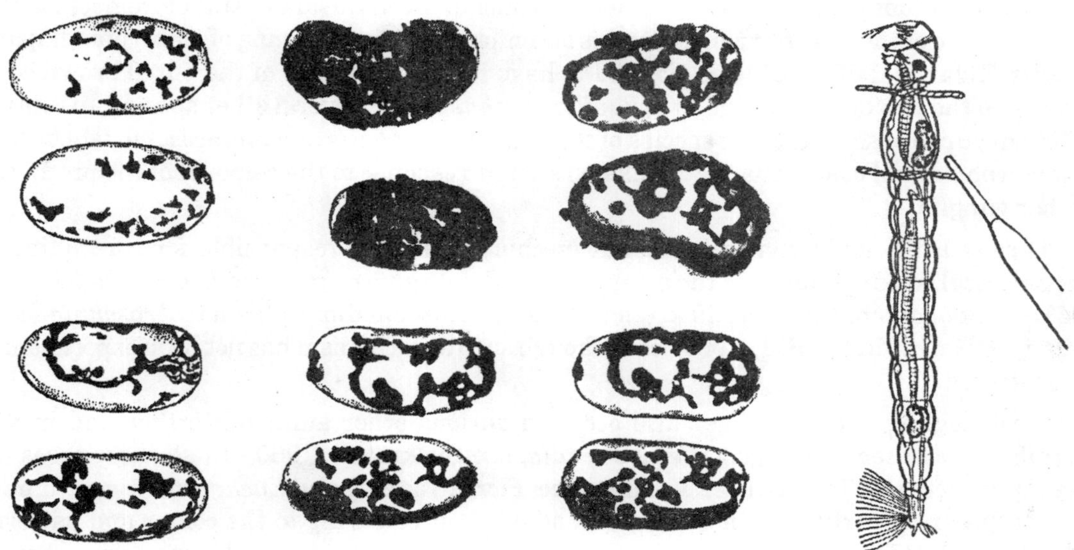

Fig. 1.6. The effect of the neurohormones on the melanophores of the tracheal sacs in *Corethra* larvae. *Above*—diagram of the larva, into the ligatured thorax of which the hormone is injected, the untreated hind pair of sacs serves as a control (right pair in each series). Upper series-insect-Ringer control; middle series-neurohormone C (left pair of bladders); lower series-neurohormone D. (After Gersch, 1957.)

Some examples of morphological colour change have already been mentioned in connection with the metamorphosis hormones which control them indirectly. For example, the complicated colour change preceding metamorphosis in *Cerura vinula* caterpillars is a part of morphogenesis and controlled indirectly by the activity of MH. Similarly, the colour changes connected with the phase polymorphism of migratory locusts are conditioned by JH.

A quite different mechanism appears to be involved in the adaptive colouration of some butterfly pupae. This phenomenon takes the form of a partial adaptation to the background colour. Thus it has been known for a long time that the colour of the pupal cuticle in some species of the families Papilionidae and Pieridae is dependent on the background colour experienced by the individual immediately prior to pupation. Ohnishi and Hidaka (1955) have recently shown that a similar phenomenon in several East Asiatic species of the genus *Papilio* is Immorally conditioned. For example, pupae which formed on young green twigs are also green whereas those which formed on older twigs with a brown bark (on the plant *Pcncirus trifoliatd)* are brown.

In a series of experiments involving ligaturing, extirpation of individual ganglia and nerve interruptions with caterpillars before pupation Hidaka (1956) found that colour adaptation was conditioned by optical sensations requiring an undamaged cerebral-prothoracic-suboeso-phageal ganglionic complex.

The basic colour of the pupa is green, which may or may not be supplemented by brown depending on the colour of the background. If any of the aforementioned ganglia are removed or their connections broken, the brown colour fails to develop even in pupae determined for brown.

Both a nervous (optical) impulse and a neurohormonal mechanism are therefore necessary for the appearance of the brown colouration; this is similar to the conditioning of embryonic diapause. Recently Hidaka (1960, 1961) has claimed to have found the source of the pupal brown colour hormone in the prothoracic ganglion of *Papilio xanthus* L. and several other species. In contrast, the brown pupae of several other species of the genera *Nymphalis* and *Malacosoma* which do not show morphological colour changes, make no similar response to the removal of the prothoracic or other ganglia.

There is little doubt that a hormonal mechanism is also responsible for controlling the morphological colour changes in the caterpillars of *Laphygma exigua* and *L. exempta* (cf. Faure, 1943) and *Spodoptera* (Mathee, 1945) as well as the seasonal dimorphism in *Araschnia levana* (Suffert, 1924a, b; Muller, H. J., 1960), even though such a mechanism has not been experimentally demonstrated.

Morphological colour changes also occur in various other adult butterflies and in some caterpillars (of other species, e.g. *Cerura vinula,* see Buckmann, 1960). In all these cases they are adaptive (cryptic) in character. For instance, *Pieris, Vanessa* and *Luehdorfia japonica* pupae and caterpillars of certain other species can adapt their colouring to the colour and pattern of the background (Hidaka *et al.*, 1971). Similar adaptation has been observed in many grasshoppers (Wigglesworth, 1971). In some cases, the hormonal mechanism of the colour change has almost been demonstrated (Raabe, 1964); in many others, e.g. the spring and summer forms of the adult butterfly *Araschnia levana,* this mechanism can feasibly be presumed (cf. Wigglesworth, 1970). This does not apply to the newly described colour change in the *Hercules* Beetle *(Dynastes hercules)* where an endocrine mechanism is not probable.

A connection has been shown to exist between JH activity and the shape and size of the ventral black spots on the abdomen of the lygaeid bug *Oncopeltus fasciatus* (Novak, 1959d). Symmetrical black pigment spots develop on the ventral side of the abdomen of adult insects round two foci which correspond in position to the centres of the paired gonads. The differences between male and female in the shape and size of these spots correspond to the differences in the shape and size of the gonads. The size of the spots decreased in insects kept in a high temperature and their change in shape was consistent.

The spots were largest at room temperature and practically disappeared at 30°C. Low doses of JH, or supplying the JH by the implantation of corpora allata at the last larval instar, were found to produce the same effect as the high temperature. The effect both in increased temperature and the hormone can therefore be explained as a very weak form of progressive metathetely (a delay in development of the imaginal characters produced by the positive influence of other larval parts of the body, including the product of the prothoracic glands). The hypothesis was suggested that the black spots are controlled by the two sources of a determining substance, which spread out from the parts of the gonads where ripe sex cells first appear. However, castration experiments showed that removal of the gonads at the beginning of the last larval instar had no effect on the development of the spots.

Influences on Water Balance

As suggested by van der Kloot (1960), the close analogy between the insect neurosecretory system of the pars intercerebralis-corpora cardiaca and the vertebrate neurosecretory centres of the hypothalamus-neurohypophysis make it probable *a priori* that the neurosecretory cells of the insect brain have a role in water metabolism. This has been shown by a number of authors and from our present knowledge there is little doubt that AH itself influences water balance.

It has been suggested that the action of AH on membrane activity, which is manifested in its diuretic effect on water metabolism, may also be the cause of its stimulation of, perhaps, all the glands of the insect organism. Augmentation of the membrane activity of various cellular differentiations, including the nucleus, in other parts of the organism, may likewise explain its general activating character, as in the termination of diapause.

The products of other nsc, occurring both in the cerebral ganglion and in other parts of the nervous system, were also shown to affect water balance. On the basis of a histological investigation of the neurosecretory system of the cockroach *Blaberus giganteus,* Wall and Ralph (1962) concluded that a diuretic factor was released by the type A-cells of the pars intercerebralis and the thoracic ganglia. Under conditions causing dehydration, these cells produced greater than normal amounts of stainable granules. A diuretic principle was also reported by Nunez (1963) in *Rhodnius prolixus,* which he supposed was produced by the nsc of the complex thoracic ganglion.

Maddrell (1963, 1966) carried out a very detailed study of water metabolism in the blood-sucking bug *Rhodnius prolixus.* Water excretion in

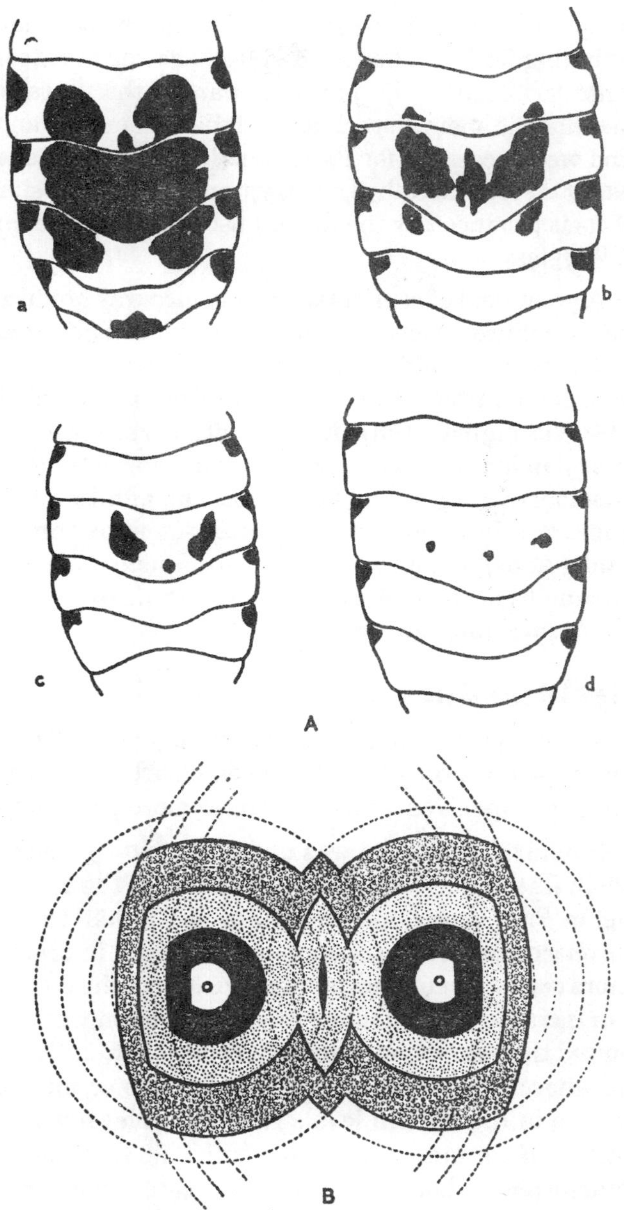

Fig. 1.7. A, The black pattern of the abdominal sternites in *Oncopeltus fasciatus* females in four different rearing temperatures: (*a*) room temperature; (*b*) 25°C; (*c*) 28°C; (*d*) 32°C. B, Diagram of the geometrical pattern produced by two factors spreading from two foci with equal and proportionate velocity, inhibited laterally. The areas with equal concentration of the factor are marked in the same way. Compare with the figure above. (From Novak, 1955d.)

this species increased markedly after each blood meal, starting a few minutes after feeding and continuing for 3 to 4 hours. Maddrell showed that this was caused by a neurohormone secreted by nsc localized in the posterior part of the thoracic ganglion mass formed by fusion of the mesothoracic, metathoracic and all the abdominal ganglia. Distension of the abdomen by imbibed blood was necessary for its production, and its release was inhibited if the ventral nerve cord was cut. Serotonin (5-hydroxytryptamine) induced similar activity. A neurohormone with similar effects is produced by the medial nsc of the cerebral ganglion in the pyrrhocorid bug *Dysdercus* (cf. Wigglesworth, 1970).

A great deal of confirmatory evidence was produced by different authors from various other insects. A diuretic neurohormone was found in *Schistocerca, Locusta, Gryllus, Anisotarsus* larvae, etc. The removal of the nsc of the pars intercerebralis resulted, in these species, in distension of the abdomen from water retention (Highnam *et al.,* 1965; Girardie, 1965; Cazal and Girardie, 1968). Vietinghoff (1967) made a detailed study of water metabolism and its regulation by the rectal glands in *Carausius morosus,* and Mordue (1961, 1970, 1971) in locusts. Mordue succeeded in demonstrating both a diuretic and an antidiuretic neurohormone in *Locusta migratoria.* An antidiuretic hormone, causing water retention through increased reabsorption of water by the rectal glands, was also found in *Periplaneta americana* (Wall, 1965, 1967) and an antidiuretic hormone from the metameric neurohaemal organs was reported by Besse and Cazal (1968) in cockroaches and stick-insects, etc.

Effects on Gut Proteinase Activity

The importance of the neurosecretory products from the brain for normal activity of the gut proteinases in *Calliphora* has been clearly demonstrated. It is not known whether a special hormone is involved or, as is perhaps more probable, this is again an effect of AH.

Similar dependence was demonstrated by Dadd (1961) in *Tenebrio molitor* and by Slama (1964), who studied the effects of removal of the brain nsc and the cc on the haemolymph protein level in *Pyrrhocoris apterus* adults. Hill (1962) and Highnam (1962) made a detailed study of neurohormonal control of protein metabolism in *Schistocerca gregaria.* It was shown that various factors causing neurosecretion release from the nsc (e.g. hyperactivity, copulation, etc.) resulted in an increase in the haemolymph protein concentration, while conditions inhibiting its release reduced the haemolymph protein concentration. This occurred after cauterization of the medial nsc, whereas the reimplantation of active cc into such insects led to a rapid, though brief, increase in the protein level. The protein level was likewise raised by removal of parts of the ovaries (Highnam, Lusis and Hill, 1963). Protein metabolism and the influence of the metamorphosis hormone was investigated in a series of studies by Janda and Slama (1965), Janda (1970), Janda and Socha (1970), Janda and Sehnal (1970).

Activation of the Endocrine Glands

Perhaps the most familiar and most fully analysed effect of neuro-hormones in insects is the effect of AH on the activity of the prothoracic glands. The findings of Gersch (1961) have produced a new view on this effect. By injecting purified *Periplaneta americana* brain neurohormone extract into ligated *Calliphora* larvae, as in the *Calliphora* test, simultaneously implanting inactive ring glands, Gersch (1962) was able to induce pupariation. Without simultaneous ring gland implantation this did not occur. A later analysis led the author and his co-workers to distinguish between neurohormone D and the two AH components. Brauer (1973), in the same laboratory, showed that AH I (the high molecular weight component) strongly stimulated RNA synthesis in the pgl in the second half of the last larval instar of *Periplaneta*. AH II was inactive in this respect. At first, mRNA production was stimulated, while in later phases ribosomal RNA was mainly synthesized. A number of other authors had previously demonstrated an increase in RNA synthesis in the pgl of other insects, in the corresponding periods, by means of autoradiography (Oberlander and Schneiderman, 1963; Krishnakumaran *et al.*, 1965; Kobayashi *et al.*, 1968).

Much less has been published on the way in which the brain neurohormones act on the corpora allata, or possibly inactivate them, as assumed by some authors, e.g. Engelmann (1965), etc. (cf. Highnam, 1968). In principle, however, it does not seem to differ from the mechanism of activation of the pgl.

Effect on the Gonads

That AH is necessary for the development of ovaries in *Calliphora*, and no doubt also in most other insects, has been shown by E. Thomsen. An increase in the amounts of neurosecretory granules in the neurosecretory cells of the pars intercerebralis in the period shortly before and during egg-laying has been observed in several insect species, e.g. the stick-insects *Carausius morosus* and *Clitumnus* (Dupont-Raabe, 1951, 1952), *Bombyx mori* (Arvy, Bounhiol and Gabe, 1953), *Calotermes flavicollis* (Noirot, 1957) and in *Iphita limbata* (Nayar, 1958). The role of a brain hormone in the development of the gonads therefore seems to be quite definite (cf. Raabe, 1957).

Another interesting relationship between the neurosecretory cells and the ovaries has been demonstrated by Scharrer (1955a, b) in the cockroach *Leucophaea maderae*. A striking change in the appearance of the type B-cells occurred in adult females which were castrated between the Illrd and Vllth instars. In these cells the granules increased in size and number and stained green with paraldehyde fuchsin and Masson-Foot. The neurosecretory material in the other cells remained unchanged. Scharrer concluded that there was a humoral relationship between the type B-cells and the ovaries, and therefore named them castration cells or ovariectomy-cells. No corresponding effects of castration were observed in males.

The effects of castration were also analysed by Gersch, Fischer, linger and Koch (1960) in *Periplaneta americana* and *Blatta orientate* (cf. Gersch, 1960). Following castration, neurosecretory cells of type B with granules staining green were also found in other ganglia of the ventral nerve cord. Similar cells were also found in the ventral nerve cord of *Chaoborus* larvae. There remains little doubt that they are of general occurrence in insects. The observed effects are not necessarily a proof of a specific endocrine principle of the ovaries influencing the neurosecretory cells concerned. It is much more probable that the changes observed in the 'castration' cells are the result of a falling off in the consumption of their secretion by the ovaries. The phenomenon is therefore probably similar to the observed hypertrophy of the corpora allata following ovariectomy. In this connection it is very interesting that Noirot (1957) has observed the appearance of new neurosecretory cells in the suboesophageal ganglion of functional reproductives in termites. Another possibility is that the green colouration observed might be connected with an increase in the RNA content of neurons with injured axons, as described by Cahn (1962). However, this finding requires further analysis.

New data on the Hormonal Control of Metabolism by Neuro-hormones

New data have been obtained by a number of authors in the last three years. They were recently summarized by Gersch (1964). Some of them are discussed in connection with the activation hormone. Among the most important, mention should be made of the finding by Steele (1963) of a diabetogenic factor in the corpora cardiaca of *Periplaneta americana,* reminiscent of that produced by the sinus gland in crustaceans. The injection of corpora cardiaca extracts (even 0.002 of the amount of one pair of corpora cardiaca) resulted in a significant increase in trehalose in the haemolymph. It is assumed that this effect depends on activation of the phosphorylase responsible for the decomposition of glycogen.

Ralph and McCarthy (1964) found an increase in the trehalose content in the haemolymph of adult *Periplaneta americana* after injecting them with saline extracts of both corpora cardiaca and brain. They conclude that the active factor of the corpora cardiaca is produced by the neurosecretory cells of the brain.

Effect on Saccharide Metabolism

A number of papers on the action of neurohormones affecting the haemolymph sugar concentration has been published. In a study of hyperglycaemic activity in *Periplaneta americana,* Steele (1963) showed that two hormones present in the corpora cardiaca were responsible for raising the haemolymph sugar concentration. Their effect seems to consist in conversion of the enzyme phosphorylase in the fat body from an inactive to an active form, which then catalyses glycogen decomposition, thereby increasing trehalose production. Similar investigations in locusts (Highnam, 1969; Cazal, 1971) showed that two hormones released from the corpora cardiaca had the same effect on phosphorylase activity. One (the less active) was produced in the cerebral ganglion and the other was produced by the glandular components of the corpora cardiaca. Some evidence of their interaction was found. Saccharide metabolism and the influence of various hormones was studied in detail by Nemec (1971), Divakar and Nemec (1973) and Nemec and Rohdendorf (1973).

Effect of Neurohormones on Electrical Activity

This effect on the nervous system in insects was studied by Gersch and Richter (1963). They showed that extracts of the brain and the ganglia of the ventral nerve cord, like those of the corpora cardiaca, release specific electrical impulses in the phallus nerve of *Periplaneta americana.* The separate application of the neurohormones C_1 and D_1 isolated by paper chromatography showed that D_1 possessed strong activity, while C_1 was inactive. The effect of these two substances on the autorhythmic electrical activity of the brain *of Periplaneta americana in situ* and of the isolated mesothoracic ganglion was studied by Strejikova, Servit and Novak (1965). It was found that neurohormone D_1 evoked hyperautorhythmia in both cases, while neurohormone C_1 inhibited the existing electrical activity after first inducing a transient rise.

THE PHYLOGENETIC ORIGIN OF NEUROHORMONES AND NEUROSECRETION

The common occurrence of neurosecretion in all multicellular animals (except perhaps Coelenterata) and the striking analogies with insect neurosecretion to be found in the neurosecretory system of the higher animals, leave little doubt of the common principle and phylogenetic origin of this function.

Several theories have been formulated regarding the phylogenetic relationship between the neurosecretory cell and the nerve cell. Hanstrom (1954) emphasized the basic similarity of the neurosecretory cells and the brain cells. From his detailed comparative research on the anatomy and histology of the nervous system in both invertebrates and vertebrates, he proposed the following sequence for the phylogeny of the typical neurosecretory cell: (1) typical neurone; (2) neurone with all the characteristics of a nerve cell but containing a limited number of neurosecretory granules; (3) a neurosecretory cell with axon, dendrites and all other characteristics of a typical nerve cell; (4) a typical neurosecretory cell with axon but without dendrites and the other properties of a nerve cell; and (5) a secretory cell of neurosecretory origin without any of the characteristics of a nerve cell from the medulla adrenalis of the higher vertebrates.

Clark (1956) based his theory on his investigations of the brain in the polychaete families Nephthiidae and Nereidae. He supposes that secretory activity was phylogenetically the primary function of the neurosecretory cells, and that these were only secondarily included into the mass of the brain. He regards the cerebral organs of the Nemertini, the frontal organs of Phyllopoda (Copepoda) and the X-organs of Malacostraca as evidence supporting his theory.

Gersch (1960), on the basis of his co-worker, Uhde's, work showing the existence of neurosecretory cells in Platyhelminthes (*Dendrocoelum* sp.), agrees with Hanstrom that the nervous system is primary. The neurosecretory cells developed only secondarily from the nerve cells and parallel with the development of the gland cells from the original ectodermal cells. Only after this did they gradually acquire the functional importance they now posses in the neurosecretory system of higher animals.

Important evidence of the secondary character of the neurosecretory cells is their occurrence in very varied parts of the nervous system and in different histological units.

A theory reconciling the two contradictory concepts of Hanstrom and Gersch on the one hand and Clark on the other has recently been put forward by Novak and Gutmann (1962). They have shown that the secretory granules found in the glia cells in the nervous system of cockroaches (earlier described as gliosomata by B. Scharrer, 1939) agree in all their staining properties with the typical neurosecretory granules of the type A-cells. The granules in the glia cells, of course, lack the other physiological features of neurosecretion and do not show any signs of cyclicity and hormonal activity. The authors suggest the term 'gliosecretion' for this particular type of secretory activity. They assume that the gliosecretory granules have the same chemical composition as the carrier substance in neurosecretion but lack the physiologically active neurohormones. Both gliosecretion and neurosecretion are assumed to have developed from the secretory activity of the primary ectoderm cells. The same origin is shared by the so-called trabeculae, a network produced by a particular layer of glia cells round the neuropile of the ganglia, as well as by the perineural lamella, and by the perineural lamella produced by the perineurium cells. The primary secretory product inside the cytoplasm of particular nerve cells then became the carrier substance for the phylogenetically later neurohumoral factors (specific to nerve cells) which thus developed into true neurohormones. A detailed cytochemical and electromicroscopic study of gliosecretory granules (gliosomata) has been made by Pipa *et al.* (1961, 1962).

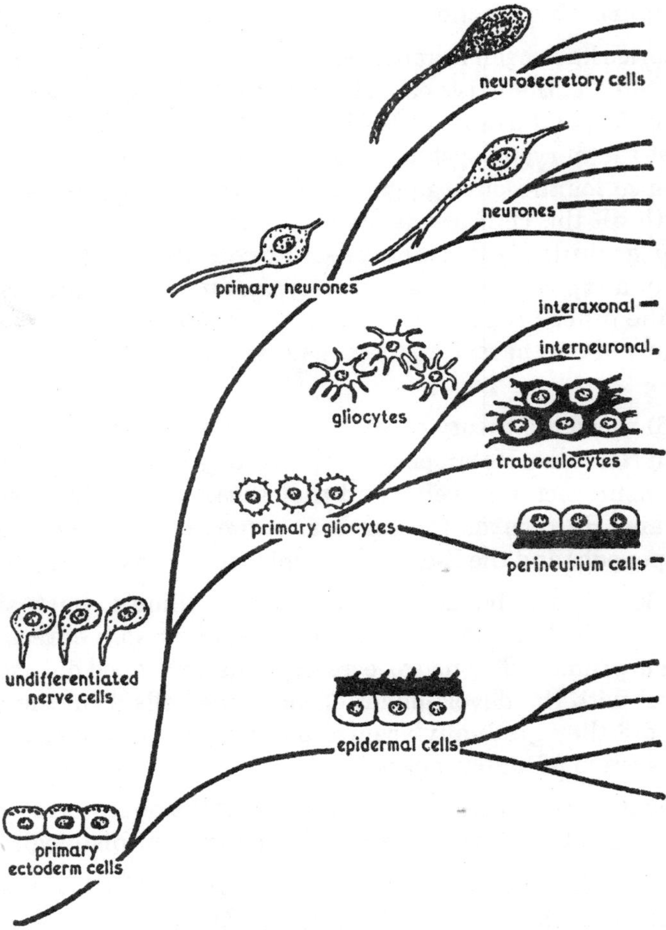

Fig. 1.8. Supposed phylogenetic relationship of the Gomori-positive structures in insects.

One component of the neurosecretory material, the carrier substance, would therefore be phylogenetically earlier than the generalized nerve cells, whereas the other, i.e. the neurohormones themselves, would be phylogenetically secondary having developed from the neurohumoral factors of the nerve cells as supposed by Hanstrom and Gersch.

2 The Protohormones

One of the most characteristic features of hormones is that they produce their effects in parts of the body other than those from which they are secreted, and reach their site of action by means of the body fluid. This should, however, be regarded as phylogenetically secondary and the result of a long and gradual process of evolution. Apart from the various true insect hormones so far considered, there have been found in arthropods a series of physiologically active substances which have their site of action at the place where they are produced and which are only occasionally carried in the body fluid when they may exert their effects elsewhere.

These substances no doubt represent a stage through which all the present hormones in every animal group had to pass in their phylogeny. It therefore seems useful to classify them as a special group of physiologically active substances, giving them the term protohormones or prehormones. They may be regarded as hormones *in statu nascendi.* Most of the present protohormones are substances whose evolution has stopped at this level, and which carry out their particular functions in the body of present-day animals, just as, for example, rudimentary light-sensitive spots instead of complex eyes are found contemporaneously with the very complicated compound eyes of arthropods, or as the primitive protozoans appear alongside the most complicated and phylogenetically advanced vertebrates.

The known protohormones may be divided into two groups: *the neurohumoral factors* which have been characterized together with the neurohormones in the previous chapter, and *the gene hormones,* which may be defined as substances which depend on particular genes and which condition particular genetic features, but can produce their effect by means of the circulation of the body fluid, even when introduced into animals which lack the appropriate genes.

NEUROHUMORAL FACTORS

The most widely occurring and the earliest known type of secretory products from nerve cells are the so-called neurohumours, i.e. physiologically active substances such as acetylcholine and noradrenaline in the nerve endings. The definition of hormones applies to these substances in that they are produced by specific cells, they are physiologically active and they produce their effect inside the organism. The definition does not apply, however, in that the substances work at their site of origin (in the nerve endings and synapses) without entering the body fluid under natural conditions. They thus correspond to the concept of protohormones as used in this book. Some of them are closely related, if not identical, with some of the neurohormones in their chemical composition. Also most, if not all, of the neurohormones will have probably passed through such a stage in the course of their phylogenetic development. In some instances it is difficult to decide whether a particular active substance isolated from nervous tissue (e.g. acetylcholine) has the character of a neurohumoral factor or perhaps that of a neurohormone produced and discharged in the form of neurosecretion.

The function of most of the known neurohumoral factors is their share in the transmission of nervous excitation in the synapses. This seems to apply to acetylcholine, which is known from the nerve endings (synapses) of the parasympathetic nerves and the preganglionic parts of the sympathetic nerves of vertebrates. Bykow and Kibjakov have demonstrated that acetylcholine is also present in nerve endings in the central nervous system. In the postganglionic parts of the sympathetic nerves, the function of acetylcholine is replaced by that of noradrenaline (sympathin). The reactions in the synapse can be regarded as consisting of the following stages: the electric current reaching the synapses from nerve processes starts specific chemical reactions. These result in acetylcholine being liberated, which produces a similar change in the electric potential in the nerve fibre (axon or dendrite) of the next nerve cell which, in turn, is transmitted to the other pole of this nerve cell in the form of an electric current.

Similar, but more complicated, changes take place through the action of other neurohumours at the neuromuscular junctions of other nerve endings, resulting in stimulation of the muscular or glandular tissue concerned or in the induction of excitement (centripetal electric current) in sensory endings. The following neurohumoral factors have been found in insects.

Acetylcholine

This has occasionally been found in very high concentrations in insects, e.g. in similar weights of tissue, up to fifteen times more acetylcholine was found in the nervous system of the cockroach *Periplaneta americana* than in the central system of vertebrates (Mikalonis, 1941; Tobias, 1948). The question remains whether acetylcholine occurs not only in the nerve endings but also as a true neurohormone inside the neurosecretory cells. Besides acetylcholine, acetylcholinesterase has also been found in large quantities within insects, again in *Periplaneta americana* (Richards and Cutkomp, 1945) and also in *Apis mellifica* and *Melanoplus differentialis* (Means, 1942).

Acetylcholinesterase

This is most abundant in the neuropile, where most synapses occur. It is also present in the neural lamella of each ganglion and in the glial membranes, where its function seems to be to protect the central nervous system from the entry of acetylcholine from outside (Treherne and Smith, 1965). Its first appearance in insect embryos was found to concur with the appearance of the neuroblasts (Treherne, 1966). On the other hand, as distinct from the condition in vertebrates it is absent from neuromuscular junctions, suggesting that, in insects, it is not the chemical transmitter in this type of nerve ending.

Serotonin (5-hydroxytryptamine)

This has also been found in considerable amounts in insect ganglia and in corpora cardiaca (Treherne *et al.,* 1967). It seems to be produced by specific neurosecretory cells (Hinks, 1967) and to have various excitatory and inhibitory effects in different tissues and species, e.g.

stimulation of the heart rate and gut peristalsis in *Periplaneta,* stimulation or inhibition of neuromuscular transmission and enhancement of night flight in Lepidoptera, stimulation of secretory activity in isolated salivary glands of *Calliphora,* stimulation of the action of the diuretic hormone in *Rhodnius,* etc. (cf. Wigglesworth, 1970).

Hiripi and Salanki-Rosza (1973) studied the presence of 5-hydrotryptamine (5HT) and catecholamines in the central nervous system and heart of *Locusta migratoria* with an Aminco Bowman spectrophotofluorimeter, according to their excitation and emission spectra. 5HT and dopamine were found in considerable amounts in both the ganglia and the heart.

Noradrenaline was present in the cerebral ganglion only, and in only about 10 per cent of the dopamine concentration. Adrenaline was found neither in the ganglia nor in the heart. The three positive substances are regarded as transmitter or modulator substances in the insect CNS and 5HT and dopamine in the heart also.

Catecholamines

Catecholamines have been found in insects in relatively large amounts. Dopamine (3-hydroxytyramine) seems to be the most abundant, while adrenaline and noradrenaline occur in smaller amounts (Treherne, 1966). Among their various important effects, adrenaline, noradrenaline and dopamine were shown to stimulate the activity of the central nervous system, followed by blocking of the giant fibre synapses, in cockroaches (Hodgson and Wright, 1963). They produce pharmacological effects on the heart rate, gut peristalsis, the Malpighian tubules and the oviducts (Wigglesworth, 1970). An adrenaline-like transmitter affects the luminous organ in the fire-fly *Photuris,* where the application of reserpine (known to counteract adrenergic transmitters in vertebrate nerve endings) abolishes the light produced by nerve stimulation. Catecholamines can be demonstrated histologically by the characteristic yellow-green fluorescence in lyophilized ganglia, specific areas in the cerebral ganglion of *Periplaneta* (Frontali and Norberg, 1966) and the ventral nerve cord of *Gryllus* and other insects (Cech and Knoz, 1971).

GABA (γ-aminobutyric acid)

This is another neurohumoral factor common to insects. It has been shown to inhibit impulse conduction in caterpillars of *Dendrolimus pini.* It is formed by decarboxylation of glutamic acid. The corresponding decarboxylase is also of common occurrence in the insect nervous system (Treherne, 1966). L-glutamate was shown to induce depolarization in the synapses in this way (Usherwood, 1968).

Certain other Pharmacologically Active Substances

Such substances have been found in insect tissues, but their chemical structure and possible function in the insect organism have not yet been determined. A substance which is produced in the opaque male accessory glands of male insects, and stimulates the nerves controlling peristalsis in female reproductive tracts, merits special attention. It seems to be a pharmacologically active amine which stimulates the heart rate, similar to the one produced by the pericardial cells in *Periplaneta* through the action of a peptide neurohormone from the corpora cardiaca (Davey, 1964; Wigglesworth, 1970).

GENE HORMONES

Gene hormones have so far been demonstrated only in insects, although they can be presumed to exist in other groups of animals also (at least in other arthropods). They are clearly of a protohormone character. The messenger RNA responsible for induction of the enzyme synthesis mentioned below can probably be regarded as being the actual gene hormone. Thus the term 'gene hormone' is well justified. This also follows from the existence of morphogenetic gene hormones as mentioned further below. At present, the following types of gene hormones are known.

Gene Hormones and the eye-colour of *Ephestia Kuhniella*

Perhaps the first gene hormone discovered was the one which conditions the development of the black colour in the compound eye of *Ephestia kuhniella* (Caspari, 1933; Plagge, 1936a, b). A red-eyed mutation has been isolated in cultures of this moth, whereas the eyes of the wild type are black. Genetical analysis has shown that red eye-colour is recessive to black. The wild form AA is further characterized by dark pigment spots on the testes and black pigmentation of the larval epidermis and of the ganglia of the nervous system. These features are all missing in the aa form.

It has been shown that the colouration peculiar to the wild form AA may equally develop in the red-eyed mutant if testes from normal males are implanted during a particular sensitive period before pupation. Transplantation of the ovaries, the brain and even of individual eggs from the AA form, produce the same effect as transplanting the testes. This effect is not species specific: implantation of ovaries from the moth *Acidalia virgulata* and transfusion of haemolymph from pupae of *Galleria mellonella,* and even from the flies *Drosophila melanogaster* and *Calliphora erythrocephala* have the same effect (cf. Plagge, 1936a, b; Ephrussi, 1942a, b). It is interesting to note that the effect is not directly proportional to the amount of gene A. It has been shown that the tissues of the hybrid race Aa have the same effect as those of the AA form. From this it is concluded that the effect is not due to the gene itself but to a substance produced in the tissue when the gene is present. It has further been shown by Plagge that the presence of AA tissue for 24 hours is sufficient to produce the effect whether the implant is then removed or not. A possible explanation is that the pigment may catalyse its own further synthesis once it has developed in the new medium (the aa host).

Gene Hormones in the eye-colour *of Drosophila Melanogaster*

Even more detailed investigations have been carried out on the development of eye-pigmentation in *Drosophila,* where a similar protohormone effect has been demonstrated. Genetical experiments with *Drosophila* have produced numerous mutations differing in slight degrees of tone in their eye-colouration, from the dark red of the wild type to pure white. Some of these colourations are effected autonomously by genes, while in others a hormonal effect very similar to that described for *Ephestia kuhniella* has been found. The difference between these two groups of colouration is evident in gynandromorph individuals of *Drosophila* from the fact that, whereas in the former group each eye can have a different colour tone, in the latter group, if one eye is of the wild type (w), the other necessarily shows the same colouration. Of the many mutations described the following have been shown to be hormonally controlled: vermilion

(v^+), cinnabar (cn^+), cardinal (cd^+) and scarlet (st^+). The substances controlling these colour tones are again present in various parts of the body such as the Malpighian tubules and the fat body. By transplanting these tissues the particular eye-colour may be induced in the mutants lacking the relevant gene. Similarly, if an imaginal eye disc with a particular gene missing is transplanted into a larva which has that particular gene, the eye which develops from the disc shares the colour tone of the host. It has been shown that these gene hormones produce their effects even when they are administered in food, though of course only if accepted by the larvae during the short period of about 70 to 65 hours before pupation (Beadle and Law, 1938).

The Chemistry of the Gene Hormones

Much attention has been paid to the chemical composition of the gene hormones which affect eye-colouration. The careful investigations of Butenandt, Weidel and Becker (1944) and Tatum and Beadle (1938) have shown that the gene hormone A of *Ephestia kuhniella* and the gene hormone v^+ of *Drosophila* are identical. Moreover, their effect can be repeated with 1-kynurenine, which has some physical and chemical features in common with the gene hormones. In contrast, d-kynurenine and kynurenic acid have no effect on eye-colouration.

L-kynurenine is a substance formed by the enzymatic oxidation of the amino acid tryptophane. The gene hormones concerned have been shown to work as oxidation enzymes in such a way as to continue the reaction until the particular colour substance has been formed. Table 6 shows how, according to Beadle (cf. Scharrer, B., 1942), the various eye-pigments develop.

Table 2.1.

brown pigment
↑
scarlet pigment
↑ ← st^+ gene
cardinal pigment
↑ ← cd^+ gene
cinnabar substance (chromogen)
↑ ← cn^+ gene
$C_6H_4(NH_2)$—C(=O)—CH_2—CH(NH_2)—COOH 1-kynurenine (=A-substance = y^+-substance)
↑ ← v^+ gene
(indole ring with N, OH)—CH_2—CH(NH_2)—COOH oxytryptophane
↑ ← v^+ gene
(indole ring with NH)—CH_2—CH(NH_2)—COOH tryptophane

The brown eye-pigment of *Drosophila* is a compound of the ommochrome group (Becker, 1942). The cinnabar pigment (cn^+) may be regarded as the chromogen for the black eye-pigment (Kikkawa, 1941)-Quantitative investigations on the formation of eye-pigment in *Ephestia* have shown that the amount of pigment is directly proportional to the amount of the injected pigment.

Other Gene Hormones

Apart from the gene hormones affecting the pigmentation of the eyes and various other parts of the body, a homorally transmissible factor which conditions the number of facts of the eye has been shown in *Drosophila.* In *Drosophila melanogaster* a mutation known as 'bar' occurs, which differs from the wild type in that its compound eyes have 50 to 60 facets instead of the normal 600 to 650 facets. It is said to lack the gene B^+. When mutant larvae receive an extract of wild type larvae in their food, however, the number of facets shows a significant increase (Ephrussi, Chevais and Khouvine, 1938; Chevais, 1943, 1944). An extract of *Calliphora* pupae is equally effective. The active substance concerned, known as 'anti-bar substance', seems to be very common in insects. Its chemical composition has not yet been determined, although the same effect can be obtained using imidazole derivatives such as histidine, creatinine and methylhydantion, which is especially active (Khouvine and Harnby, 1936; Khouvine and Chevais, 1938).

It has been adduced against the term 'gene hormone' that some experimental findings show that it is not the gene itself which causes the eye-colour change in red-eyed insect mutants, but that it is the absence of the enzyme converting tryptophane to kynurenine which inhibits formation of the brown ommochrome. In normal insects (the AA gene in *Ephestia,* or the w^+ gene of wild type *Drosophila),* the enzyme determines the conversion of tryptophane to kynurenine, which is then transformed to hydroxykynurenine and further to the brown ommochrome (cf. Wigglesworth, 1970). The enzyme is evidently a product of protein synthesis determined by the relevant DNA locus, so that the term 'gene-dependent' hormone is still justified. The substance produced in transplantation experiments is undoubtedly the corresponding mRNA. Findings of morphogenetic protohormones of this type testify in favour of this conclusion. As mentioned by Wiggles-worth (see above), the findings on insect gene hormones were one of the first and clearest examples in support of the 'one gene-one enzyme' hypothesis generally accepted in modern molecular biology.

3 The Exohormones

The exohormones may be considered to be hormones in that they are produced by the organism and are physiologically active at extremely small concentrations. However, they do not act solely (if at all), in the individual which produces them, but instead affect other individuals in the same colony of social insects. For this reason some authors have called them 'socio-hormones'. The term 'ectohormones' has also been used, but this is less apt because not all exohormones occur on the body surface of the individuals producing them and scarcely any are produced there or act there.

Karlson and Luscher (1959a, b) have proposed the term 'pheromones'* for this kind of active substance, the term to include substances 'secreted by an individual and received by a second individual of the same species in which they release a specific reaction, for example, a definite behaviour or developmental process. The principle of minute amounts being effective holds. They function as chemical messengers among individuals.'

Apart from the exohormones as defined above, the authors also include in the term pheromones (later changed to pherormones by Pain, 1961) sexual attractants and other olfactorily active substances which reach their site of action through the air, such as marking scents and warning substances. The authors further include orally active substances such as the alleged determining substances in royal jelly.

In another paper reviewing this category of substances Karlson and Butenandt (1959) suggest the 'possible future necessity' of distinguishing between substances which act sensorily, for which the term telomones is suggested, and those which act biochemically, to which the term pheromones would be restricted.

There are major reasons against including both the sensorily and biochemically active substances under the same heading, within which one may separate the biochemically active substances working inside the body (hormones) from those working outside the body (exohormones). First of all, a basic phylogenetic difference exists between the two groups. Those substances which act sensorily, indirectly and by means of the nervous system have secondarily acquired a physiological action on a particular function (e.g. copulatory movements) after that function had been fully developed phylogenetically. By contrast those substances which act biochemically and directly have an activity which is phylogenetically primary. From this a more complicated function (e.g. inhibition of ovary development by the queen inhibiting substance of

*From the Greek 'pherein' = to carry, and 'hormon' = to stimulate, to excite.

the honey bee, or inhibition of morphogenesis in permanent nymphs by inhibitory substances of the reproductives in *Kalotermes)* has developed secondarily. This sharp distinction between the exohormones and substances acting sensorily exists even though both types of substance seem to work together in some cases as in the control of honeybee workers by the queen. The distinction furthermore holds even though there may exist a group of substances which control neurosecretion and which seem to combine both an indirect (sensory) and a direct (biochemical) effect but are nevertheless of a definite sensory character.

Further clear evidence that there is a close relationship between exohormones and hormones in the restricted sense is provided by the extremely interesting effects that exohormones of the honey bee have on the development of the ovaries in crustaceans and the similarity of their physiological effects to those of particular neurohormones.

The fundamental affinity of exohormones with hormones is even clearer if we view the social insect colony as the beginnings of an organism of a higher order. Then one may regard exohormones as substances circulating and producing their effects inside the organism of the higher order, *i.e.* the colony.

It therefore seems preferable to restrict the use of the term pheromones to those substances which act via the nervous system (the sense organs) and to reserve the term exohormones for substances with a direct biochemical activity being transmitted within the colonies of social insects. This is the sense in which the term has been used in the earlier editions of this book.

Table 3.1. Physiologically Active Substances (Wirkstoffe)

directly (biochemically) acting substances	enzymes	digestive
		respiratory
		etc., etc.
	hormones	hormones s. str.
		protohormones
		exohormones
	vitamins	vitamins s.str.
		determining substances (?)
		milk substances
indirectly (sensorily) acting substances (=pheromones)	neurosecretorily active scents (?)	inhibitory odours
		phase determining odours (?)
		stimulating odours (?)
	telomones	behaviour influencing odours
		sex attractants
		marking scents
		territorial demarcation scents
		warning scents

Lüscher (1972) recently suggested that the metamorphosis hormones, MH and JH, together with JHa, may also act as exohormones (pheromones in his terminology, but with a different meaning from the one given to this term in the present book). His conclusion is based on the discovery that the hormones can be passed from individual to individual by trophallaxis (Lüscher, 1969), by faeces (Luscher, 1957) and by vapours. Confirmation of this is provided by the experiments of Springhetti (1969, 1970), which showed that reproductives, with their excess of JH, stimulated the production of soldiers and inhibited the production of further reproductives —all of which can be attributed to (and was actually demonstrated experimentally as being due to) JH action.

As far as the active substances contained in royal jelly are concerned, it seems justifiable to consider them as a special group of 'milk' (or nourishing) substances which are neither pheromones or exohormones as they work mainly nutritionally. In this connection the term 'milk' is used in its widest physiological sense to include the pharyngeal products of bees, i.e. the royal jelly and worker food as well as the milk of mammals (cf., e.g., Brian, 1957). The various physiologically active substances can be classified as shown in Table 3.1.

Of these substances only the hormones, including the exohormones, come within the scope of this book. However, a few of the pheromones of bees and termites have also been included, as they are produced together with the exohormones and have not yet been clearly distinguished from them.

THE EXOHORMONES OF TERMITES

Polymorphism

In spite of the low phylogenetic level of the termites, the highest level of caste polymorphism and division of labour is present in this group. The diversity of types shows a substantial increase over the social Hymenoptera with the occurrence of termites of all the castes. Not only reproductives, but also workers and soldiers of various kinds are produced in both sexes. This is, no doubt, connected with a much higher phylogenetic age of the polymorphism here than in the phylogenetically younger Hymenoptera.

A normal termite colony has one pair of reproductives, the king and queen, which were originally winged and only lose their wings during the foundation of a new colony after the mating flight. The proportion of workers and soldiers in a normal well-developed colony of a particular termite species appears to be more or less constant. In many phylogenetically primitive termite species, however, there are no true adult workers, and their position and function is occupied by the so-called pseudergates (Grasse, 1947). These are larvae whose development has been arrested at a particular morphogenetic stage prior to metamorphosis. Their prothoracic glands remain intact and active, however, so that they continue moulting at regular intervals, but without any growth or differentiation. These moults have been called stationary moults (Luscher, 1961). The pseudergates are therefore permanent larvae in the full sense of the word. If the reproductive pair is removed from the colony, the development of the pseudergates may continue and they may moult to form the so-called supplementary reproductives.

The newly hatched larvae in *Kalotermes* and other phylogenetically primitive termite species are identical, without any morphological or potential caste differences. Until the end of the IIIrd larval instar each nymph in the colony may eventually become a reproductive, a soldier or a worker (pseudergate). During the IVth instar some of the larvae differentiate in the direction of a soldier and moult to a pre-soldier, which becomes a soldier after the next moult. After the Vth instar a specimen may become either a pseudergate with stationary moults or it may moult to a supplementary reproductive or to a pre-soldier. Alternatively, it may continue differentiating towards the adult form, moulting to the 1st and IInd nymphal instars with wing pads. Further, the nymphal instars retain the potential to develop not to the winged adult but to supplementary reproductives or to pre-soldiers. Regressive moults back to the pseudergate (Grasse and Noirot, 1947; Lüscher, 1952, 1961) are also possible. Table 3.2 summarizes the possible lines of differentiation.

Table 3.2 Polymorphism in Kalotermes flavicollis (modified from Luscher, 1961). Each arrow represents a moult, the broken line indicates the possibility of several moults

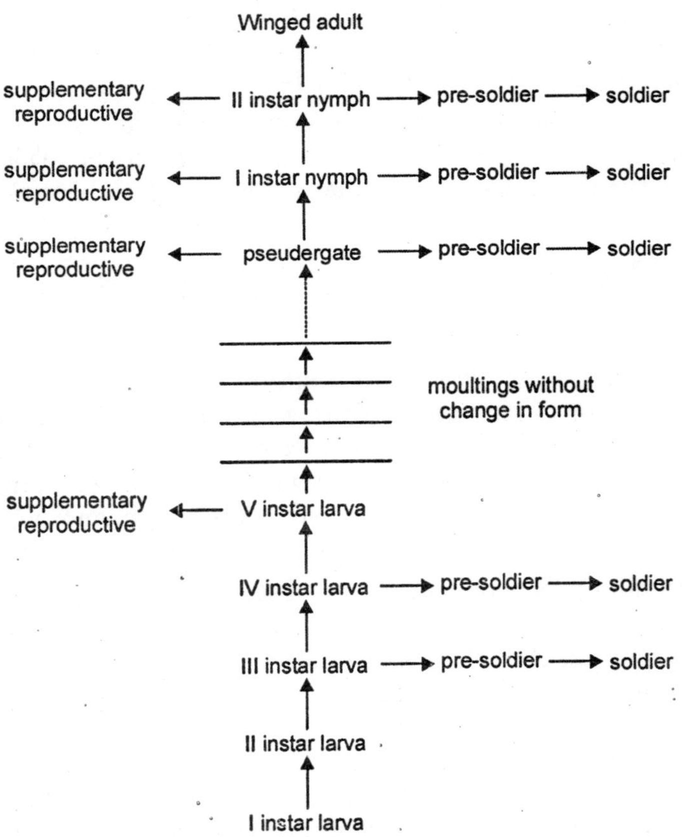

In phylogenetically advanced termites the situation is more complicated and several authors (cf. Brian, 1957) have alleged that genetic and histogenetic determination is involved. The fact that some termite species form their workers from other species has been stressed as evidence for genetic determination. Whereas both sexes are able to produce both castes in other species, particularly in the more primitive termites, each sex in the more advanced termites seems secondarily to have lost the potential to produce one or other caste. Caste determination in these species is therefore controlled by sex and in this way evidently controlled genetically.

In many species of the family Termitidae differences in internal structure between winged and apterous adults are already apparent in the first instar. In *Tenuirostritermes tenuirostris* (Desneaux) the inhibitory affect of reproductives on the formation of further reproductives has been observed at the moment of hatching, and according to Weesner (cf. Brian, 1961), may occur even earlier.

The Inhibiting Substances

Research over the last few years has shown that inhibition of the growth and differentiation of pseudergates (e.g. in *Kalotermes)* is caused by the effect of particular chemical substances produced by the functional reproductives: the so-called queen inhibitory substance, and probably also a king inhibitory substance, these substances inhibit the development of ovaries (and testes) and doubtless at the same time inhibit growth (including differentiation) of the imaginal parts of the body and of the body as a whole. Thus, by the action of the inhibitory substances the development of the majority of the larvae in a colony is arrested at the pseudergate level. When the reproductive pair dies or is artificially removed from the colony the inhibitory effect disappears and many of the pseudergates change to supplementary reproductives at the next moult. The pseudergates which change are those which are in a suitable physiological state when the inhibitory effect disappears, *i.e.* at the appropriate stage of the intermoult period. The same applies to larvae in the appropriate stage of the Vth larval instar and to those in the Ist and IInd nymphal instar. As soon as the first pair of supplementary reproductives starts to reproduce, further supernumerary reproductives are killed by the workers and the inhibitory substances produced by the new king and queen inhibit further differentiation of reproductives in the colony.

The recent papers by Luscher (1961a, b) suggest the existence of four different inhibitory substances concerned with regulating the development of the supplementary reproductives. The effect of at least two of them seems to be amplified by the workers (pseudergates) through a remarkable sex-linked effect: the male pseudergates seem selectively to inhibit the development of female supplementaries and *vice versa.* Further evidence, both experimental and comparative, is necessary before these conclusions can be confirmed and the effects fully understood (cf. Luscher, 1956).

The source of the inhibitory substances has not yet definitely been discovered. Experiments in which a wire mesh prevents pseudergates reaching either the anterior or the posterior parts of the body of living reproductives, and others involving the use of extracts from various parts of the body, suggest that the active substances originate either in the head or thorax of the specimen and that they leave the body via the intestinal tract, most probably with the faeces (Light, 1944; Luscher 1953; Hinton, 1955-1966). The inhibitory substance or substances produced by females is said to differ from that produced by males. The substances produced by the king and queen are complementary in such a way that the full effect is only obtained if both reproductives are present. The inhibitory effect of the female by itself is feeble and that of the male is virtually nil.

The chemistry of the inhibitory substances is almost unknown. The active principle can be extracted from the queens with methanol or with water, and there is little difference in the activity of the two extracts (Light, 1944). The substance remains active for several months after the death of the animal, yet the effect of the reproductives disappears within 24 hours of their removal from the colony. This has been interpreted as a rapid inactivation of the inhibitory substance (Lüscher, 1952-1956), although the substance may be quickly removed from the circulatory system by the Malpighian tubules and may perhaps be inactivated during its passage through them. Buchli (1956) has shown that the substance (substances) is only active when applied during the first third of a particular instar but that the state of the fat body (quantity of reserve materials) also seems to be important. It is not yet clear whether the substance can also be produced by specimens which are unable to reproduce (cf. Hinton, 1955-1956) although some experiments (Luscher, 1956b, c) suggest that this is so. The results of Luscher's (1955) experiments appear to demonstrate that these substances are exohormones. When a sexual pair were separated from pseudergates by a double barrier of wire gauze which prevented any contact but allowed the passage of odours, the result was equivalent to the complete removal of the reproductives: a pair of functional reproductives was produced and they survived. When only a single barrier allowing antennal contact was used, all developing supplementary reproductives were eliminated. It is not clear from the experiments whether the substance (or substances) perceived by the antennae is identical with the substance (exohormone) inhibiting the development of reproductives or whether it merely affects the behaviour of the pseudergates and causes them to eliminate the excess reproductives. In the latter case the substance would be a pheromone in the sense used in this book. Such a parallel action of an exohormone and a pheromone has been found in bees.

The Stimulating Substances

As well as the inhibitory substances in both winged and supernumerary reproductives, Luscher (1956b, c) claims to have found one or perhaps two substances which have a stimulating effect on the development of supplementaries. He was able to extract this substance from the heads of supplementaries with methanol, and strips of filter paper soaked with the extract proved to be active. Compared with control experiments, application of the extract produced a definite positive effect, though only if the reproductives were removed. This showed that the stimulating effect is weaker than the inhibitory one. Lüscher assumes that one of these different sex-specific substances is produced by the females which stimulates the production of males, and the other substance by males stimulating the production of females. In the natural colony these substances would assume importance if one of the reproductive pair were lost. The presence of the remaining reproductive would then stimulate the replacement of its mate by a supplementary of the appropriate sex. The selection value of the stimulating substances would perhaps lie in the acceleration of the replacement process and it seems to be all the more important as the loss of only one reproductive is much more frequent in natural conditions than the simultaneous loss of the pair.

Exohormones and the Metamorphosis Hormones

Several authors have tested the possibility that exohormones modify the secretion and activity of the metamorphosis hormones in termites. Kaiser (1955) has shown that differentiation from the pseudergate level in the more primitive termites involves a different number of moults for different castes, *i.e.* three for a normal winged reproductive, two for a soldier and one for a supplementary reproductive. These differences suggest corresponding differences in the production and action of the metamorphosis hormones, and histological differences were in fact found in the ventral glands and the corpora allata. In the more advanced termites, the development of workers is correlated with reduced functioning of the ventral glands while the development of soldiers is correlated with a notable increase in size of the ventral glands and corpora allata. The ventral glands degenerate following the imaginal moult. A large increase in the size of the corpora allata was observed in each instar during the development of supplementaries, but this increase was not found during the development of normal reproductives where the increase occurs at the time of egg-laying. During stationary moults, the ventral glands of the pseudergates were about twice the size of those in the intermoult period.

Lüscher (1957a) and Lüscher and Springhetti (1960) studied the influence of implanting corpora allata from different donors into pseudergates in normal colonies and colonies which had lost their reproductive pair. In most cases, the implantation of the corpora allata resulted in moults to pre-soldiers. Under some conditions moults to intercastes between pre-soldiers and supplementary reproductives and between pseudergates and pre-soldiers resulted. The authors concluded that at least three factors influence the type of differentiation which follows implantation of corpora allata. These factors are (1) the physiological state of the recipient specimen; (2) the physiological state of the colony; and (3) the physiological state of the implanted gland (*i.e.* the physiological state of the donor specimen).

The authors went on to develop a theory based on the assumption that there are two different corpus allatum hormones, one with a juvenilizing effect and the other with gonadotropic activity. According to the theory, 'juvenile hormone' favours the differentiation of supplementaries while the 'gonadotropic hormone' favours the differentiation of pre-soldiers. This theory needs to become rather complicated to explain, even in part, the various experimental results obtained by the authors. However, a more simple theory can be suggested which agrees with the GF theory. Such a theory involves the assumption of only one corpus allatum hormone, which does not disagree with anything known about JH. In the presence of the inhibitory substances (*i.e.* in the presence of the reproductives) growth (both isometric and gradient growth), as well as production of JH, are more or less completely inhibited. Under these conditions a low dose of JH favours development towards supplementaries while increasing doses (prolonged period of action) result in intercastes and eventually the differentiation of pre-soldiers. If the exohormones are absent, supplementaries will develop. This prothetelic effect seems to be entirely due to a reduced proportion of nutrients for the imaginal parts by the accelerated growth of the ovaries in the absence of the inhibitory factor, in accordance with the law of correlation of consumption. In the presence of the inhibitory factor the application of JH will again induce the formation of pre-soldiers with possible intercastes at the expense of the supplementaries. A further piece of evidence in favour of the GF theory is that the ventral glands do not degenerate until a definite level of differentiation, as in supplementaries and adults, has been reached. They remain active both in pseudergates, where differentiation is inhibited by the exohormones, and in pre-soldiers, where differentiation is delayed by JH. The dissociation in pseudergates of chronically repeated moulting from growth and differentiation is further evidence in support of the GF theory assumption that the two processes are independent and that MH acts directly only on the former.

On the basis of his experiments with Roller's JH preparation, Lüscher (1969) himself abandoned the hypothesis that there are two or more different JH, after demonstrating that the same JH can induce both effects. In place of his earlier views, he adopted the new hypothesis that these effects are caused by different doses of the same JH produced and communicated to other members of the colony by individuals belonging to different castes, so that it would thus have the character of a true exohormone (Luscher 1972). In that case, only one inhibitory hormone would be necessary, *i.e.* the one stabilizing the pseudergates and inhibiting soldier development. It has been suggested that it is a JH inhibitor, but its effect must be more general, *i.e.* inhibition of any growth and morphogenesis. This is in agreement with the findings of Springhetti (1960) and Lebrun (1967, 1970). For instance, the latter author showed that the implantation of *Periplaneta americana* corpora allata into nymphs developing into adults (1st and IInd instar nymphs) caused the development of intercastes between soldiers and sexual adults.

The Phylogeny of Social Life in Termites

The discovery of exohormones has clarified the picture of the origin of social life in the Isoptera. As Kaiser (1955) pointed out, the termites differ from the Hymenoptera in that no examples are known of a solitary way of life or of transitions between this and social life. The first signs of a development towards a social life can, however, often be seen in the orders Dictyoptera (in the suborder Blattodea) and Dermaptera which are closely allied to the ancestors of the termites. In these orders it is a frequent phenomenon that nymphs in various instars live together with the female or with the parent generation. Under such conditions the appearance during phylogeny of an inhibitory substance of an exohormone nature would necessarily prolong the situation by keeping the morphogenetic stage of the nymphs static. This would automatically keep the stage of development of the brain unchanged and thus preserve the nymphal instincts. The selection value of each of these effects as a step towards social life is clear. These first stages would prepare the way for the development of trophallaxis and other features of social life (cf. Schneirla, in Roeder, 1953). Such conclusions are confirmed by the wide non-specificity and wide distribution of the exohormones as shown by several authors (Carlisle and Butler, 1956; Hrdy and Novak, 1960).

EXOHORMONES IN THE HYMENOPTERA

Polymorphism

Polymorphism in the Hymenoptera is fundamentally quite different from that in termites. The main difference is that (apart from perhaps a few unimportant exceptions) only the female sex is involved. Usually only one caste besides the reproductive females and males is found; *i.e.* the female workers, which may in some cases be polymorphic. A second sterile female caste, the soldier, is only found in some species of ants. In the colonies of the honey bee *(Apis mellificd)* only one extra caste exists. These are females with reduced ovaries which are known as workers. The first differentiation to worker or queen larvae does not start until the third day after the larva hatches. Up to this moment any larva may develop in either direction. The determining factor involved is the quality and quantity of the food supplied by the workers (Rhein, 1933, 1954). Under normal conditions, however, the fate of each individual is already decided at the time of egg laying, as the eggs from which the queens develop are laid into special large and distinct queen cells. In the adult, the ovaries of workers in colonies with queens are reduced and functionless, whereas in queenless colonies the ovaries of the workers become functional and the workers start laying eggs. It is now established that both ovarian development in the workers and the consequent instinct of workers to construct queen cells are controlled by definite substances produced by the queen which are similar to the exohormones and pheromones already discussed in termites. Unlike the more primitive termites, however, both the bee and ant workers are adult insects in which the prothoracic glands have disappeared and are therefore unable to moult and develop further. There is, however, a major difference between the bees and ants: whereas the queen bee is slightly neotenic in comparison with the worker bees, the wingless ant workers are strongly neotenic in comparison with the winged sexual females.

Recent studies by various authors have elucidated the problem still further, although some questions still require investigation. A complex determinant system is now assumed to control caste determination in honey bee females. Environmental differences start with differences in the larval food from the very first day of larval life. The food prepared for the larvae in the worker cells (worker jelly) differs from the brood food in the queen cells; in addition to pharyngeal gland secretion, it also contains an inhibitory substance from the mandibular glands of the nurse bees (Jung-Hoffmann, 1966; Rembold, 1967; Wirtz, 1973). After about three and a half days, the worker jelly is modified by the addition of honey and pollen. If unmodified jelly is fed to older larvae, it inhibits morphogenesis. The first differences between worker and queen larvae (in the structure of the fat body cells) appear before the third day of larval life (Wirtz, 1973).

Various differences between the structure and activity of the larval worker and queen endocrine systems have been described. In queen larvae, the corpora allata seem to attain a higher level of activity, and at an earlier stage of development, than in worker larvae. Royal jelly probably activates the endocrine system in some way. The topical application of JHa to 3-day-old worker larvae in worker cells produces a number of queen characters in the treated specimens. The effect of queen jelly on caste differentiation thus seems to be mediated by modified JH production. A worker-like appearance of the fat body cells can be induced in larvae reared *in vitro* by the addition of mandibular gland secretion (Wirtz, 1973).

The Inhibitory Substance of the Honey Bee

The inhibitory substance is produced by the fertilized queen. The first effect of removing the queen from the colony is that the workers construct emergency queen cells. These queen cells are constructed above some of the original worker cells in which eggs or young larvae are still present. If neither eggs nor larvae are available to the workers in the colony when the queen is lost, the ovaries of many (of about 10 per cent of the workers within a week of removing the queen) start developing. After some time the workers begin to lay eggs, but of course these remain unfertilized and only drones develop. Both the construction of queen cells and the development of ovaries in workers are the symptoms of absence of the inhibitory substance, *i.e.* the absence of the queen. There is also a definite positive effect of the substance (or group of substances) which causes the queens (alive or dead) or extracts prepared from queens to be highly attractive to the workers. There is still doubt whether Butler's original assumption (1954) is correct that only one queen-inhibiting substance produces all the effects mentioned, or whether there are at least two substances as suggested by Pain (1952, 1954a, b), Groot and Voogd (1954) and Voogd (1955), one substance being the inhibitor of ovarian development and the other Butler's 'Queen Substance', which Brian (1957) has called a hormone and a signal.

The latest findings, including Butler's, seem to have decided in favour of the second view. On the other hand, several pheromone-type substances interfering with various kinds of social behaviour in the honey bee have been described (Velthuis, 1970; Callow, 1971; Butler *et al.*, 1973). The origin of the queen inhibitory substance in the mandibular glands now seems to be beyond doubt (Garry, 1961a, b; Butler *et al.*, 1973; Wirtz, 1973).

Pain (1961) made a detailed study of all three aspects of the effects of the queen on the workers, and suggested that the inhibitory substance is complex and a mixture of several acids.

Pheromone I. The substance isolated by Barbier (1958) from active extracts of queens and workers in colonies with queens; a substance also isolated by Callow and Johnston (1959) from the heads of fertilized queens. Both groups of authors simultaneously synthesized this acid: Barbier (1960) synthesized it from cycloheptanone and Johnston (1959) from azelaic acids. It has the following formula:

$$CH_3—CO—(CH_2)_5—CH{=}CH—COOH \quad ...(1.7)$$

Pheromone I

and is α, β unsaturated 9-oxydec-2-enoic acid. Pain called this acid pheromone I. The acid forms white flakes and has the following effects on queenless workers:

(*a*) It inhibits the construction of queen cells - at a concentration of 0.13 µg per bee according to Butler and at 0.5 µg according to Pain (1961a).

(*b*) It is an indispensable component of the queen odour but it is not itself attractive to the workers (Barbier and Pain, 1960).

(*c*) It does not inhibit the formation of eggs in the ovaries of workers when injected, or in food or when licked from the body surface of the dead queen.

Pheromone II. A second substance which Pain (1961) called pheromone II was isolated in the form of highly volatile esters by distillation from the active pentane soluble fraction (pentane acids). Two of these esters were identified by gas-chromatography as methyl phenylacetate and methyl phenylpropionate (Pain, Hugel and Barbier, 1960). This substance is not attractive to workers by itself, but when mixed with 'pheromone I' gives the characteristic queen odour (Barbier and Pain, 1960).

The substance was later identified as trans-9-hydroxy-decenoic acid, with the following formula (Butler and Callow, 1968):

$$CH_3—CH(OH)—(CH_2)_5—CH = CH—COOH \quad ...(1.8)$$

Pheromone II

It is also produced by the mandibular glands and development of the worker's ovaries is inhibited only in the presence of both substances (cf. Rembold, 1973).

Both pheromone I and II were found to occur in the mandibular glands of fertile queens. Only traces of exohormone I were present in emerging queens, and pheromone II appeared later. This agreed with the lack of attractivity of freshly emerged queens. The rate at which the pheromone complex (I and II) was produced appeared to depend on, apart from the age of the queen, the number of workers accompanying her, *i.e.* the state of nutrition of the queen seems to be the important factor.

Basing her argument on an analysis of Müssbichler's (1952) work, Pain went on to claim that the substance causing inhibition in ovarial development of the workers was not identical with the inhibitor of queen cell construction. She was certainly also right in attaching importance to the effect of what she calls nutritional castration of the workers by the queen which causes the inhibition of ovarian development. A number of earlier workers (cf., e.g., Hess, 1947) have shown that the queen influences the development of ovaries in the workers, first by herself consuming a quantity of the pharyngeal gland secretion (royal jelly) of her attendant bees, and then secondarily, and what is much more important, by producing larvae which are the main consumers of royal and/or worker jelly. This alimentary mechanism, which works parallel with the inhibitory substance, is without doubt quantitatively the main cause of suppression of ovarian development in workers.

Pain's conclusions (1961) on the olfactory character of the queen inhibitory substance, which she suggested worked by inhibiting the secretion of the corpora allata, seem to be less convincing. Her conclusions in no way invalidate the existence of a special inhibitory substance of the exohormone type, which directly and biochemically suppresses the development of the ovaries. Such a substance has been suggested in the papers of Butler (1954, 1957), Carlisle and Butler (1956) and Hrdy, Novak and Skrobal (1961). Thus the honey bee queen probably produces at least two pheromones (I and II of Pain) and one exohormone. These three substances work together to produce the inhibitory effects.

Apart from the views expressed by Pain (1961) other authors seem entirely to agree that actual contact is necessary for the substance inhibiting ovarian development to be transmitted. This has been shown by the fact that a double wire screen between the queen and the workers prevents this inhibition. That the effect of the substance is weaker when taken with food than when licked from the body surface of a dead queen is in no way contradictory of the purely direct biochemical action that Butler (1957, cf. Karlson and Butenandt, 1959) suggested. Both the amount taken per specimen as well as the concentration of the substance seem to be important in producing the inhibition, and both these were probably lower in the food than on the body surface. Thus the grounds for Voogd's 'psychic stimulation' hypothesis (1955, 1956) do not seem to be sufficient.

In contrast to Hinton's conclusions (1955b), the worker bees appear to play an important part in the transmission of the exohormone and the pheromone. It has been shown that if the queen substance, in the form of an aqueous solution of an alcohol extract, is given to worker bees, ovarian development and queen cell construction are suppressed not only in these workers, but also in others which had not originally drunk the extract.

The mode of action of the *queen inhibitory substance* on the worker bees was studied by Luscher and Walker (1963). They found that in workers separated from the queen the corpora allata increased in volume. This increase was significantly inhibited by the queen extract prepared from about 100 queen bees by Butler's method. The authors concluded that the inhibitory effect of the queen substance on the ovaries of workers was caused by the suppression of the secretory activity of their corpora allata.

The finding of a *new phenomenon in the honey bee* is reported by Butler, Callow, and Chapman (1964). They show that a synthetic substance, 9-hydroxydec-trans-2-enoic acid, caused dispersing clusters of swarming honey bees to reform and settle down again in the same way as does the odour of the queen, *i.e.* her mandibular gland secretion.

The Occurrence of Exohormones in other Hymenoptera

Little is known on exohormones in Hymenoptera other than honey bees. According to Sakagani (1954) the effect of the queen substance (or substances) in the Japanese bee *Apis Indica japonica* seems to be much weaker than in *Apis mellifica,* for fully developed ovaries were found in seventeen out of twenty workers accompanying the queen. This seems to be at least partly due to the workers taking some of the royal jelly, *i.e.* the secretion of the pharyngeal glands which is fed to the queen. In this species the stimulating effect of the jelly appears to be stronger than the inhibitory effect of the queen substance.

Cumber's (1949) observations suggest that substances with an inhibitory effect on ovarian development also occur in bumble bees. He found that the sterility of auxiliary females was not determined by the quantity of food available to the bumble-bee colony but by the presence of the queen, for removal of the latter resulted in development of the ovaries of the auxiliary females. Similarly the loss of the queen in colonies of *Polistes gallicus* resulted in egg-laying by the 'workers' within about 24 hours (cf. Hinton, 1955-1956). Similar effects have been found in some species of the genus *Halictus* (Que, 1958)*.

Stumper (1956) has found an exohormone system, similar to that in bees, in a number of ant species. This system appears to control the following behaviour pattern: (1) crowding of the queen by the workers; (2) extensive licking of the queen; and (3) carrying the queen.

Similar effects are caused by the same substance in various Hymenoptera, and are especially pronounced in *Lasius alienus* and *Pheidole pallidula.* Petroleum ether proved the most suitable solvent for extracting this exohormone. A dead *Lasius* worker impregnated with an extract of a *Pheidole* queen was accepted as a queen by an orphan *Pheidole* colony. Bier (1956, 1958; cf. Karlson and Butenandt, 1959) has shown that the presence of a functional queen inhibits ovarian development in workers of the species *Leptothorax unifasciatus* and *Formica pratensis.*

The influence of the metamorphosis hormones on caste differentiation has been studied by Brian (1958a) in the ant *Myrmica rubra.* He has shown that there is a critical period in the last larval instar of all female larvae during which the functioning of the corpora allata may be prevented by the neurohormone (AH?) of the neurosecretory cells of brain. When this happens only sterile workers develop. In the absence of this inhibition the increase in secretory activity of the corpora allata promotes the growth of the imaginal wing and ovarial discs and the ovaries, so that sexual females (queens) develop. Development, however, is also dependent on the state of nutrition. Starvation causes a more or less limited development of the wing discs at a critical period, resulting in intercastes of various grades.

Brian (1961) later suggested that when the MH (ecdysone) concentration increases quickly the relatively small ovaries and wing discs do not have enough time to respond, and workers develop. On the other hand, when the MH concentration increases gradually, as in diapausing larvae, the corpora allata are active and the ovaries and the imaginal discs of the wings grow and develop normally so that winged sexual queens are produced. The determination of diapause which in its turn influences the type of neurosecretory activity of the brain cells thus also affects determination of the female caste.

Brian (1959) has also discussed the possible existence of a special substance with the character of an exohormone which would simultaneously suppress the development of ovaries and other female features and influence the instincts of the workers. He assumed that it would also suppress the secretory activity of the larval corpora allata at the same time.

---

* Those female wasps and bumble bees which do not lay eggs are not true Workers as they are in honey bees and ants. They are merely unfertilized females of reduced size and with undeveloped ovaries, and are morphologically identical the 'queen' in every respect, including the activity of gonads.

The idea was developed further on the basis of the author's later findings on the effect of starvation on queen determination and the effects of the presence of a fertile queen on worker development (Brian, 1963, 1965). It was found that the *Myrmica* queen produced (also in the mandibular gland) a substance related to, but not identical with, the honey bee queen inhibitory substance (Brian and Hibble, 1963) and that 9-oxodec-trans-2-enoic acid was likewise partly effective in *Myrmica* (Butler and Paton, unpublished, cf. Brian, 1965).

The existence of two active principles in the mandibular glands of the *Myrmica rubra* queen was suggested by Brian and Hibble (1963). One, probably 9-oxodecenoic acid, is supposed to produce a stimulatory effect on brood growth similar to that caused by the presence of the queen, the other, which probably corresponds to a lipoid fraction of queen mandibular gland extract, has an inhibitory effect. In another paper, the authors analysed the influence of the queen on the growth of the brood in connection with the size of the larvae in Myrmica (Brian and Hibble, 1963b). The influence of temperature, food and the presence of queens on caste differentiation was studied by Brian (1963) in female larvae.

CONCERNING THE PHYLOGENETICAL ORIGIN OF EXOHORMONES

The occurrence of hormonal substances inhibiting ovarian development and other growth processes in arthropods is not limited to the social insects. It has been known for a long time that neurohormones produced by the eye stalks of many Decapoda have an inhibitory effect on moulting and growth (Scudamore, 1942; cf. Carlisle and Knowles, 1959). It is therefore extremely interesting that Carlisle and Butler (1956; cf. Carlisle and Knowles, 1959) found that the ovary-inhibiting hormone in the eye stalks of Decapoda produced an effect similar to that of the queen substance on the ovaries of honey bee workers in the absence of the queen. The same result was obtained on ant and termite workers. Similarly, the queen substance has been shown to suppress moulting and the functioning of the ovaries in prawns. It has been suggested (Carlisle and Butler, 1956) that these substances are steroids.

The wide non-specificity of the queen-bee inhibitory substance has been further confirmed by Hrdy, Novak and Skrobal (1961; cf. Hrdy and Novak, 1961). An aqueous solution of an evaporated alcohol extract of fertile honey bee queens was tested on groups of *Kalotermes flavicollis* workers. The extract caused a significant delay in the development of supplementaries compared with control groups. Moreover, whereas the workers in the controls allowed one pair of supplementaries to remain, the workers in the experimental series destroyed the reproductives as they appeared.

Another inhibitory humoral effect is that of the so-called diapause hormone. This is produced by the suboesophageal ganglion of female pupae in those Lepidoptera which show embryonic diapause.

It is a rather tempting hypothesis that such Immorally active substances were the phylogenetic origin of the various inhibitory exohormones of social insects mentioned above. Such substances probably attained the position of exohormones in one of several ways: (1) by removal from the blood via the Malpighian tubules and then leaving the body with the faeces as in termites; or (2) by accumulating in glands such as the mandibular glands of the honey-bee; (3) by reaching the alimentary canal or perhaps the salivary fluid (cf. Noirot, 1955).

It is therefore very interesting that, according to Noirot (1955), various groups of neurosecretory cells occur in the cerebral and suboesophageal ganglia of the ventral nerve cord of termites. A paired group of neurosecretory cells in the posterior part of the pars intercerebralis, which are not identical with the cells producing AH, are of particular interest in this respect. These cells stain more deeply in adult insects than in larvae.

An alternative possibility has been suggested by Novak (1961), who emphasizes the similarity in the mode of action of the inhibitory exohormones and the so-called pseudo-juvenilizing effects of some substances like the unsaturated fatty acids.

This conclusion was largely confirmed by recent findings (discussed above) on the role of JH and the effects of JHa in caste determination in both termites and Hymenoptera.

The findings on the leading role of the metamorphosis hormones, and of JH in particular, on caste differentiation in social insects show the dominant importance of JH in most mechanisms concerned with morphogenesis on the one hand and the broad unity of morphogenetic differentiation mechanisms on the other. The same hormonal mechanism which modifies cell differentiation at organism level is utilized by natural selection to ensure morphological and functional differentiation at the animal society level. This reconfirms the general experience that 'nature prefers simplicity wherever possible'.

4 Neurosecretion and Neuroendocrine Mechanisms in the Invertebrates

Remarkable progress has been made in elucidating the neuroendocrine systems of invertebrates and lower vertebrates, and certain concepts of great importance have emerged. While members of the animal kingdom are widely divergent in genetic endowments, body structure, and living habits, they are all forced to adjust to the same general environmental conditions in order to thrive as individuals and as species. Since protoplasm is fundamentally the same, irrespective of the species in which it is found, all organisms must obtain from their surroundings about the same materials for growth, maintenance, and repair of their bodies. In other words, all organisms have about the same needs and face very similar problems in their efforts to survive. It is not surprising, therefore, to find that invertebrates and vertebrates often employ comparable functional (analogous) arrangements in meeting their basic requirements. The comparative study of analogous systems has often been more rewarding than the study of homologous structures.

The traditional distinctions between endocrine and nervous systems appear to be largely artificial. The two systems function together to transform a loosely bound collection of cells, tissues, and organs into a cooperative enterprise, an organism that can adjust internal processes to changing conditions in the external world. Even the simplest organisms are complicated machines consisting of a multitude of parts that must be integrated in order to make life possible. During the course of evolution there has been an increase in the number of cells comprising organisms and also differentiation of cells into tissues and organs, *i.e.,* into groups of cells specialized for rather specific functions. With evolutionary advancement the problem of knitting diverse tissues and organs into a smoothly operating unit assumed increasing importance. In a multicellular individual like man, the billions of cells are structural and functional units in a very real sense, but each cell has become an interacting part of the whole organism and hence has lost the capacity of being self-sustaining.

THE CONCEPT OF NEUROSECRETION

In addition to neurohumors, such as acetylcholine and norepinephrine, which function in the transmission of nervous impulses, it is known that certain neurons secrete long-range, long-acting neurohormones. These latter substances are more stable than the neurohumors and may act at many distant points within the organism over a period of hours rather than milliseconds.

Neurosecretory Cells

Some neurons have developed secretory functions to such an extent that they become morphologically distinguishable from other nerve cells. These are called "neurosecretory cells," since they are neurons that elaborate and discharge special products after the manner of glands. Glandular neurons of this type are almost ubiquitous among metazoan animals. Though some of the smaller invertebrate phyla have not been adequately studied, such cells appear to be present wherever a central nervous system exists and some degree of cephalization has occurred.

Since neurosecretory cells may have dendrites, axis cylinders, and Nissl bodies, they are modified neurons rather than glia cells that have migrated into the nervous system. Visible granules appear in the perikarya, undergo various transformations, and are eventually discharged from the axons. The neurosecretory substance is transmitted distally, probably by axoplasmic flow, and often accumulates in terminal bulbs. There are indications that in special cases the neurohormone may be released from other points along the axon or even from the perikaryon itself. The axons of neurosecretory cells often end in close association with blood vessels or hemocoels. This complex, formed of axonal bulbs and vascular structures, serves as a storage and release center for the neurohormones, and is called a *neurohemal* organ. The best known neurohemal organs are the neural lobe of the vertebrate pituitary, the caudal neurosecretory system (urophysis) of certain fishes, the sinus gland of crustaceans, and the corpus cardiacum of insects. Neurohemal organs do not always occur in connection with neurosecretory systems, but the absence of true synaptic terminations is quite characteristic of neurosecretory axons.

Very little is known concerning the form in which the active ingredients are contained within the neurosecretory granules. Electron microscopic studies indicate that the granules are surrounded by distinct membranes, and these are presumed to be semipermeable. The neurosecretory droplets present in the rat's neurohypophysis decrease in density after the animals are dehydrated in order to deplete the stores of antidiuretic hormone. When tested for effects on pigment cells, homogenates of the sinus gland of the crab *(Uca)* are almost inactive in isotonic sucrose, but are activated by dilution with water, heating, freezing, and other procedures that are known to damage or disrupt biologic membranes. Physicochemical studies suggest that the neurosecretory granules found in sinus gland extracts contain neurohormones that are polypeptide in nature and that they are probably bound to large protein carriers. The Gomori and other staining techniques probably do not actually stain the neurohormones within the granules, but rather carrier molecules with which the active principles are associated.

Though the neurosecretory axons do not synapse with other neurons, muscles, or exocrine glands, there are indications that the glandular neurons can conduct impulses. The intra-axonal transport of secretory droplets has been observed in fresh preparations of the hypophysial stalk of the goosefish, *Lophius.* The hypophysial stalk of this fish is exceptionally long and has been useful in the demonstration of action potentials. Direct electrical stimulation of these fibers near the proximal end of the stalk leads to activity that may be recorded at the distal end. Delayed responses result from stimulation of the hypothalamic region near the base of the stalk.

The *urophysis* is a neurosecretory structure located at the posterior end of the spinal cord in fishes. Neurosecretory cells within the cord (Dahlgren cells) send axons ventrally and their enormous end-bulbs cause this region to bulge slightly. While it is agreed that the urophysis is analogous to the neurohypophysial division of the pituitary gland, its function remains obscure. There are suggestions that its products may be involved in osmore gulation and buoyancy, but others do not confirm these findings It has been shown that the giant, neurosecretory cells contributing to the urophysis of the eel are capable of conducting impulses.

If neurosecretory cells can receive and conduct impulses, as is the case in certain instances, the significance of this action remains problematical. Two possibilities might be considered: (1) the impulse could conceivably trigger the release of secretion from the cell itself or (2) the impulse might activate pituicytes or other glia-like cells which are often closely applied to axonal endings within neurohemal organs. Some investigators have suggested that these glia-like elements may effect changes in the neurosecretions to form the active principles which reach the body fluids.

NONARTHROPOD INVERTEBRATES

Neurosecretory cells have been identified in all invertebrate phyla that possess well-defined nervous systems, with the possible exception of lower coelenterates and ctenophores. Evidence that the products of neurosecretory cells perform functional roles in certain invertebrate species has not been pursued far enough to be convincing. It is often difficult to correlate cytologic changes in the neurosecretory cells with prevailing physiologic states in the organism. Furthermore, classical deficiency experiments are often difficult or impossible, and as a substitute, extracts of the nervous system have been used in injection experiments. When extracts containing all elements of nervous tissue are employed, there is the danger of misconstruing what may be a pharmacologic effect of the tissue for a physiologic response to a neurohormone. Unfortunately, none of the invertebrate neurohormones are yet available in chemically pure form.

Some Lower Invertebrate Phyla

Neurosecretory cells are known to be present in the cerebral ganglia of flatworms such as *Dugesia* and *Polycelis,* A water-soluble factor, prepared from planarian brain, induces the regeneration of excised eyespots. Since the number of neurosecretory cells in the brain of *Polycelis* increases while posterior regeneration is in progress, it is likely that neurohormones are involved in this process.

The circumoral ring and radial nerves of starfishes are said to contain neurosecretory cells, and a number of physiologic effects have been ascribed to their secretions. Neurosecretory cells have been reported in the motor ganglia of the Ophiuroidea. Materials obtained from nerve extracts are found to effect colour change, locomotor activity, water content, and gamete shedding. Extracts of radial nerves from starfishes contain a principle (shedding substance) that induces the release of gametes from the ripe gonads, and another (shedhibin) that inhibits the shedding of eggs. The exact cellular source of these agents acting upon the gonads is obscure, but they are thought to be neurosecretions.

NEUROENDOCRINE MECHANISMS IN ANNELIDA

Neurosecretory cells are present in the central nervous systems of all three major classes of annelids, and neurohemal organs have been identified in some species. The neuroendocrine complexes of polychaetes and oligochaetes appear to be involved in the control of three processes: (*a*) maturation of the gonads, (*b*) somatic transformations related to reproduction, and (c) the regeneration of posterior segments. The presence of true, non-neural endocrine glands in this phylum has not been established.

Polychaeta

As many polychaetes become sexually mature, they undergo somatic modifications which equip them to swim at the surface of the sea and engage in spawning activity. This transformation of an immature individual (atoke) into a reproductive individual (epitoke) is called *epitoky,* and involves changes in the parapodia, chaetae, musculature, eyes, size of the segments, etc.

Much evidence indicates that neurosecretory cells within the brain of nereids are the source of a neurohormone that inhibits gonadal maturation and epitoky. By analogy with insects, this is sometimes called "juvenile hormone" since it acts to keep the animals sexually immature. Surgical removal of the cerebral ganglia, in nereids with or without natural epitoky, results in premature and accelerated development of the gametes. In species in which epitoky occurs, decerebration of the atokous worms causes the premature attainment of the same metamorphic transformations as normally occur when the worms become sexually mature. All these changes that follow decerebration can be prevented by implanting ganglia from atokous donors; ganglia contributed by epitokous donors do not prevent the precocious epitoky and sexual maturation. The onset of normal epitoky can be delayed in intact individuals by the implantation of ganglia from atokous donors. Production or liberation of the inhibitory brain neurohormone apparently ceases after the onset of sexual maturity. Certain species and races of nereids normally reproduce in the atokous condition (without epitoky) and, in these forms, decerebration does not induce epitoky; the operation, however, does result in the precocious maturation of male germ cells. It is not known what factors operate to shut off the production of brain neurohormone as the nereids approach sexual maturity.

Ablation and implantation experiments indicate that neurosecretory cells in the posterior portions of the cerebral ganglia of nereids are the source of a neurohormone that controls the regeneration of posterior segments. Little or no posterior regeneration occurs if the cerebral ganglia are removed at the time of segment amputation. Intracoelomic implants of the ganglia into decerebrate hosts restore the ability to replace excised posterior segments. Studies on *Nereis* have shown that both normal growth and regenerative capacity decline as the worms age, and that neurosecretions from the cerebral ganglia are apparently involved in both processes. It is not known whether the growth-promoting and regeneration-promoting neurohormones are identical or distinct secretions.

Oligochaeta

Neurosecretory cells within the supraesophageal ganglia of *Lumbricus* are the probable source of a secretion that governs the maintenance of external sex characters *(e.g.,* clitellum), and possibly the differentiation of gametes. This neurohormone has an inhibitory effect on gonad maturation and is probably involved in the process of egg laying. As in polychaetes, the brain is essential for the regeneration of posterior segments.

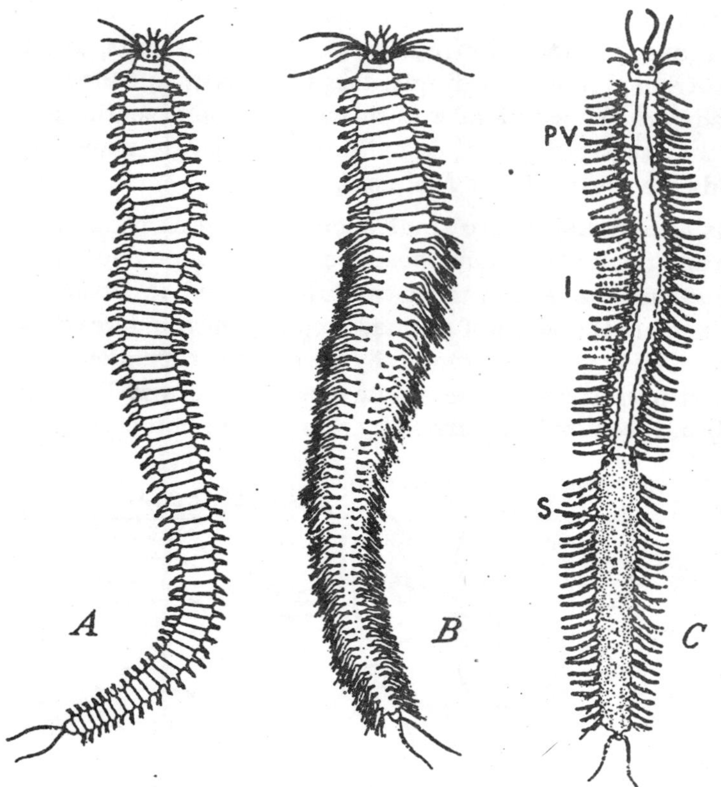

Fig. 4.1. Metamorphosis in polychaete annelids. A, the immature or atokous stage of *Nereis; B*, the adult or epitokous stage of *Nereis; C*, stolonization in a syllid polychaete. I, intestine; PV, proventriculus; S, stolon. (From Charniaux-Cotton and Kleinholz: *In* Pincus, Thimann, and Astwood [eds.]: The Hormones, Vol. 4. New York, Academic Press, 1964; after Durchon, 1960.)

Hirudinea

The central nervous system of the leech is rich in cells that are obviously neurosecretory in nature, but no physiologic role for their products has been established. The giant chromaffin cells in each ventral ganglion of *Hirudo* have been shown to contain an epinephrine-like substance. Other neurosecretory cells of the nerve cord and subesophageal ganglion are reported to give a positive chromaffin reaction.

NEUROENDOCRINE MECHANISMS IN MOLLUSCA

Neurosecretory cells are found within the brains of all mollusca that have been studied, and, in addition, appear in connection with many of the ganglia. The epithelial vesicles of cephalopods (squids and octopuses) show some structural evidence of neurosecretory competence, though this has not been proved. The epistellar body of the octopus is such a structure, and the indications are that it is a rudimentary photoreceptor.

The genital ducts and accessory glands of slugs *(e.g., Limax)* undergo functional development at the time of gonadal maturation. Precocious development of these sex accessories may be induced by transplanting pieces of mature gonads into young recipients. The accessory sex organs involute following the castration of adults, and these changes are repaired by gonadal transplants. Accordingly, there is reason to believe that the gonads of pulmonate gastropods are the source of a hormone which is essential for the development and functioning of the accessory genital complex. In addition to neurosecretory complexes, true endocrine glands are present in the Mollusca.

The optic glands of cephalopods are small *endocrine* organs lying on the optic stalks on either side of the brain. Such glands have been found in all cephalopods examined, with the exception of *Nautilus,* and have been most carefully studied in Octopus. They contain no neurosecretory cells and are the source of a gonadotrophin which induces ovarian and testicular enlargement. The production of gonadotrophin by the gland is regulated by an inhibitory nerve supply extending to it from the subpeduncular lobe of the brain. The inhibitory centers in this Portion of the brain seem to be governed by changes in photoperiod.

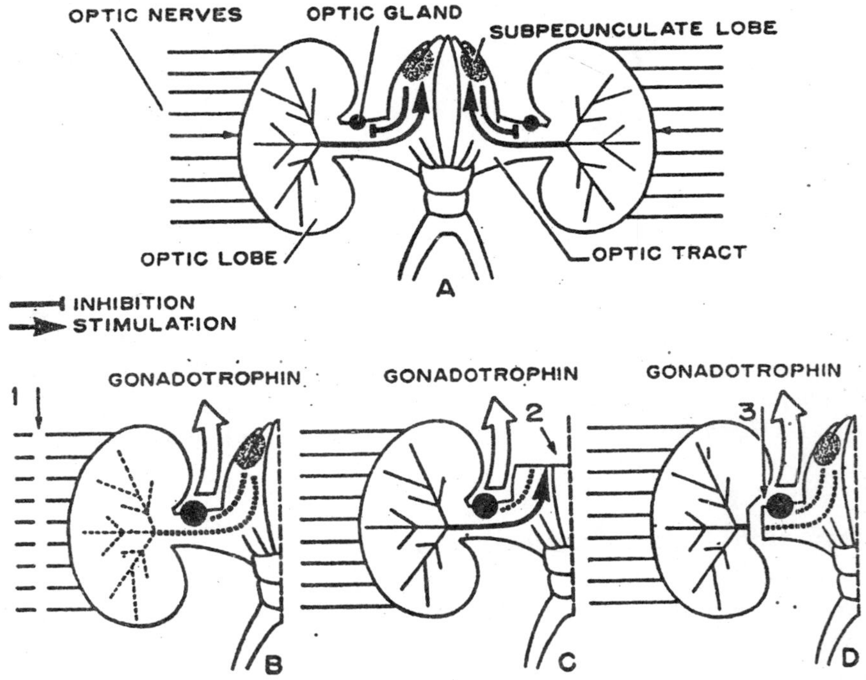

Fig. 4.2. Neuroendocrine control of gonad maturation in the octopus. *A,* In an immature animal, unoperated upon, the production of gonadotrophin by the optic glands is held in check by an inhibitory nerve supply originating in the subpedunculate lobe of the brain. Activation of this brain center appears to depend upon changes in photoperiod. *B,* Section of the optic nerves (point 1) prevents activation (broken lines) of the inhibitory nerve center; the optic glands enlarge and secrete gonadotrophin, which induces hypertrophy of the gonads. The same result may be accomplished by ablation of the subpedunculate lobes (C, at point 2) or by section of the optic tract *(D,* point 3). (Modified from figures by Wells and Wells.)

Thus, light stimuli received by the eyes activate nerve centers in the brain and these hold the optic glands in check and the gonads are inhibited through a lack of gonadotrophin. After blinding or excision of the subpedunculate lobes, the optic glands enlarge and the ovaries develop precociously. These procedures have no effect on the gonads after removal of the optic glands. Functioning of the brain-optic gland-gonadal system of *Octopus* is suggestive of the regulation of sexual maturity in vertebrates by the hypothalamic-pituitary-gonadal axis. It should be noted, however, that neurosecretory cells seem not to be involved in the case of *Octopus*. After excision of the optic glands *of Octopus*, the oocytes develop normally but follicle cells do not, and yolk deposition fails to occur. The gonadotrophin from the optic gland of the male appears to promote spermatogenesis. Ablation of the subpedunculate lobes of very young males results in enlargement of the optic gland, increased testicular weight, and the precocious appearance of spermatophores in the testis.

There is no information on the chemical nature of the optic gland hormone. Experiments on the transplanatation of optic glands and the effectiveness of fractionated organ extracts would probably yield interesting results.

NEUROENDOCRINE MECHANISMS IN THE CRUSTACEA

Location of Structures

The regulatory mechanisms of crustaceans are extremely complex and are very closely related to the central nervous system. The principal endocrine structures of a generalized crustacean. Neurosecretory cells are abundant in the brain and in practically all of the other ganglia. The endocrine organs in crustaceans, as in insects, fall into three categories: (*a*) aggregations of neurosecretory cells that produce neurohormones and discharge them from their axonic terminals, (*b*) neurohemal organs for the storage, possible modification, and liberation of neurohormones, and (*c*) true endocrine glands (non-neural) that release hormones into the blood.

Important neurosecretory centers are found in connection with the optic ganglia which lie within the eyestalks. Best known of these are the "X organs" which are present in the eyestalks of most stalk-eyed species, or in the head when stalked eyes are absent. Two kinds of X organs are now recognized: the *ganglionic X organ* and the *sensory pore X organ*. The two X organs fuse to form a single structure on the medulla terminalis of brachyurans but, in natantians, they are separated groups of neurosecretory cells. Clusters of secretory neurons are also found within the brain, the thoracic ganglia, the esophageal connective ganglia, and the tritocerebral commissure.

The sinus gland was early recognized as a potent source of hormones, and it was assumed initially that the secretions arose from glandular elements comprising the structure. Since the sinus glands do not show the histologic characteristics of a typical gland of internal secretion, it became apparent that their products probably arose elsewhere. As surgical procedures improved, it was possible to remove this small organ alone without resorting to ablation of the entire eyestalk. It became apparent that these two surgical procedures result in different physiologic effects, and this led Bliss and Welsh (1952) to conclude that the sinus glands are essentially reservoirs for the storage and discharge of neurohormones derived from the axons of neurosecretory neurons. This concept has been amply confirmed and is now generally accepted. Like other neurohemal organs, the sinus glands consist chiefly of axonic

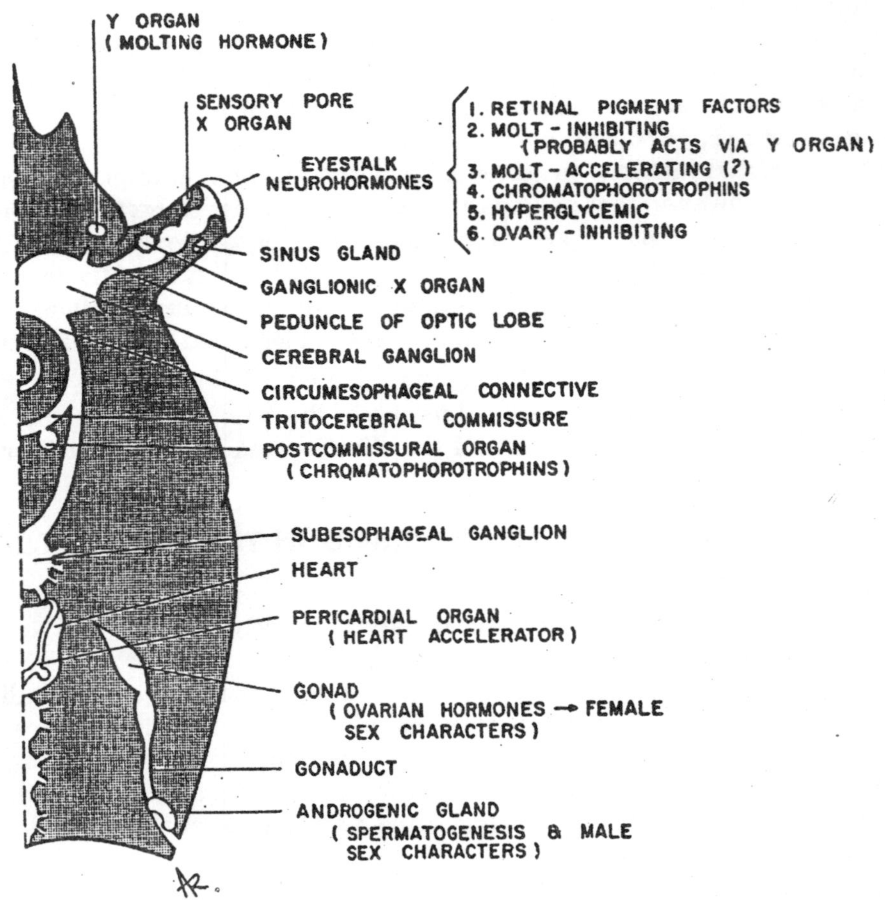

Fig. 4.3. Summarizing diagram of the cephalothorax of a generalized crustacean, showing the locations of the best-known endocrine glands and neuroendocrine areas. The active principles from these structures are listed or shown in parentheses. Neurosecretory cells are present in all components of the central nervous system, and no attempt has been made to show them.

terminals, and are closely associated with rich vascular channels. In addition to discharged secretions, they contain nonglandular cells that seem comparable to the pituicytes present within the pars nervosa of the vertebrate neurohypophysis. These glia-like elements often appear to be wrapped around the bulbous terminals of the secretory axons. While it is possible that the sinus glands may possess some inherent secretory ability, like the corpora cardiaca of insects, they are chiefly storage-release centers for neurohormones.

The postcommissural and pericardial organs also appear to be essentially neurohemal in nature. The secretion-charged axons from cells within the brain and esophageal connective ganglia terminate in the postcommissural organs which store and discharge the neurohormones. Neurosecretory cells, having their perikarya within the ganglia of the ventral chain, deliver their products to the pericardial organs located near the openings of the large veins into the pericardial sac.

Three endocrine glands, not composed of secretory neurons, are found in the Crustacea: the Y organs, the androgenic glands, and the ovaries. The Y organs are located in the antennary or maxillary segment and resemble, in some respects, the prothoracic, molt-regulating glands of insects. They appear to be devoid of direct innervation, and are probably regulated by neurosecretions derived from the eyestalk complex. Androgenic glands have been found in a variety of crustaceans, and in a few species of insects. They are generally located outside the testes and are typically found along the vas deferens. Although rudimentary androgenic glands are present in females, they develop only in males. These masculinizing glands are probably controlled by neurohormones from the X-organ sinus gland complex. The crustacean ovary, unlike the testis, serves in an endocrine capacity.

Retinal Pigment Migration

The compound eyes of higher crustaceans are composed of a large number of units called *ommatidia.* Three functionally distinct groups of pigments are possessed by each ommatidium: distal retinal pigment, proximal retinal pigment, and reflecting white pigment. The distal and proximal pigments screen the sensory component of the ommatidium, the rhabdome, in bright light, and migrate away from the rhabdome in darkness. The white pigment increases the effectiveness of dim light as a stimulus by reflecting the light that enters the eye over adjacent receptors. When the eye is adapted to darkness, the distal pigment migrates distally to enclose the dioptic apparatus, the proximal pigment moves to a position below the basement membrane, and the reflecting pigment surrounds the rhabdome. In the light-adapted eye, the distal pigment disperses proximally as far as the retinula cells, the proximal pigment moves distally to meet the distal retinal pigment, and the reflecting pigment moves away from the rhabdome and assumes a position below the basement membrane. Eyestalk extracts, prepared from light-adapted prawns, result in light adaptation of the distal and reflecting pigments when injected into dark-adapted recipients; the proximal pigment remains unchanged.

It is probable that the responses of these retinal effectors are controlled by two neurosecretions, one light-adapting and the other dark-adapting. There are many species variations but, in general, the two factors can be extracted from the eyestalk and various portions of the nervous system. The regulation of retinal pigments is a very complex process, and available extracts do not duplicate exactly the normal actions of variable illumination.

Chromatophorotrophins and Colour Change

The chromatophores, or pigment cells, of crustaceans may contain white, red, yellow, black, brown, and blue pigments. By appropriate condensation or dispersal of particular pigments within the chromatophores, crustaceans can approximate rather accurately the coloured backgrounds upon which they rest. It is known that crustacean pigment cells are devoid of nerves and are adjusted by regulatory agents present in the blood (chromatophorotrophins). Studies involving eyestalk or X-organ ablation, coupled with the testing of extracts made from different regions of the nervous system, have demonstrated that the chromatophorotrophins are products of neurosecretory cells. Some of these factors are abundant in the tritocerebral commissure and the postcommissural organs. Multiple chromatophorotrophins have been extracted from the nervous systems of decapod crustaceans, but the exact number has not been agreed upon. It is possible that the effective agents obtained through the extraction of neurosecretory aggregations *(e.g.,* ganglionic X organ) are not of the same chemical form as the finished products liberated by neurohemal organs *(e.g.,* sinus gland).

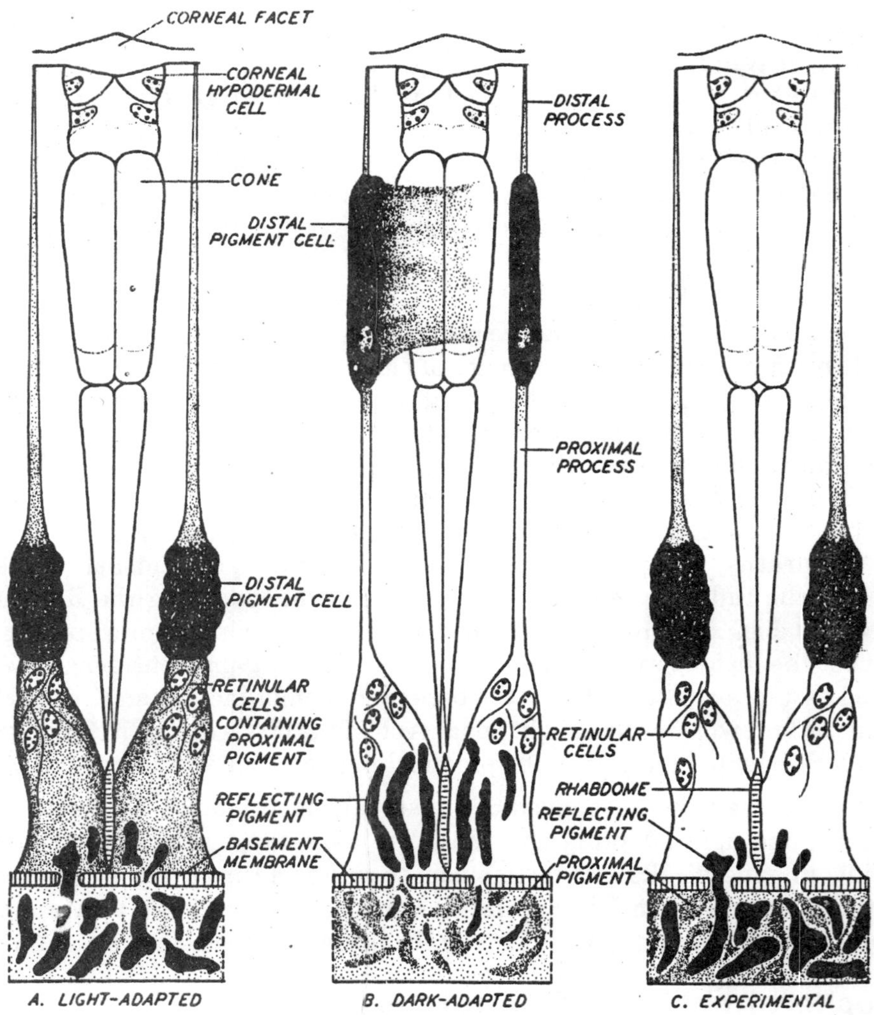

Fig. 4.4. Ommatidia from the eyes of the shrimp *Palaemonetes vulgaris,* showing general morphology and the positions of the pigments under different conditions. A, The light-adapted condition. The distal pigment has migrated inwardly and is concentrated near the retinular cells. Most of the proximal pigment, contained within the retinular cells, lies distad of the basement membrane. The reflecting pigment moves through the basement membrane, *i.e.,* away from the rhabdome. B, The dark-adapted state. The distal pigment surrounds the distal end of the cone; the reflecting pigment surrounds the rhabdome, and the proximal pigment migrates into the portions of the retinular cells which lie below the basement membrane. C, An ommatidium from an animal which had been adapted to darkness and then had an injection of an extract of the eyestalks (sinus glands). The extract causes the distal and reflecting pigments to assume the positions characteristic of light adaptation; the proximal pigment apparently is not influenced in this species. (Modified after Kleinholz, 1936.)

Though barnacles have neither compound eyes nor chromatophores, extracts of their nervous systems have chromatophore-activating properties when tested on decapod crustaceans. Blind cave crayfishes possess no retinal pigments and no chromatophores, yet their nervous systems contain both the distal pigment, light-adapting principle and another affecting integumentary chromatophores.

Molting

Growth is a discontinuous process in arthropods, increase in size being restricted to the period between loss of the old exoskeleton and the production of a new one. Molting may be seasonal or continuous, and is influenced by a great variety of environmental conditions. Loss and replacement of the exoskeleton is an outward expression of a whole complex of major metabolic adjustments occurring within. Four periods are recognized in the crustacean molt cycle: (1) During *premolt,* inorganic constituents of the old exoskeleton are resorbed and stored in gastroliths or hepatopancreas. Oxygen consumption increases, glycogen is deposited in the hypodermis, and lost limbs are regenerated rapidly. (2) *Molt* is the sloughing of the old cuticle, and this is accompanied by a marked increase in size. The immediate increase in size results from a rapid absorption of water, which thus reserves enough space to permit growth even after the new cuticle has hardened. (3) *Postmolt* is the period during which a new exoskeleton is formed through the redeposition of chitin and inorganic salts. (4) *Intermolt* is characterized by relative quiescence, but there is generally some storage of reserves in preparation for the next molt. Some crustaceans eventually cease molting and con-sequently undergo no further growth, a condition called *anecdysis.*

In most species of crustaceans, the ablation of both eyestalks results in accelerated molting and precocious growth. Extracts of the ganglionic X-organ sinus gland complex act to prevent molting. These and other observations led to the conclusion that neurosecretory cells in the ganglionic X organ secrete a moltinhibiting neurohormone that is stored and released from the sinus gland. Studies on the formation of gastroliths in crayfish indicate that the moltinhibiting factor is derived from components of the eyestalk. These concretions normally form in the stomach wall during premolt, but may be induced at any period of the molt cycle by ablating both eyestalks or by excision of the X-organ sinus gland complex. Gastrolith formation in eyestalkless animals can be prevented by the implantation of sinus glands. Profound quantitative shifts in the metabolism of carbohydrate, protein, lipid, and inorganic materials occur in relation to the molt cycle. As might be expected, these biochemical adjustments are facilitated by the endocrine factors that control molting.

Whether the components of the eyestalk produce a second factor, molt-accelerating principle, remains unresolved. Under certain conditions, and in certain species, moltinhibition may follow eyestalk abalation; certain workers feel that this can be explained on other grounds, and hence there is not sufficient evidence for a molt-accelerating neurohormone. On the other hand, it has been suggested, from studies on the crayfish *(Orconectes virilis),* that both molt-inhibiting and molt-accelerating factors from the eyestalk are involved in regulating the cyclical changes which characterize metabolism of the hepatopancreas.

The Y organs were discovered by Gabe (1953), and it is certain that they produce a hormone which performs a positive role in the molting process. Young crabs are prevented from molting by bilateral removal of the Y organs, but molting cycles may be restored in these animals by implanting several Y organs. These organs normally degenerate after puberty in the crab *Maia,* and no further molting occurs. It seems probable that the moltinhibiting neurohormone from the ganglionic X-organ sinus gland complex may exert its effects by regulating the functional status of the Y organs. It would follow that the moltinhibiting neurosecretion is continuously produced during postmolt and intermolt periods to stall the production of molting hormone by the Y organs; during premolt, the production of neurosecretion ceases, or diminishes, and the Y organs are allowed to secrete molting hormone.

The Y organs are analogous (possibly homologous) to the prothoracic glands of insects. It is interesting that both endocrine glands secrete hormones that control molting, and that both are regulated by neurosecretions from the central nervous system. The prothoracic glands degenerate in adult insects, following the last molt, and the Y organs are greatly reduced in crabs during anecdysis.

Reproduction

There is more evidence for the existence of sex hormones, in the malacostracans, the higher crustaceans, than in the other invertebrates. Hormones regulating the differentiation of male and female sexual characters arise from the ovaries and the androgenic glands; the testis itself probably has no endocrine function. Neurosecretions of the ganglionic X-organ sinus gland complex have inhibitory effects on ovarian maturation and secretory activity of the androgenic glands. There is evidence that the molting hormone from the Y organs is also essential for normal differentiation of both the ovary and the testis. After bilateral ablation of Y organs in very young crabs *(Carcinus),* mitotic processes are impaired in both the ovary and testis. In the ovary, oogonial mitoses cease and follicles are not formed around the oöcytes; vitellogenesis, or yolk deposition, does not occur in the oocytes that lack follicles. Spermatogonial mitoses are arrested, and the testes become depleted of mature germ cells.

Bilateral removal of the ovaries in certain crustaceans *(e.g., Orchestia)* results in the loss of specific secondary sex characters. Secondary characters of the female type may be induced in genetic males by removing the androgenic glands and implanting ovaries. Two ovarian hormones have been postulated on the basis of good evidence, but neither is known chemically.

It has been shown in many species of crustaceans that ablation of the eyestalks, during periods of sexual quiescence, is followed by enlargement of the ovaries and by a precocious deposition of yolk in the oocytes. Extracts prepared from eyestalks, ganglionic X organs, or sinus glands effectively prevent ovarian enlargement when administered to females that are entering the period of reproductive activity. Such extracts are without effect when given to very immature females or to females immediately after the end of the reproductive period. It appears that the primary action of this ovarian-inhibiting neurohormone is to prevent vitellogenesis. The production of this factor is apparently regulated by environmental stimuli that impinge on sensory receptors.

The important androgenic glands were discovered by Charniaux-Cotton (1954) in the amphipod *Orchestia gammarella,* and have since been found in quite a variety of crustacean species. The following observations indicate that the androgenic gland hormone regulates spermatogenesis and the secondary sex characters of the male: (1) In *O. gammarella,* ablation of the androgenic glands causes spermatogenic activity in the testis to wane, and regenerated appendages develop neither the male nor the female sex characteristics; in *O. montagui,* oogenesis ensues in the testis. (2) If androgenic glands are grafted to intact female hosts, the ovaries eventually transform into testes and proliferate spermatocytes, spermatids, and functional spermatozoa. The appendages progressively assume male form, and masculine sexual behaviour is acquired. The bipotentiality of the primordial germ cells is clearly indicated. (3) Castration of normal males, the androgenic glands remaining intact, has no effect other than to remove the source of germ cells. (4) If ovaries are grafted into males that have been deprived of androgenic glands, they persist as functional ovaries. When grafted into males possessing androgenic glands, with or without host testes, the ovaries quickly acquire testicular structure. (5) Aqueous extracts of androgenic glands, injected into males deprived of these organs, effectively prevent changes in the testes and secondary sex characters. In certain species, blood plasma from males, injected into female recipients, causes the ovary to assume testicular structure and function.

It follows from these observations that the differentiation of germ cells of crustaceans is reversible, regardless of their genetic constitution: in the absence of androgenic gland hormone the gonads can become ovaries, but testicular differentiation requires the presence of this hormone. Several species of decapod crustaceans are protandric hermaphrodites, and it is known that androgenic glands are present during the male phase and are lost before the onset of the female phase. The androgenic glands of some species are closely applied to the testes, and it is probable that the sex reversals formerly attributed to destruction of the gonads by parasites actually result from destruction of or damage to the androgenic glands.

Heart Acceleration

The pericardial organs are neurohemal in nature and release a neurohormone which increases the frequency and amplitude of the heartbeat. Although extracts of pericardial organs contain 5-hydroxytryptamine, it is believed that the major activity of the extracts is due to another substance not yet identified.

Metabolism

Striking variations in tissue metabolism occur during the molting cycle in Crustacea, and eyestalk factors, presumably neurosecretions, are known to be involved. Identity of the active principles and the manner in which they interact are problems that require further elucidation. Molt-inhibiting and Molt-accelerating factors from the eyestalk, and molting hormone from the Y organs are apparently implicated in these cyclical changes in metabolism. After removal of the eyestalks, the blood sugar concentration decreases, whereas the glycogen content of the hypodermis increases. The metabolic changes following eyestalk removal are comparable to those that occur in the intact animal in preparation for a molt. Hyperglycemia may be induced by administering extracts of the eyestalk or X-organ sinus gland complex to normal animals. Since the same effect is induced by subjecting intact animals to stressful stimuli, it appears that the response is normally mediated by the neuroendocrine system of the eyestalk. The importance of carbohydrate metabolism in the molting process is indicated by the fact that chitin is one of the major constituents of the exoskeleton and is derived from glycogen.

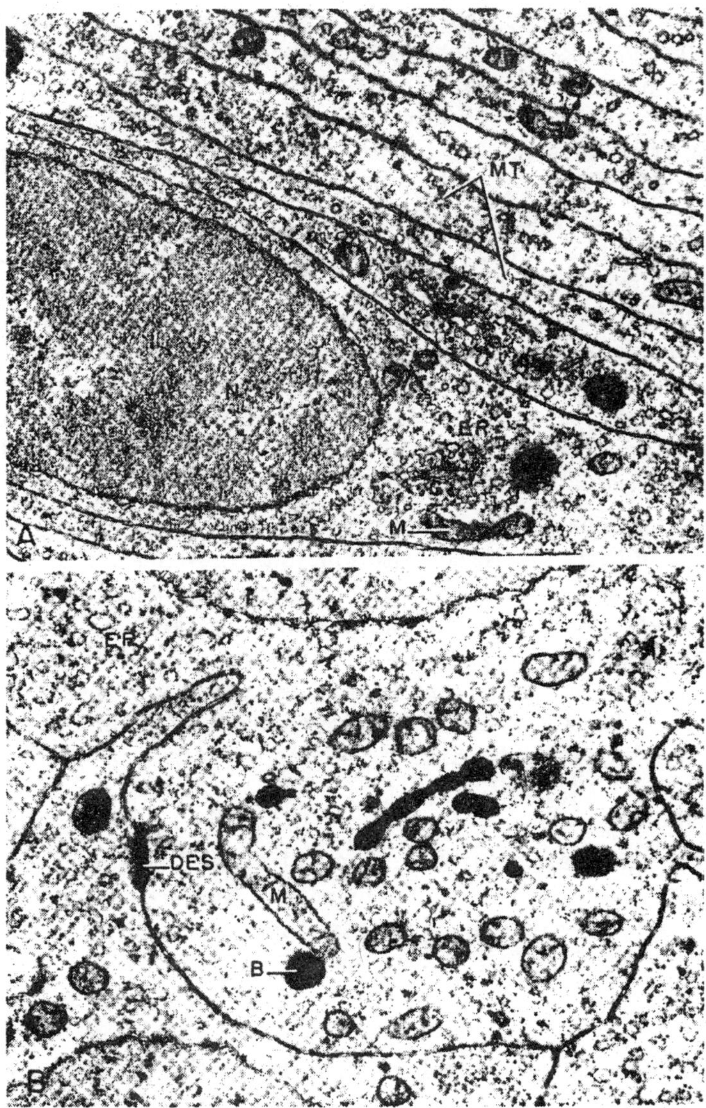

Fig. 4.5. Electron micrographs of the androgenic glands of the amphipod *Orchestia gammarella*. The cells of these glands are characterized by a well-developed granular endoplasmic reticulum, long mitochondria with transverse cristae, and frequent Golgi bodies. These ultrastructural features make it very probable that the androgenic gland hormone is a polypeptide or protein, rather than a steroid. A, × 15,300; B, × 21,780. B, probably a lysosome; DES, desmosome; ER, endoplasmic reticulum; G, Golgi body; M, mitochondrion; MT, microtubules; N, nucleus. (Courtesy of Mr. Jean-Jacques Meusy, Centre National de la Recherche Scientifique, Paris.)

Evidence for a hyperglycemic neurohormone has been adduced. It appears to be a protein and thus differs from the chromatophoro-trophins and the light-adapting retinal·pigment principle.

NEUROENDOCRINE MECHANISMS IN THE INSECTA

The experiments of Kopec (1917 to 1922) provided the first clear evidence that hormones are of functional importance in the invertebrates. The original view that pupation in Lepidoptera is influenced by a factor coming from the brain has been confirmed and extended to other orders of insects. Among the Insecta, as in the Crustacea, the key position in the endocrine system is held by neurosecretory centers that elaborate neurohormones acting directly on target tissues or indirectly on endocrine glands. Since insects are a very diversified group, many kinds of microsurgical procedures have been employed for the study of hormonal mechanisms. Among these may be mentioned transplantation, ablation of organs, parabiosis, isolation of larval parts by constriction, tissue cultures, the fusion of individuals by capillary tubes, and the isolation of pupal parts by fusion to glass plates. Methods are being perfected for the bioassay of insect hormones, and one hormone has apparently been isolated in pure form. Certain unifying concepts of hormone action have emerged, making it clear that the same general principles operate in both hemimetabolous and holometabolous insects. It has become apparent that the various hormones interact with one another, just as in vertebrates, and this makes it difficult to assign definite actions to a specific hormone. What a particular hormone accomplishes often depends upon other hormones that are present with it, upon the hormone titer in the body fluids, and upon the competence of the target tissues to respond.

Insect Life Cycles

In hemimetabolous development, illustrated by the locust, the cockroach, and the blood-sucking bug *Rhodnius,* the newly hatched young resemble the adults in many respects; the most important differences are size, the absence of wings, and immature genitalia. The young nymph, 'as it is called, undergoes a series of molts during which it gradually acquires adult characters, the most profound metamorphic changes occurring at the final molt. The number of nymphal instars varies with the species and may be altered by environmental conditions. All instars are capable of feeding, and there is generally no cessation of development during this type of cycle; however, growth occurs mostly between the nymphal molts.

Moths and butterflies are examples of holometabolous insects. Their eggs hatch into wormlike larvae that bear practically no resemblance to the adult. These caterpillars feed and undergo a series of molts, and, since growth occurs at molting, each succeeding instar is somewhat larger than the preceding one. The larval instars, though differing in size, usually do not undergo profound structural changes. Eventually, the last instar molts and transforms into a chrysalis or pupa, a quiescent stage which is incapable of feeding. During pupation the individual undergoes a rather complete structural reorganization, so that it resembles the adult much more closely than did the preceding larval stages. Eventually, the pupal case is burst, and an imago emerges; it may feed, but it undergoes no further growth or profound modification in structure.

Though certain organs may be transmitted from the larva to the adult without much change, many must undergo a complete reorganization. These changes involve the destruction of larval tissues (histolysis) and the differentiation of adult tissues and organs (histogenesis and organogenesis) from larval anlagen called imaginal discs. All these postembryonic changes must be timed and coordinated properly in order to produce the normal adult.

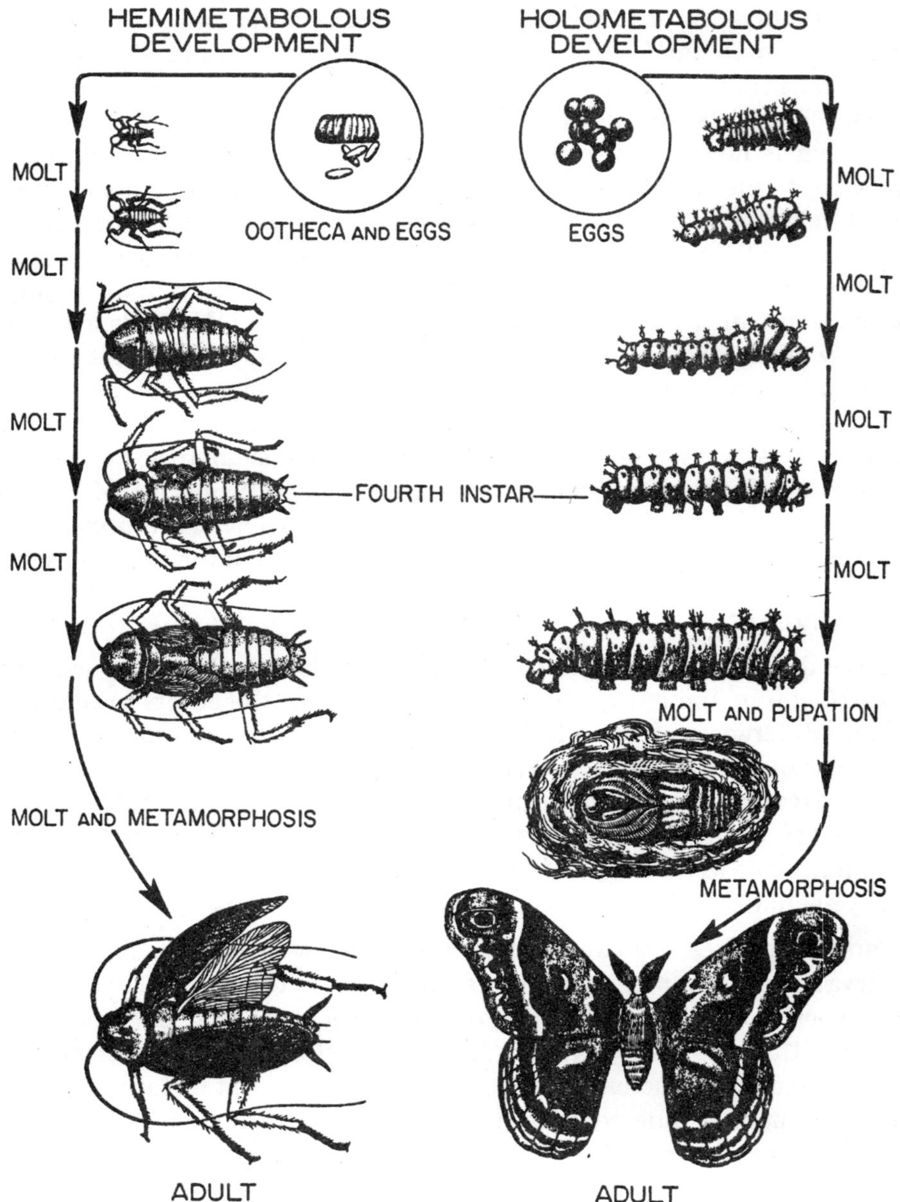

Fig. 4.6. Two contrasting types of life cycles among insects. In those exhibiting direct or hemimetabolous development, the last nymphal instar molts and metamorphoses into the adult. In forms undergoing indirect or holometabolous development, the last larval instar passes through a pupal stage before metamorphosing into the adult. The cockroach *(Periplaneta americana)* is taken as an example of hemimetabolous development; the giant silkworm moth *(Hyalophora cecropia)* is used as an example of holometabolous development. The instars are drawn to scale.

Metamorphosis includes the changes in form normally undergone by an animal during its life cycle, and these events are genetically determined. Since the cells of the larva, pupa, and adult contain the same genetic information, it appears that different sets of genes are sequentially brought into action during the life cycle. There is increasing evidence that the genetic switch, or activation of an alternative set of genes, is triggered by neuroendocrine adjustments. The establishment of differential hormone environments, known to prevail during particular periods of the life cycle, may often be effected by external changes such as photoperiods, temperature, food, etc.

Diapause

The life cycles of many insects include a period of dormancy, called *diapause,* during which growth and differentiation are almost in complete abeyance. The diapause may result from the direct effects of adverse environments, or it may appear regularly under the most favourable conditions as a phase of the life cycle. Depending upon the species, such periods of suspended development may occur at any time in the life cycle-egg, nymph, caterpillar, pupa, or adult. These periods of dormancy may have the net effect of producing young at a time when their chances for survival are maximal, of enabling the individual to survive hazardous environmental conditions.

Periods of developmental arrest in the egg are characteristic of certain insects. Some silkworm *(Bombyx mori)* races are single brooded or univoltine, each generation appearing in the spring following a period of arrested embryonic development during the winter. Other races are divoltine or multivoltine, the nondiapausing summer eggs producing one or more generations before the winter generation of diapausing eggs is produced. The diapausing eggs can be distinguished by the appearance of pigment in the serosa. Whether a female produces diapause or nondiapause eggs depends upon the photoperiod to which she was exposed while an egg or very young larva. In case of the divoltine race, the diapause eggs survive the winter and produce adults in spring. Females of the first generation produce nondiapause eggs because of their exposure to short days. Females of the second generation, derived from nondiapause eggs of spring, produce their eggs during mid-summer, under the influence of long days and high temperatures. These eggs are of the diapause type and do not develop until spring. The necessary information for the production of diapause eggs by these moths is derived from the long hours of daylight, and this photoperiodic mechanism provides the basis for winter survival.

It has been shown that the immediate stimulus for the production of diapause eggs by *Bombyx* is a neurohormone derived from neurosecretory cells in the subesophageal ganglion of the mother, and this secretion is released only in mothers that were exposed to long periods of daylight when they themselves were eggs, or more precisely, developing embryos. The neurohormone of the mother acts upon the eggs (embryos) before they are released from the genital tract. When the ovaries of a univoltine race (diapause eggs only) are transplanted into larvae of a divoltine race (diapause or nondiapause eggs), the eggs produced by the grafted ovary, after the host attains adulthood, always show the voltinism of the host. There are indications that the brain exerts an inhibitory effect, by way of the esophageal connectives, on the release of diapause neuro-hormone by the subesophageal ganglion.

In certain butterflies *(e.g., Araschnia levana),* the photoperiod to which the young larvae are exposed determines whether or not a pupal diapause occurs. The adults emerging from diapausing and nondiapausing pupae may be quite different in appearance, and occasionally, spring and summer varieties have been mistaken for different species.

The Cecropia moth *(Hyalophora cecropia)* undergoes a pupal diapause in nature, and the individual passes the winter in this condition. Mating occurs in the spring, and in about two weeks, the eggs hatch to give rise to the first larval instars. There are four larval molts, and the fifth instar spins a silken cocoon in which it pupates. With the rising temperatures of spring, the diapause ends and the adult emerges from the old pupal cuticle and escapes from the cocoon. Adult Cecropia do not feed, and between a quarter and a third of the pupal weight is utilized in the production of about 300 yolk-laden eggs." Pupal dormancy in this species normally lasts for about eight months, but it has been found that the length of the diapause may be materially shortened by chilling the pupae and then returning them to room temperatures.

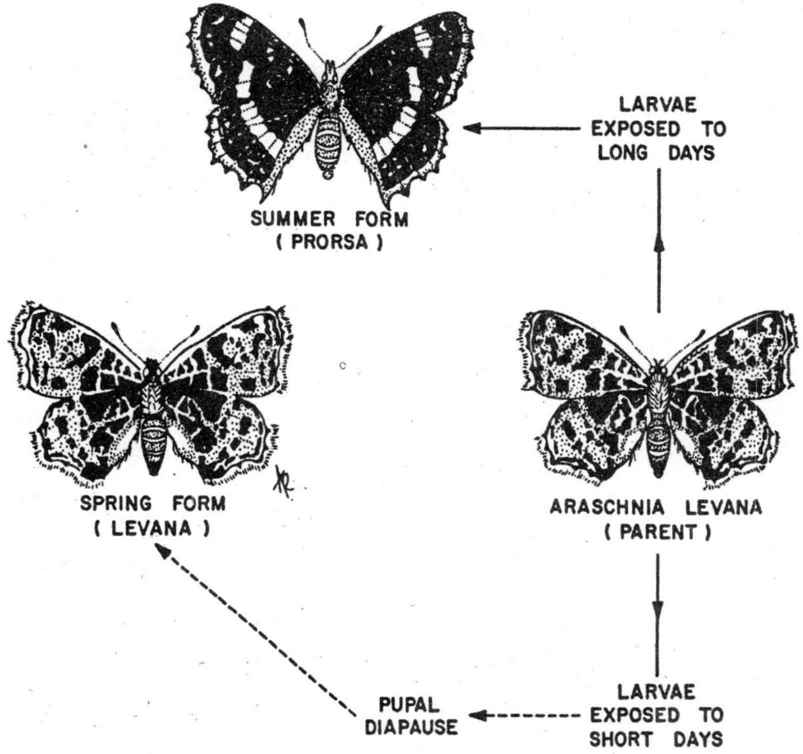

Fig. 4.7. In the butterfly *(Araschnia levana)*, the photoperiod to which the young larvae are exposed determines whether or not a diapause will occur at the pupal stage. Larvae exposed to short days undergo a pupal diapause; those exposed to long days do not. Diapausing pupae produce the spring form and nondiapausing pupae produce the summer form. The spring and summer varieties are so different in appearance that they have been mistaken as separate species. (Redrawn and modified from Wigglesworth, V. B.: Endeavour, 1965).

With the onset of pupal diapause, there is a sudden and complete cessation of growth, and few if any mitotic divisions occur during this long dormancy. The heart continues to beat and the animal utilizes oxygen and produces carbon dioxide, but all metabolic processes are reduced to a very low level. The rate of oxygen consumption in diapausing Cecropia is only 1.4 per cent that of the mature larva, and 5 per cent that of the adult moth just before the last molt. Just prior to the spinning of the cocoon, the larva ceases the intake of substances other than atmospheric oxygen; henceforth until its death as a mature moth 8 to 10 months later, the materials of its own body are reworked as a source of energy. The metabolism of an adult moth in flight is approximately 2000 times that of a diapausing pupa.

The destructive larvae of the European corn-borer *(Ostrinia nubilalis)* enter pupal diapause in response to falling temperatures and short daily photoperiods.

Some insects undergo diapause in the adult stage, and this generally takes the form of arrested reproduction in the female. In the Colorado potato beetle *(Leptinotarsa),* for example, the females begin to burrow into the soil during autumn and become quiescent; the metabolic rate falls, thoracic muscles degenerate, and egg maturation ceases. All these diapausal changes can be reversed in this beetle by implanting active corpora allata into their bodies. This suggests that the reproductive arrest is brought about by environmental factors which act via the corpora allata to inhibit the production of juvenile hormone.

Control of Postembryonic Development

The endocrine mechanism regulating growth and differentiation in all insects centers in the neurosecretory system of the brain. The extrinsic stimulus varies from species to species: in the blood-sucking bug *Rhodnius,* stretching of the larval abdomen by a meal of blood provides a nervous stimulus to the brain; in locusts, the act of chewing and swallowing may accomplish the same effect; in Cecropia, temperature changes activate the brain. The activated neurosecretory cells of the brain produce a brain hormone (BH) which their axons discharge into the corpora cardiaca where it is stored and from whence it is liberated into the blood. The BH specifically stimulates the prothoracic glands to produce a hormone that has been variously named molting hormone (MH), growth and differentiation hormone, prothoracic gland hormone, and ecdysone. The MH acts directly upon the tissues of the body causing them to differentiate in the direction of adult structures. A second hormone, called juvenile hormone (JH), is secreted by the corpora allata, and its action is to encourage the laying down of larval or nymphal structures. In other words, JH suppresses pupal and imaginal differentiation. The secretory activity of the corpora allata is also regulated by the nervous system, and the amount of JH present in the blood diminishes progressively with successive molts. Imaginal differentiation occurs under the influence of MH, when very little if any JH is present.

The body form attained by organisms depends upon the genetic control of developmental processes. It appears that multiple sets of genes are present in insect species, and that these sets can be called into expression successively during the life cycle. As these genie complexes are activated (or suppressed), larval, pupal, and imaginal characters differentiate.

Studies on the salivary gland chromosomes of *Chironomus* show that molting hormone (ecdysone) acts promptly to cause puffing of only a few gene loci, but this is followed by sequential puffing of other loci. Since molting hormone seems to activate only several gene loci, it has been proposed that RNA messages produced by these genes, and the protein synthesis occurring as a consequence, set in motion certain cytoplasmic reactions which activate other genes.

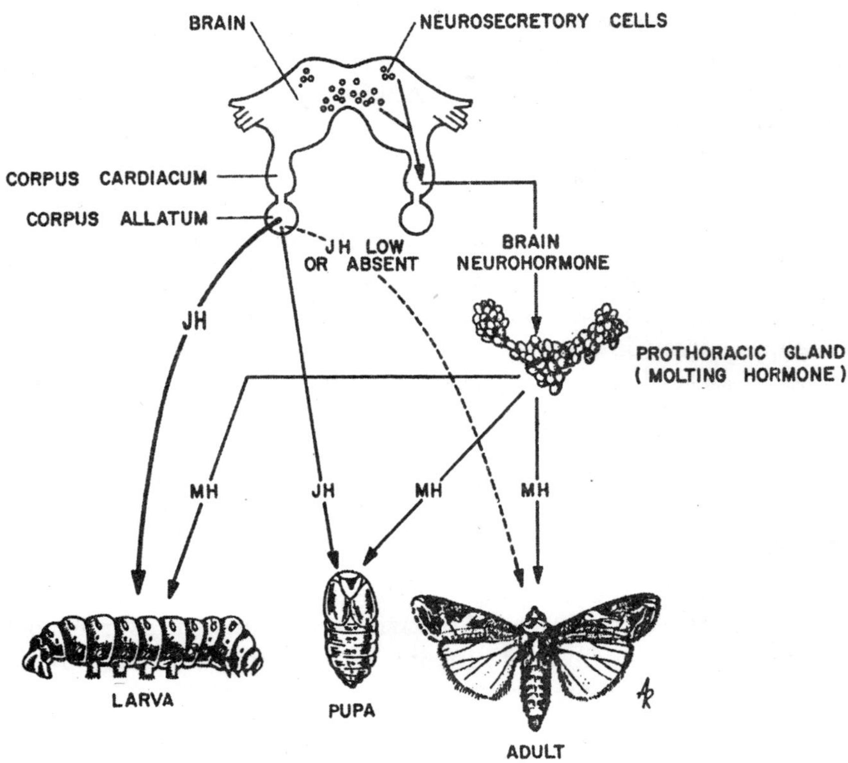

Fig. 4.8. A diagram illustrating the neural and endocrine control of growth and molting in a moth. Neurosecretory cells of the brain release a principle which stimulates the prothoracic glands to secrete molting hormone (MH or ecdysone). Juvenile hormone (JH) arises from the corpora allata and promotes the retention of larval characters. Adult differentiation occurs when MH acts in the absence of JH.

ENDOCRINE STRUCTURES AND THEIR SECRETIONS

Organs of Nervous Origin

Neurosecretory cells are numerous in insects and have important functions. Medial and lateral groups of such cells are present in the brain (protocerebrum); other aggregations are found in the subesophageal ganglion, and in all other ganglia of the ventral chain.

The corpora cardiaca arise from the nervous system and are situated behind the brain in close association with the dorsal aorta. They are paired in some species, but are fused in others. The corpora cardiaca are neurohemal organs and are composed of four cellular elements: (*a*) the bulbous endings of neurosecretory axons whose perikarya are located in the dorsum of the brain, (*b*) the perikarya of neurosecretory cells that send axons into nerves that supply various peripheral organs, (*c*) glia-like cells, and (*d*) intrinsic corpus cardiacum cells. Although the corpora cardiaca are storage-release centers, there is increasing evidence that their own cells are capable of producing secretions.

Organs of Epithelial Origin

The best known endocrine glands of the insect originate from ectodermal cells proliferated from the surface epithelium in the vicinity of the mouth parts. Aggregations of such cells approach the posterior margin of the brain and become the corpora allata. These glands are commonly paired and laterally placed *(Periplaneta),* or they may fuse to form a single structure *(Rhodnius).* Other clusters of cells, derived in the same manner, may come to rest in the

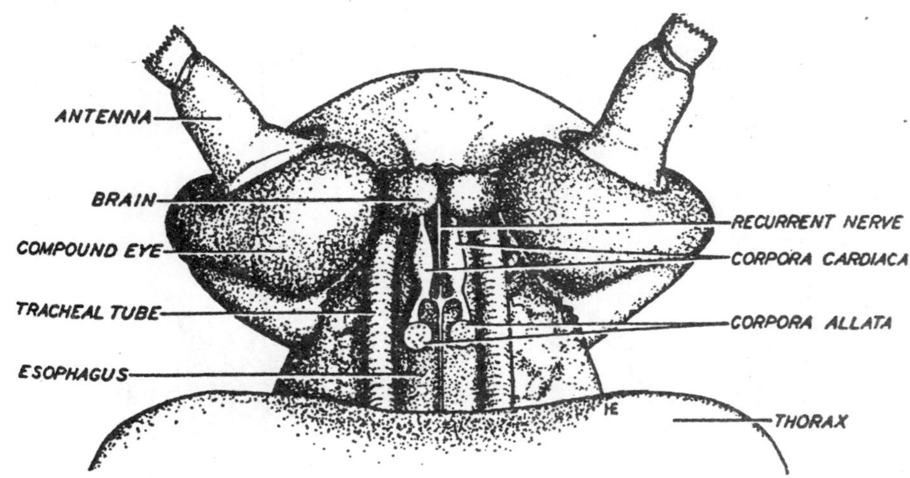

Fig. 4.9. A dissection of the head of the roach *(Periplaneta americana),* showing the paired corpora allata and corpora cardiaca and the relations of these glands to other structures of the head.

head or be carried farther back into the thorax. These anlagen form the *ventral glands* of the head in certain species, or the thoracic or *prothoracic glands* in others. The ventral and prothoracic glands are structurally and functionally comparable, differing only in location. The prothoracic glands are circumscribed and can be surgically removed in some species such as the roach, but in others, the tissue is so scattered that ablation is difficult or impossible. These glands undergo autolysis shortly after completion of metamorphosis, and hence are not present in adult insects.

The prothoracic glands were described in the goat moth by Lyonet as early as 1762, and Toyama followed their embryonic development in the silkworm in 1902. The functional significance of these ductless glands was not fully appreciated until the classic studies of Fukuda on the larvae of *Bombyx mori* in 1940. Through ingenious ligation and transplantation experiments, he proved that the prothoracic glands are essential for the larvae of this moth to undergo pupation. The clear-cut experiments of Williams on the isolated pupal abdomens of the Cecropia moth demonstrated that the prothoracic glands are incapable of secreting molting hormone unless they are activated by a principle from the brain.

In some of the Diptera, the corpus cardiacum, corpus allatum, and prothoracic gland fuse to form a structure surrounding the aorta. This retrocerebral complex is called the *ring gland* of Weismann.

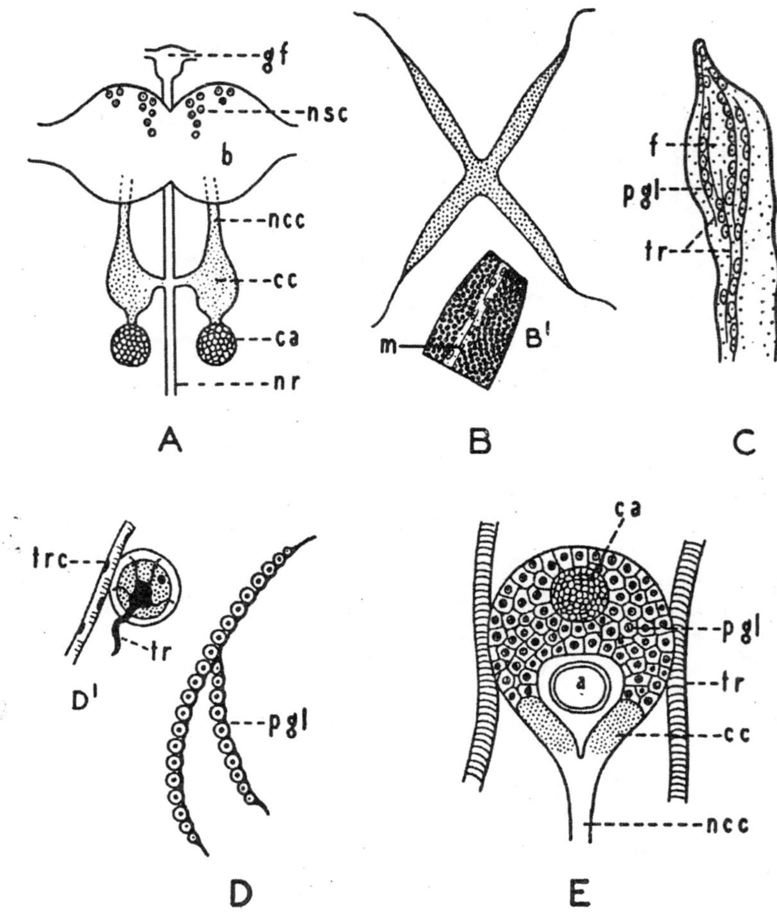

Fig. 4.10. The endocrine glands of insects. *A*, The brain-corpus cardiacum-corpus allatum complex. *B*, Prothoracic gland of the roach *Periplaneta*. *B'*, Section showing small cells in relation to the muscle core. *C*, Prothoracic gland cells in the fat body of *Rhodnius*. *D*, Prothoracic gland of Lepidoptera. *D'*, A single gland cell showing its tracheal supply *(tr)*. *E*, Ring gland of *Drosophila* showing its three components: prothoracic gland *(pgl)*, corpus allatum *(ca)*, and corpus cardiacum (cc). Abbreviations: *a*, aorta; *b*, brain; *ca*, corpus allatum; cc, corpus cardiacum; *f*, fat body; *gf*, ganglion frontale; m, muscles; *ncc*, nervus corporis cardiaci; *nr*, nervus recurrens; *nsc*, neurosecretory cells; *pgl*, prothoracic gland; *tr*, trachea; *trc*, tracheal cells. (From Bodenstein, D.: Rec. Prog. Horm. Res., 10, 1954.)

Epithelial cells of the hind gut are known to be the source of a hormone (proctodone) which is required to activate the neurosecretory system of the brain. The hormone has been found in two species of Lepidoptera in which it plays a role in photoperiodism and diapause.

Neurohormones of the Brain

Kopec (1917) used ligatures in order to divide the larva of the gypsy moth into blood tight compartments. He correctly concluded that the brain is the source of a blood-borne factor which is required for pupation. Wigglesworth (1940), working on *Rhodnius,* traced the source of this brain principle to collections of neurosecretory cells present in the pars intercerebralis. He accomplished this by excising the pars intercerebralis regions of brains from nymphs that had fed a few days previously (during the critical period), and implanting these into nymphal hosts that had been decapitated immediately after feeding (before the critical period). The decapitated hosts could live 6 to 10 months, but never molted; the activated neurosecretory cells contained in the brain implants induced them to molt. He found that no other part of the nervous system would stimulate molting when implanted into the same recipients. For the first time, it was possible to assign a functional role to particular neurosecretory cells.

Pupae of the Cecropia moth remain in diapause for 5 to 6 months when they are kept at a temperature of 25° C. After surgical removal of the brain, such pupae never metamorphose, though they may live for approximately a year. The intact pupae may be induced to metamorphose precociously by chilling (3-5° C.) them for about 6 weeks and then returning them to room temperatures. Implanted brains from chilled pupae induce metamorphosis

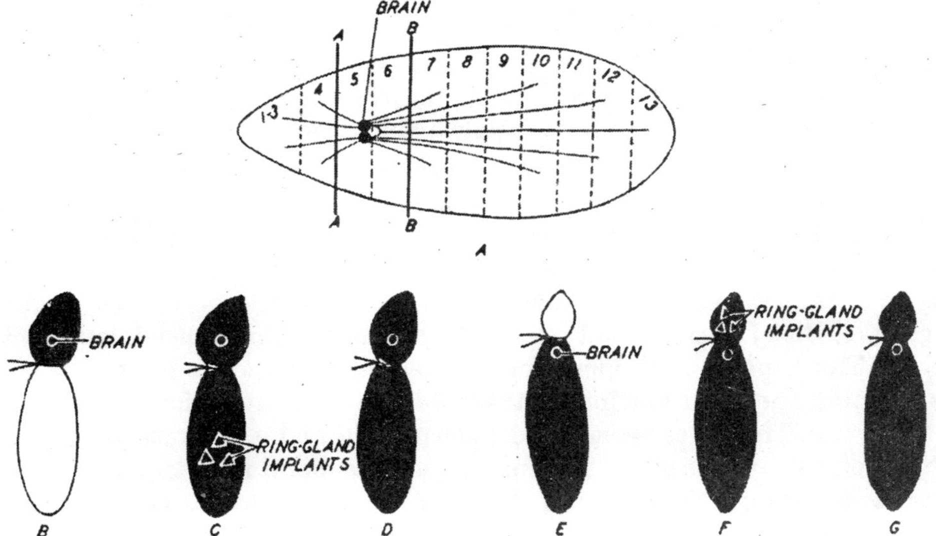

Fig. 4.11. Ligation experiments upon the larvae of dipterans. Studies of this type indicate that the brain, or some closely associated structure, releases one or more blood-borne agents that are essential for pupation. A, Diagram of a muscid larva, showing the position of the brain. Ligatures may be placed in front of the brain (line A—A) or behind it (line B —B). In diagrams B to G the pupated regions of the body are indicated in black. B, Larva ligatured behind the brain before the critical period. Only the portion of the body containing the brain undergoes pupation. It is presumed that the posterior part of the body fails to pupate because the ligature prevents the dissemination of an essential substance (ecdysone) contained in the body fluid. C, Same as *B* with ring glands implanted into the body posterior to the ligature. The ring glands induce pupation in that portion of the body isolated from the brain. *D,* Same as *B,* but the ligature was placed after the critical period, *i.e.,* after the essential endocrine substances had attained effective concentrations in the body fluid. The entire body pupates. *E,* Larva ligatured in front of the brain before the critical period. Pupation is accomplished only by the segment of the body containing the brain. F, Same as *E,* but with ring glands implanted into the fragment lacking the brain. This procedure produces pupation of the anterior region. G, Same as *E,* but the ligature was placed after the critical period. All the body pupates.

when implanted into unchilled pupae. When chilled pupae are parabiotically united with unchilled pupae, both molt synchronously in a little less than 2 months. These experiments indicate that pupal diapause in Cecropia results from a failure of the brain to produce BH, brain activation depending upon a temperature stimulus. Different mechanisms are undoubtedly involved in other species of Lepidoptera.

The chemical nature of brain hormone (BH) is still a subject of controversy. Kobayashi and Kirimura claimed that it is indistinguishable from cholesterol, whereas Ichikawa and Ishizaki found it to be water soluble and lipid insoluble. Brainless, diapausing pupae of the Cecropia moth provide a sensitive system for testing materials that are capable of activating prothoracic glands. Neither pure cholesterol nor any other lipid-soluble extract has been found effective in this system. It is well known, however, that many steroids, sterols, and steroid precursors can mimic the actions of BH in certain test systems. Even epinephrine can activate the prothoracic glands in certain insect tests, though it is not found in insects.

While it is common practice to refer to the prothoracic glandstimulating factor as "brain hormone" or "prothoracotrophin," it must be recognized that the neurosecretory cells of the brain produce additional secretions that have other actions. The medial neurosecretory cells in the brain of the blowfly *(Calliphora)* supply a neuro-hormone that stimulates egg maturation in this species. This is accomplished by evoking the formation of proteases in the gut, thereby facilitating the breakdown of proteins to supply amino acids for the synthesis of egg proteins.

Another neurohormone arising from the neurosecretory system of the insect brain is the "darkening" or "tanning" factor. It is also present in two groups of neurosecretory cells in the thoracic ganglia. In newly emerged flies *(e.g., Calliphora)*, darkening and hardening of the cuticle (tanning) starts after about 15 to 20 minutes, but this effect can be abolished by ligaturing the head. It can also be prevented by section of the nerve cord posterior to the thoracic ganglia. Blood from a 15-minute-old fly induces tanning when injected into a head-ligatured fly. The tanning neurohormone is released in response to appropriate sensory impulses reaching the brain via the ventral cord a few minutes after the fly has emerged from the puparium. This factor seems to be a polypeptide, and is different from brain hormone and molting hormone (ecdysone). The tanning factor is found in a variety of insect species. A substance originating in the eyestalks of crayfishes plays some role in tanning of the newly formed exoskeleton.

Hormones of the Corpora Cardiaca

Aqueous extracts of the corpora cardiaca of the roach *Periplaneta* elevate the blood concentration of trehalose, the main circulating carbohydrate of insects. The effective substance, reported to be a polypeptide, acts upon the enzyme system of the fat body (where glycogen is stored) to promote glycogenolysis.

Efferent nerve activity in *Periplaneta* is inhibited by the subesophageal ganglion. Extracts of the corpora cardiaca mimic the surgical removal of this inhibitory system, but it is not known whether the effective substance arises *in situ* or is merely stored there.

The corpora cardiaca of the cockroach appear to secrete a peptide hormone that acts indirectly to increase the heart rate. It apparently stimulates the pericardial cells, scattered along the heart, to produce a pharmacologically active factor which in turn acts upon the heart to increase its rate of beating. The differential centrifugation of extracts from corpora cardiaca of the roach has yielded neurosecretory particles that accelerate the heart. Studies on the roach indicate that many cardioregulatory substances, probably small peptides, can be isolated from various tissues such as the nervous system, gastrointestinal tract, heart, utricles, and hemolymph.

Molting Hormone (Ecdysone)

In saturniid moths, activation of the brain and consequent termination of pupal diapause depends upon exposure to low temperatures. The classical experiments of Williams on Cecropia moths demonstrated conclusively that brain hormone (BH) does not act directly upon the tissues, but rather, that it exerts a trophic effect upon a specific target—the prothoracic glands. Brainless pupae survive for long periods, but never terminate diapause. When a brainless, diapausing pupa was grafted to a chilled pupa, the two animals metamorphosed simultaneously. The pupae of these moths could be transected anterior to the sixth abdominal segment, and the cut surfaces sealed over with coverslips; these parts were viable for 8 months or longer and made some instructive experiments possible. Chilled anterior parts, or brainless anterior parts that had received implants of chilled brains, metamorphosed into adults. The isolated abdomens (lacking prothoracic glands) did not metamorphose even after receiving multiple implants of chilled brains. The isolated abdomens could be induced to metamorphose by introducing both chilled brains and prothoracic glands; unactivated prothoracic glands alone were not adequate. It has been shown in a great variety of insects that metamorphosis depends upon two factors: brain hormone and molting hormone from the prothoracic glands. The prothoracic glands quickly regress after the final molt that gives rise to the adult stage.

What appears to be the molting hormone has been isolated and crystallized; it is a steroidal compound and its structural formula is partly known. The crystallized hormone has been called *ecdysone.* When tested on various insects, it is found to duplicate all the known effects of the natural hormone. Since it is quite certain that insects are unable to synthesize steroids from acetate, it is likely that they form ecdysone through the degradation of steroids consumed in their diets. By using labeled materials, it has been shown that *Calliphora* can form ecdysone from injected cholesterol. Five distinct fractions having ecdysone activity have been isolated from *Bombyx* material, and this indicates that present information is tentative.

The life cycle of the flagellate *Trichonympha,* living symbiotically in the gut of the wood-eating roach *Cryptocercus,* undergoes modifications in response to endocrine fluctuations in the insect. The administration of ecdysone to adult roaches, themselves incapable of molting in response to the hormone, induce, encystment and gametogenesis in the flagellates. It is probable that this is a direct effect of the host hormone upon the protozoa, and that it has adaptive value in their survival.

Juvenile Hormone

The tissue reactions to molting hormone (MH) are modulated by juvenile hormone (JH), which is a product of the corpora allata. The tropical bug *Rhodnius,* used so extensively by Wigglesworth and his colleagues, was a convenient insect for the experimental elucidation of the factors involved in molt and metamorphosis. There are five nymphal instars in *Rhodnius,* and molting occurs a definite number of days after a meal of blood; abdominal distention is the stimulus that activates the neurosecretory complex of the brain. The elongated head of *Rhodnius* made it possible to cut transversely at different levels and obtain animals deprived of brain (corpus allatum intact), or animals deprived of both brain and corpus allatum. The decapitated animals survive six to 10 months. It was possible to join decapitated animals by means of capillary tubes, or to graft instars and adults in many telobiotic and parabiotic combinations.

Two grafting experiments in *Rhodnius* may be mentioned to illustrate the action of JH. If a fourth stage nymph is decapitated after the critical period and is telobiotically grafted to a fifth nymph also decapitated after the critical period, both individuals molt, but the fifth stage nymph becomes a giant, supernumerary nymph instead of an adult. Though adult *Rhodnius* normally does not molt, it may be induced to do so by grafting nymphal stages. When fourth stage nymphs, possessing their corpora allata as a source of JH, are united with an adult, the latter molts and shows a partial return to the nymphal condition.

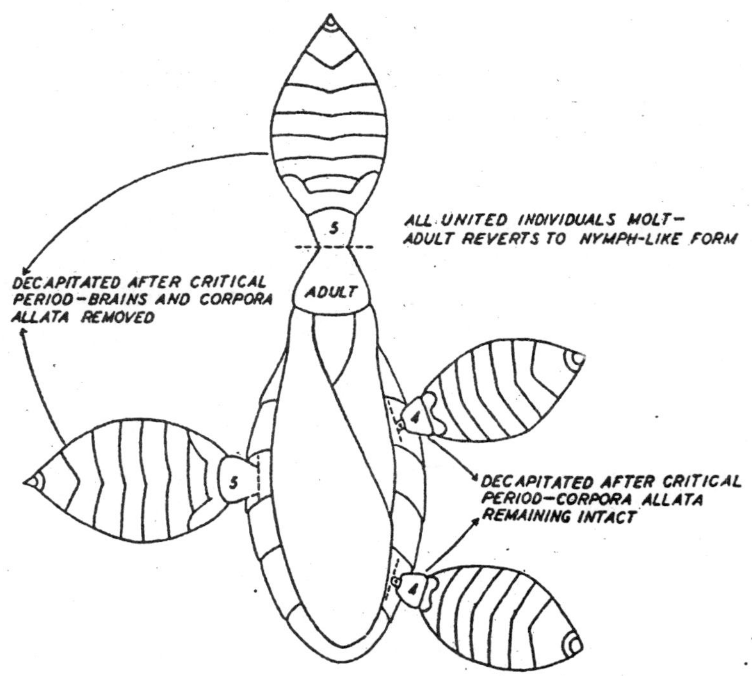

Fig. 4.12. The induction of molt in adult *Rhodnius*. The broken lines indicate the level of decapitation; the blackened dot in the proximal end of the head represents the corpus allatum, and the number of the nymphal instar in indicated on the thorax. The decapitated adult is coupled with two fifth-stage nymphs, totally decapitated after the critical period, and with two fourth-stage nymphs partially decapitated after the critical period. The fifth nymphs provide the adult with ecdysone, whereas the fourth nymphs, having their corpora allata, provide juvenile hormone in addition to ecdysone. All the combined individuals molt; the fifth nymphs become supernumerary nymphs (sixth-stage), the fourth nymphs become fifth instars, and the adult reverts to a nymphlike form.

In the roach *Leucophaea maderae,* the adult stage is preceded by an average of eight nymphal instars. Removal of the corpora allata (allatectomy) of the fifth, sixth, or seventh instars produces an abbreviation of development, the final molt being accomplished at an earlier stage than in the unoperated controls. Fifth and sixth instars, deprived of JH by allatectomy, develop characters that are intermediate between nymph and adult (preadultoids) and must molt again before attaining adult characters to a conspicuous extent (adultoids).

Walking sticks *(Dixippus)* are neotenic insects, the females beginning to reproduce parthenogenetically before they are fully adult. If third stage larvae are allatectomized, two more molts ensue before egg laying begins. By implanting corpora allata into the semi-mature adults of *Dixippus,* they may be caused to molt two more times and to become giant insects. After implanting corpora allata from young donors into sixth stage larvae, as many as four extra molts may be produced, and the insects become about twice the normal size.

Allatectomized caterpillars of the commercial silkworm undergo no further molts; they metamorphose prematurely into normally formed but miniature adults. The earlier the corpora allata are removed, the smaller the resulting adult. On the other hand, implantation of corpora allata produces supernumerary larval molts and giant caterpillars that metamorphose into giant adults.

When pupae of the Cecropia moth are grafted to a decapitated adult male moth, the pupae undergo a second pupal molt rather than metamorphosing into adults. In other words, these pupae develop as though they had received a heavy dose of JH. Studies on these moths have shown that the corpora allata of adult males produce large quantities of JH and that it is stored in the abdomen. The tissues of the adult female abdomen contain only traces of this hormone.

It is apparent from experiments of this type that three endocrine factors are involved in the growth and metamorphosis of insects: BH from neurosecretory cells of the pars intercerebralis initiates the secretion of MH by the prothoracic glands; MH acts upon the cells to promote growth and differentiation to the adult stage; if the third hormone, JH, is present, MH and JH act in concert to retard metamorphosis. When high titers of JH are present with MH, the animal grows but remains immature; imaginal differentiation results when MH is unopposed by JH.

Purified extracts of JH have been prepared and highly sensitive assays developed. The hormone is extremely stable and extracts with comparable activity have been prepared from museum insects following many years of drying. Substances having JH activity have been detected in human placenta, thymus, and adrenal glands; in higher plants, yeasts and some bacteria; and in a great variety of invertebrates.

It is known that certain parasites are capable of influencing the growth and development of their insect hosts by liberating a substance that has the same effects as JH from the corpora allata. The larvae of *Tribolium* (Coleoptera), when infected with the sporozoan parasite *Nosemba,* undergo supernumerary molts to produce giant larvae which weigh twice as much as the uninfected controls. These infected larvae seldom pupate: the infection mimicks JH in prolonging larval life at the expense of pupal and imaginal differentiation. Studies indicate that the parasite itself produces a substance with JH activity, rather than acting upon the host's corpora allata to stimulate the production of intrinsic JH.

The exact chemical nature of JH remains undetermined. Chemical studies indicate that effective extracts contain farnesol and farnesal. This is interesting because farnesol is an intermediate step in the biosynthesis of cholesterol and carotene. A number of steroids, terpenoids, and related compounds are known to copy the activities of JH when tested in appropriate ways. There are indications that all the insect hormones are chemically related: This is suggested by the fact that prothoracic glands can be activated by MH (ecdysone), BH, JH, cholesterol, and such compounds as farnesol. There are strong indications that ecdysone is a steroid. Schneiderman and Gilbert have pointed out that "Rather than being a recent innovation of the vertebrates, steroid hormones may prove to have a far more ancient lineage."

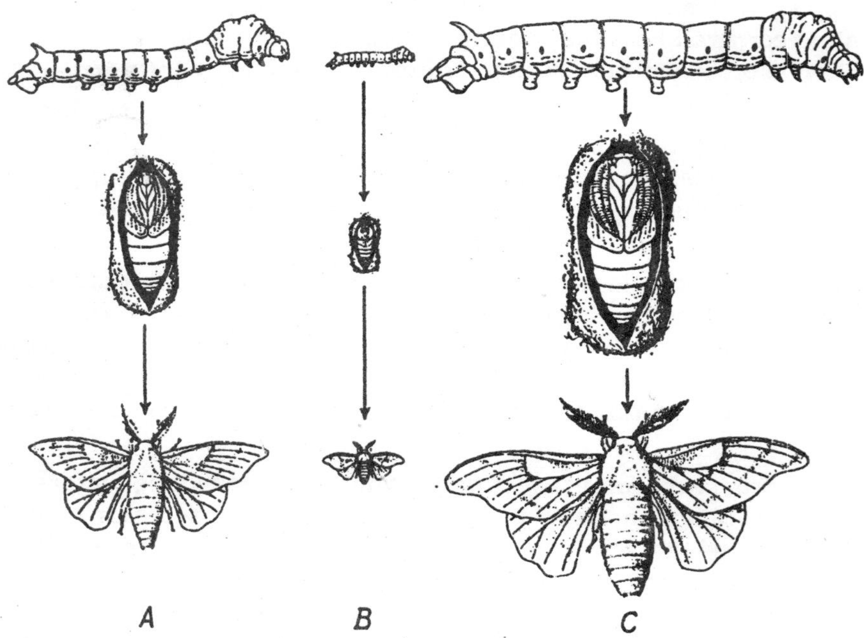

Fig. 4.13. Effects of the corpora allata on postembryonic development of *Bombyx mori*. *A*, Normal fifth instar, pupa, and adult. *B*, Allatectomized third instar, diminutive pupa, and diminutive adult. C, Supernumerary larva produced from fifth instar by transplanting corpora allata of young larvae, giant pupa, and giant adult.

Proctodone

Larval diapause in the European corn borer *(Ostrinia nubilalis)* is induced by short-day photoperiods and terminated by long-day photoperiods. When diapausing larvae are ligatured between the sixth and seventh abdominal segments, thus cutting off circulation in the two terminal segments of the abdomen, exposure to long-day photoperiods fails to terminate diapause. Cutting the ventral nerve cord at the level of the sixth abdominal segment does not prevent diapause termination in response to long days. These observations indicated that some endocrine mechanism, necessary for brain activation and consequent termination of diapause, might be located in the terminal abdominal segments. Glandular cells were found in the proctodeal epithelium, and because of its source, the name *proctodone* was applied to the hormone. It is blood-borne and participates in diapause termination by activating the neurosecretory complex of the brain.

Endocrines in Insect Reproduction

Experiments on castration and gonad transplantation have failed to yield convincing evidence that insect gonads are the source of hormones. Gonadal functions, however, are conditioned by several internal secretions. Juvenile hormone from the corpus allatum is essential for yolk deposition in the eggs and for formation of spermatophores, in many species of insects. In *Rhodnius,* for example, removal of the corpus allatum prevents the production of ripe eggs: the oocytes grow as long as they are attached to the nurse cells, but degenerate at the period when yolk deposition should occur. Allatectomy and ligation experiments on the milkweed bug, *Oncopeltus fasciatus,* indicate a similar relationship, corpora allata being necessary for yolk deposition and secretion by the oviducts. Full production of eggs in *Oncopeltus* also depends on the medial neurosecretory cells of the brain. Among certain flies *(e.g., Calliphora),* the medial neurosecretory cells are of special importance inasmuch as their products promote the formation of amino acids to be used by the eggs for the synthesis of proteins.

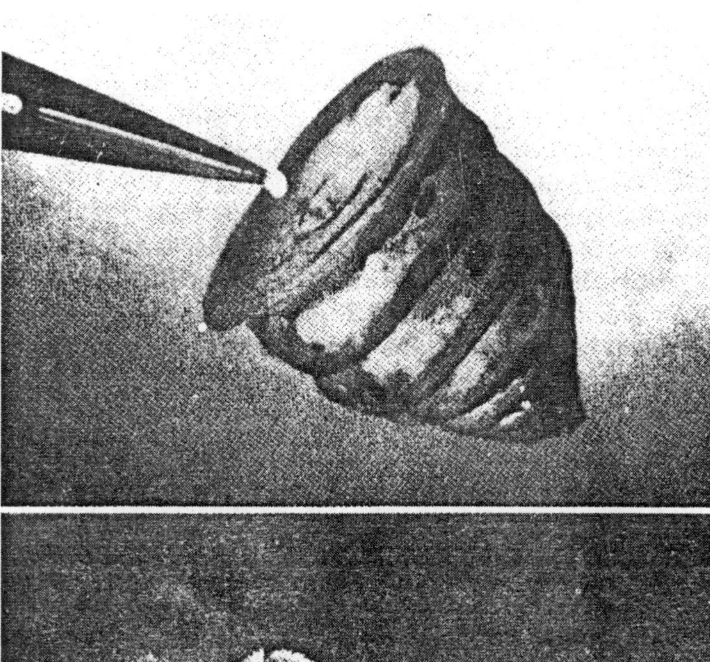

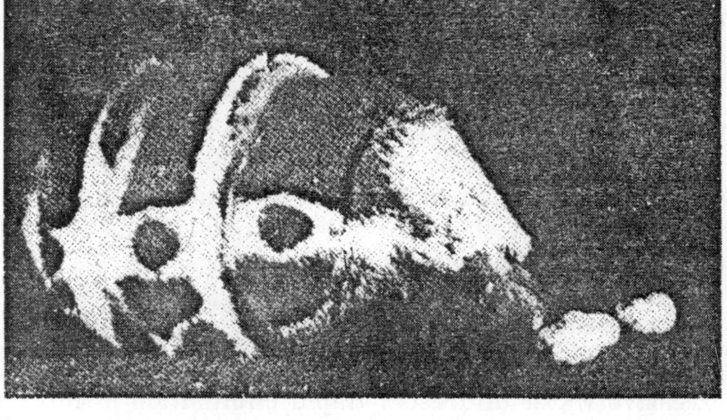

Fig. 4.14. ***Above,*** **Isolated abdomen of a diapausing Cecropia pupa sealed to a plastic cover slip. Prothoracic glands from postdiapausing pupae are being introduced into the abdomen through a central hole in the slip.** ***Below,*** **The same abdomen after differentiating adult characters; eggs are being deposited. Corpora allata are not necessary for gamete maturation in these moths.**

Reproduction in the cockroach *Byrsotria* depends upon the release of a volatile sex attractant by the female, this pheromone being essential for attracting males and releasing their courtship behaviour. Gonadectomy does not impair the mating behaviour of either sex. Allatectomy of the female, shortly after the imaginal molt, prevents the production of sex pheromone; the production of pheromone is reinstated by the implantation of corpora allata.

In this species of roach, secretions from the corpora allata are not only essential for oocyte maturation and accessory gland secretion, but also for production of the sex attractant which makes mating behaviour possible. Allatectomy of the male does not impair reproduction behaviour.

In *Bombyx* as well as the Cecropia silkworm, the corpora may be removed from the pupae of either sex without disturbing the sexual functions of mature moths; the organs are not necessary for the maturation of gametes and the production of normal young. Molting hormone of the prothoracic gland appears to be the only one necessary for sexual maturation of both sexes in these moths. Pupae deprived of their brains, corpora cardiaca, and corpora allata become sexually mature adults after the administration of crystalline ecdysone. Thus, in these Lepidoptera, the corpora allata are highly active in the adults, but no function has as yet been ascribed to them at this stage of the life cycle. It may be significant, however, that the silkworms that have been studied most carefully are short-lived moths which lack functional mouth-parts in the adult. The adult stage is greatly abbreviated and ripe eggs are ready for oviposition soon after the individual emerges from the pupa. It may be that the lack of functional mouth-parts in the adult correlates with a precocious maturation of the gonads which can occur in the absence of any gonadotrophic function of the corpora allata. The corpora allata of feeding, long-lived species of adult Lepidoptera may possibly exert gonadotrophic effects similar to those described for numerous other orders of insects.

In *autogenous* mosquitoes of the genus *Culex,* ovarian development does not depend upon the consumption of food in the adult stage, but the ovaries of *anautogenous* species do not develop beyond a resting stage until a blood meal of adequate size has been consumed. The implementation of corpora allata from autogenous donors into anautogenous recipients enables the ovaries to mature without the consumption of a meal. When anautogenous ovaries are transplanted to autogenous hosts, egg maturation occurs in the grafts even though the host does not feed. Ovaries from autogenous donors do not mature after being transplanted to anautogenous hosts. In addition to a hormone from the corpus allatum, a factor from the brain is also involved in these phenomena.

Studies on several species of insects show that the functional status of the corpora allata is regulated by the brain. In the milkweed bug and some of the roaches, the corpora allata are supplied by Inhibitory nerves from the brain: section of these nerves releases the corpora allata from inhibition, and egg maturation and ovulation occur. Distention of the gut appears to be the effective signal for egg development and ovulation in anautogenous mosquitoes; in certain roaches, it is the act of mating.

Water Balance in Insects

Some insects that feed upon fluids have evolved mechanisms for the quick elimination of excessive water (diuresis). Rapid elimination of fluid occurs in *Rhodnius* following the consumption of a blood meal, and it has been reported that the fused thoracic ganglia are the source of a diuretic neurohormone which acts upon the malpighian tubules to promote the loss of fluid. The response can be evoked in isolated malpighian tubules by exposing them to blood taken from recently fed *Rhodnius.*

Evidence for an antidiuretic principle in the roach *Periplaneta americana* has been obtained through studies on the rate of indigo carmine uptake by malpighian tubules, tested *in vitro,* following subjection of the animals to different osmotic conditions. Since the diet of these insects consists of solids or semisolids, there is little likelihood that excessive fluid intake would constitute a problem; however, periods of desiccation might require a diminished excretion of fluids and withdrawal of water from the gut. The malpighian tubules from dehydrated and salt-loaded animals showed lower than normal levels of dye uptake, indicating that they had been exposed to a factor inhibiting excretion. Extracts of corpora allata from normal animals, in contrast to those from dehydrated animals, reduced the rate of dye uptake by the tubules. Brain extracts, prepared from dehydrated subjects, also reduced dye uptake. The experiments were interpreted as indicating that an antidiuretic factor is produced by the brain and conveyed to the corpora allata where it is stored and released when the organism needs to conserve water.

Colour Changes in Insects

A few insects are capable of modifying the coloration of their bodies in response to various environmental changes. Both morphologic and physiologic colour changes may occur, and the indications are that both are effected by hormonal agents in the body fluids.

A morphologic colour change occurs in the locust *(Locusta migratorid)* that is hormonally regulated by the corpora allata. Animals of the migratory type are orange-black and live together in swarms, but individuals of the solitary phase are green. When corpora allata are implanted into nymphs of the migratory type, the green pigment characteristic of the solitary type may appear after the next molt. These results are obtained even when the corpora allata are taken from nymphs or images of the migratory type.

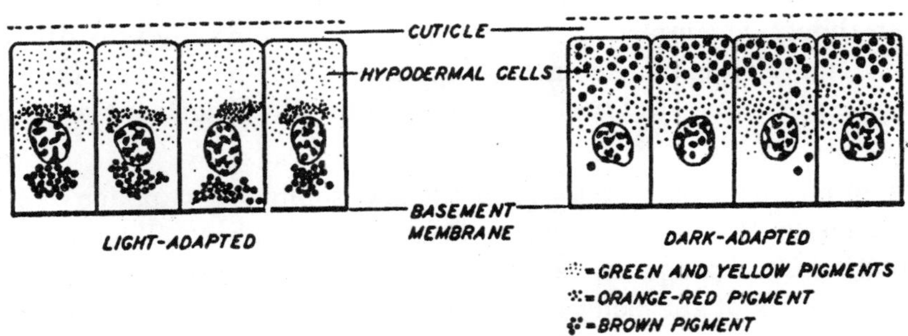

Fig. 4.15. Physiologic colour change in the walking-stick, *Dixippus morosus*. The pigments that condition the coloration of the body are contained within the hypodermal cells. In the light-adapted animal the brown and orange-red pigments are concentrated, but in the dark-adapted animal the same pigments are dispersed throughout the supranuclear regions of the cells. The green and yellow pigments have a permanent position in the cells and are not changed by light. (Modified from Hanstrom, after Giersberg.)

Factors originating from neurosecretory cells in the brain and adjacent ganglia seem to be responsible for the physiologic colour adaptations that occur in insects. Colour changes in *Carausius* are predominantly under the control of a chromatophorotrophin present in the brain and in the corpora cardiaca. Extracts from these two structures lead to a darkening of the integument, whereas extracts of the corpora allata have little if any effect.

Examples of Rhythmic Activity in Insects

As fruitflies *(Drosophila)* reproduce in nature, the adults emerge from their pupal cases in the early morning. If the eggs, larvae, and pupae are maintained in complete darkness, the adults emerge at all hours of the day and night. Exposure of these metamorphosing flies to a single flash of light (as short as 1/2000 sec.) causes a resetting of the biological clocks, and the adults emerge synchronously at an hour corresponding to the hour of light exposure. These adjustments in response to illumination are probably mediated by the neuroendocrine system.

A diurnal rhythm of locomotion and feeding activity occurs in the cockroach *Periplaneta:* the animals are relatively quiescent during the day, but become active with the onset of darkness. By ingenious surgical methods, it was demonstrated that the circadian rhythm in this insect results from the action of blood-borne materials produced by neurosecretory cells. Cockroaches, taken from continuous light, and showing no measurable activity rhythm, were parabiotically united with others which exhibited normal diurnal rhythms. The arrhythmic member of the pair assumed a circadian rhythm in which the phase times were the same as those shown by the rhythmic parabiont. The decapitation of animals, kept in alternating periods of light and darkness (12 hours each), abolished their rhythms.

The subesophageal ganglion in the head of the roach is the source of a neurosecretion which is associated with, and at least partly responsible for, the initiation of locomotor activity. Exposure of the subesophageal ganglion to low temperatures (3° C.) *in situ,* the remainder of the body not being chilled, stops the biological clock contained in the ganglion. If the ganglion is chilled for 2 or 3 hours, and left within the individual, the clock is "reset" and timing of running activity is not altered. When the ganglion is chilled for as long as 8 hours, it does not "reset" even though left intact, and the onset of locomotor activity is delayed by a period of 8 hours. If the ganglion is transplanted to an arrhythmic host immediately after chilling, there is a time lag in the onset of host activity equaling the length of time the ganglion was chilled.

The timing of neurosecretory activity in the subesophageal ganglion can be adjusted to inverted light cycles, so that secretion occurs 12 hours out of phase. When ganglia from donors secreting 12 hours out of phase are transplanted to hosts whose ganglia are set to normal time of day, the tissues of the body are exposed to two cycles of neurohormone secretion per day. Under these conditions, the animals soon die from cancerous growths in the wall of the mid-gut. The tumors metastasize and are transplantable. Growths of this type seem comparable to those produced by B. Scharrer in the roach *Leucophaea* by cutting the recurrent nerve.

The biological clock regulating and locomotor rhythm of the cockroach is set by the diurnal cycle of light and darkness, and this stimulus is perceived through the ocelli (simple eyes). The corpora cardiaca are indirectly activated and a neurosecretory stimulus from them is conveyed along nerves to the subesophageal ganglion. The neurosecretory cells in the latter ganglion normally release their products every 24 hours to initiate locomotor activity.

Much remains to be learned about the operation of biological timepieces, and this simplified version of the clock system of the roach only serves as an illustration. The essential concept is that the central nervous system produces and rhythmically releases blood-borne agents, in response to external stimuli, which organize the behavioural patterns of the animal in adjusting to its particular habitat.

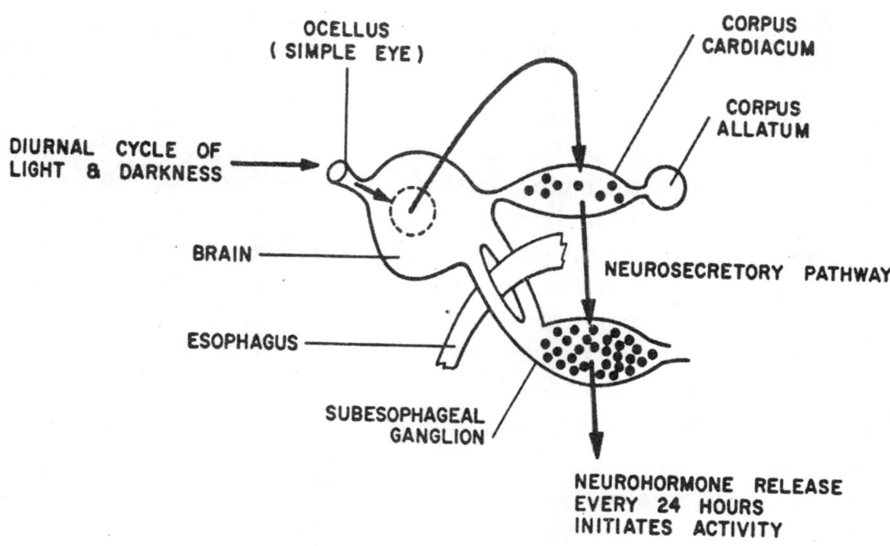

Fig. 4.16. A diagram of the brain and retrocerebral glands of a cockroach to illustrate how diurnal cycles of activity may be regulated. Black circles represent neurosecretory cells. The change from light to darkness is perceived through the ocelli; impulses are integrated in the brain and these result in activation of the corpora cardiaca; a neurosecretory stimulus is conveyed from the latter organs to a biologic clock, consisting of neurosecretory cells, in the subesophageal ganglion; the clock is set to discharge neurosecretions into the blood at regular intervals to initiate locomotor activity.

ANALOGOUS NEUROSECRETORY SYSTEMS OF INVERTEBRATES AND VERTEBRATES

From the foregoing description of endocrine mechanisms in invertebrates, it is apparent that neurosecretory cells are a distinct cell type performing definite roles in the organism. They do not occur haphazardly in the nervous system, without rhyme or reason, but form definite groups that are always found in particular locations characteristic of the species. In invertebrates and vertebrates alike, external and internal stimuli are received and integrated by the central nervous system. This system has evolved its own endocrine mechanisms, aggregations of neurosecretory cells, which serve to translate nervous impulses into diffusible chemical messengers. The timing of biologic phenomena, such as hibernation, migration, estivation, molting, metamorphosis, reproduction, diapause, locomotor rhythms, etc., is possible because organisms possess both clocks and calendars. Many, if not all, of these devices in animals are controlled by chemical messengers of one kind or another.

The simplest pituitary glands are found in the agnatha, and there is no certainty that structurally comparable organs exist in the invertebrates. Nevertheless, analogous neurosecretory systems are present in organisms as distantly related as crustaceans and primates. The neurosecretory systems of crustaceans and insects consist of two chief components: clusters of neurosecretory cells in the brain and related ganglia, and neurohemal organs that serve principally as depots for the storage and release of neurohormones. The ganglionic X organ and other neurosecretory aggregations of the crustacean brain, as well as the medial and lateral neurosecretory cells of the insect brain, probably correspond to the supraoptic and paraventricular nuclei of the vertebrate hypothalamus. The sinus gland of crustaceans and the corpora cardiaca of insects are neurohemal organs and are analogous to the neural lobe of the vertebrate pituitary gland. Non-neural glands of internal secretion arise among the invertebrates: the optic glands of cephalopod molluscs, the Y organs, ovaries, and androgenic glands of crustaceans, and the prothoracic glands and corpora allata of insects have been mentioned. The Y organ is probably a crustacean analog of the insect prothoracic gland. The corpora allata of insects are comparable, in many ways, to the vertebrate adenohypophysis. The insect glands arise out of ectodermal proliferations from the ventral side of the head; the adenohypophysis develops from an ectodermal anlage (Rathke's pouch) which pushes upward from the stomodeum.

The endocrine systems of annelids, molluscs, crustaceans, insects, and vertebrates must have evolved independently. The caudal neurosecretory system of fishes (urophysis) illustrates how such comparable patterns can develop independently even within the same individual. The urophysis consists of large neurosecretory cells which send axons into a neurohemal structure that causes a bulging below the terminal end of the spinal cord. The function of the urophysis is uncertain, but, structurally it is strikingly comparable to the cephalic neurohypophysis.

The neurosecretory cells and endocrine glands of arthropods govern such processes as growth, reproduction, colour change, metabolism, and water balance, thus covering about the same functional categories as the vertebrate pituitary gland. The exact manner in which the various hormones of the adenohypophysis are regulated remains a mystery, but they are clearly under hypothalamic control. The neurosecretory cells of vertebrates tend to be concentrated into one small area of the brain, the hypothalamus; whereas in invertebrates, such cells are much more widely distributed within and near the central nervous system.

Neurohypophysis: Neurohormonal Peptides

The oxytocins and vasopressins are neurosecretory products of the neurohypophysis, and a large body of information is available on their chemical structures, physiologic and pharmacologic actions, and their phyletic distribution among the vertebrates. The extremely rapid progress made in the elucidation of these peptides since 1953 stems largely from the fact that their molecular structures are known and the natural secretions, as well as many analogues not found in nature, may be prepared synthetically. Information is accumulating on relations between chemical structure and biologic effectiveness, and on possible mechanisms whereby they may act at the cellular level.

Five functions have been ascribed to these neurohormones: (1) contraction of the smooth muscle of the uterus (oxytocic effect), (2) contraction of the myoepithelial cells which surround the mammary alveoli (milk ejection), (3) actions upon the kidney to prevent excessive loss of water (antidiuretic effect), (4) contraction of the smooth muscle in the walls of blood vessels (vasopressor effect), and (5) regulation of the release of adenohypophysial hormones, especially in fishes in which the neurohypophysis and adenohypophysis are intimately associated. An examination of these actions reveals that the effectors employed are contractile elements and semipermeable membranes such as the kidney tubules and, in anuran amphibians, the skin and urinary bladder. The action on the kidney promotes the reabsorption of water from the glomerular filtrate, thus resulting in the excretion of a concentrated urine. The effect in amphibians is not only upon the kidney, but also upon the skin to increase its permeability and allow water from the environment to pass into the body. The mammalian vasopressins are often referred to as antidiuretic hormones (ADH) since their main action is to conserve water. Although vasopressin elevates the blood pressure of mammals, and has the reverse effect in birds, there is no satisfactory evidence that it normally plays any role in the regulation of vascular tone or blood pressure. The effect may be pharmacologic, rather than physiologic, since the dose required is seemingly in excess of the amounts normally released by the intact neurohypophysis.

CHEMISTRY AND PHYLETIC DISTRIBUTION

Chemistry

The secretions that are stored in the mammalian pars neuralis are oxytocin and vasopressin. Oxytocin is very potent in causing uterine contractions, in lowering the blood pressure of birds, and in promoting milk ejection. One milligram of pure oxytocin contains about 500 USP units of each activity. Vasopressin produces all of these effects to a much slighter degree. Vasopressin is unquestionably the main pressor and antidiuretic principle. Oxytocin produces the same effects, but only to about 1 per cent and 0.5 per cent, respectively, of the activity possessed by vasopressin.

All of the neural lobe secretions are octapeptides; that is, they consist of eight different amino acids, the two sulfur-linked cysteine molecules generally being counted as one cystine molecule. Three of the amino acids are present in the form of amides. The sulfur linkage forms a pentapeptide ring (five amino acids) to which is attached a side chain composed of three amino acids. All of the neurohypophysial principles whose structures are known show this pattern. Substitutions occur in the natural secretions at positions 3, 4, and 8, producing

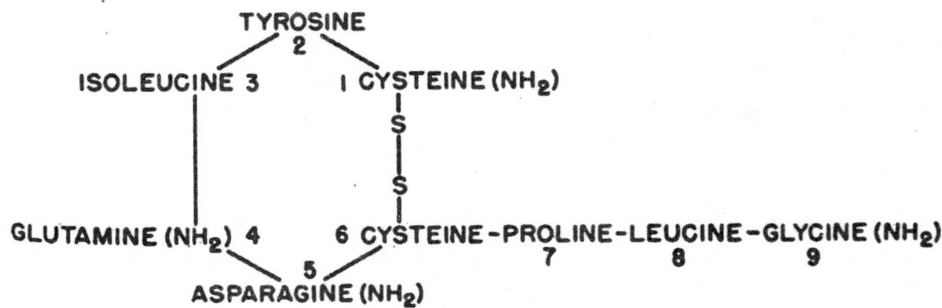

Fig. 5.1. The arrangement of amino acids in a molecule of oxytocin.

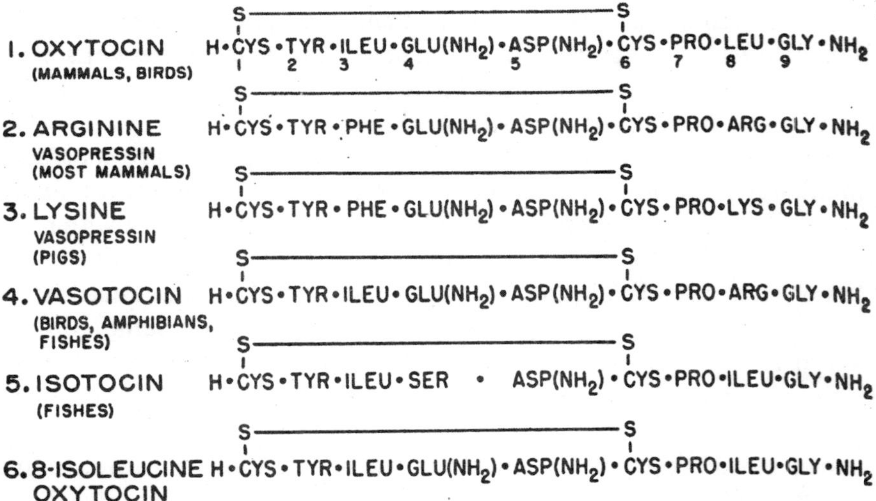

Fig. 5.2. Amino acid sequences in the natural neurohormones of the posterior lobe (neurohypophysis).

peptides with different biologic potencies. The six naturally occurring principles known at present are arginine vasopressin, lysine vasopressin, arginine vasotocin, oxytocin, isotocin, and 8-isoleucine oxytocin. Two oxytocin-like peptides have been found in pituitaries from different species of elasmobranchs, but their structures have not been determined.

Oxytocin was first identified in extracts prepared from the pituitary glands of cattle and swine, and was synthesized in 1953 by du Vigneaud and his colleagues. Shortly thereafter, the molecular structures of the vasopressins were determined and synthetically duplicated. These were brilliant chemical achievements, since these were the first peptides to be synthesized. The mammalian neuro-hypophysis is the source of three principles; oxytocin, arginine vaso-pressin, and lysine vasopressin. It may be seen that oxytocin has isoleucine at position 3 and leucine at position 8; the vasopressins have phenylalanine at position 3, and either arginine or lysine at position 8. While these differences in molecular structure might seem to be slight, they have profound effects upon biologic potencies.

Arginine vasotocin was first identified in the chicken. It has isoleucine at position 3 (like oxytocin) and arginine at position 8 (like arginine vasopressin). This analogue had been synthesized and named "arginine vasotocin" before it was found to be present in the pituitaries of a large variety of nonmammalian vertebrates. An analogue of oxytocin, called *ichthyotocin* or *isotocin,* has recently been obtained from the pituitary glands of holostean and teleostean fishes. This peptide is the same as oxytocin, except for serine at position 4 and isoleucine at position 8. Chemically, the compound is 4-serine, 8-isoleucine oxytocin. Another oxytocin-like principle, 8-isoleucine oxytocin, has been identified in the pituitaries of the primitive ray-finned fish *Polypterus,* lungfishes, and several species of Amphibia.

Phyletic Distribution

Neurohypophysial octapeptides are found in all classes of vertebrates, but they have undergone changes in amino-acid composition which make them pharmacologically different. The natural peptides vary in potency, when tested in the same or in different species, and this suggests that the receptor sites have also changed during the course of evolution.

Arginine vasotocin is the most widely distributed neurohormone. It is the only principle present in cyclostomes, the most primitive of all living vertebrates. Very small amounts of an octapeptide, closely resembling arginine vasotocin, have been found in elasmobranch pituitaries. Even though it is not known to occur among mammals, it must be regarded as a very ancient molecule. It is present in the four major lines of evolution that lead respectively to cyclostomes, the chondrichthyes (shark-like fishes), the actinopterygians (ray-finned fishes), and the tetrapods. Arginine vasotocin, together with oxytocin, has been identified in the pituitaries of teleostean fishes, amphibians, reptiles, and birds.

Two vasopressins are found among mammals, the arginine variety being most widely distributed. Arginine vasopressin has been identified in marsupials and also in *Echidna,* an egg-laying monotreme, and this suggests that it appeared very early in the evolution of mammal. Lysine vasopressin has only been identified in the neural lobes of surviving Suiformes, a suborder of the Artiodactyla or even-toed ungulates. Lysine vasopressin alone occurs in the domestic pig and hippopotamus, whereas both kinds of vasopressin are found in certain wart hogs and peccaries. Arginine vasopressin could have evolved from arginine vasotocin, present in the reptilian ancestors of mammals, by a single mutation causing isoleucine at position 3 to be replaced by phenylalanine. Lysine vasopressin could have arisen from arginine vasopressin through another mutation causing arginine in position 8 to be replaced by lysine.

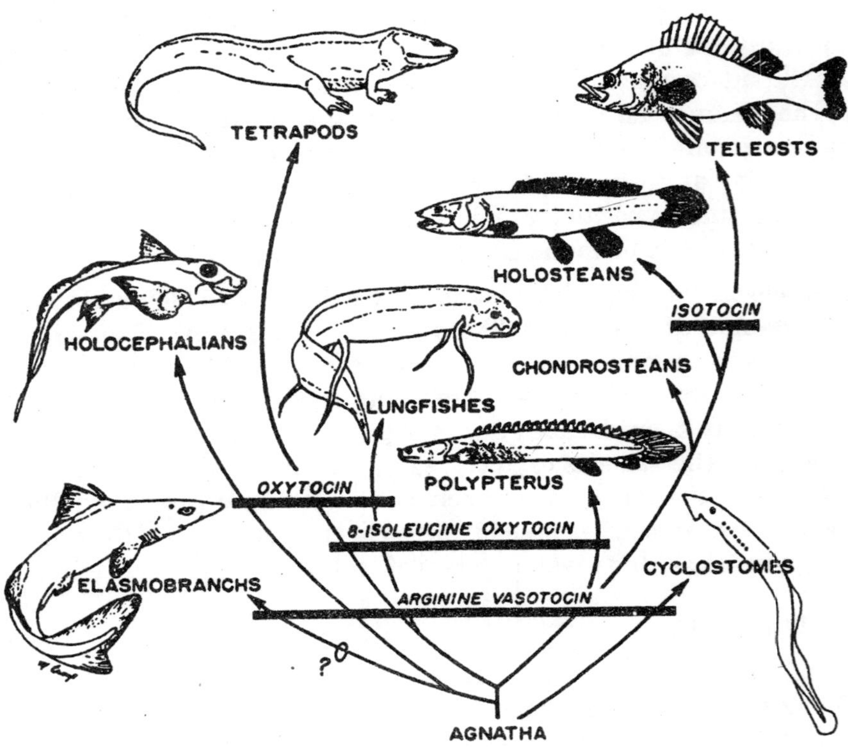

Fig. 5.3. Schematic representation of the evolution of vertebrates. The distribution of the chief neurohypophysial principles is indicated by the heavy bars which cut across the arrows. Observe the wide distribution of arginine vasotocin.

Oxytocin and similar compounds must also be very ancient molecules. Oxytocin is present in two evolutionary lines; the one leading to lungfishes and tetrapods, and the one giving rise to cartilaginous fishes. Oxytocin-like principles are present in the pituitaries of elasmobranchs, though their specific structures have not been determined, and oxytocin itself is presumably present in the chimaeras or ratfish (holocephalians). 8-Isoleucine oxytocin appears to be present in lungfishes and in *Polypterus,* a primitive ray-finned fish. Isotocin (4-serine, 8-isoleucine oxytocin) has been identified only in the pituitaries of holostean and teleostean fishes. Since arginine vasotocin seems to be the most primitive octapeptide, one could postulate that oxytocin arose from it by a mutation which caused leucine to be substituted for arginine at the 8 position. 8-Iso-leucine oxytocin could have arisen from vasotocin by substituting isoleucine for arginine at position 8. Isotocin could have been de-rived from 8-isoleucine oxytocin by substituting serine at position 4 for glutamine.

It is apparent from the above discussion that the octapeptide neurohormones are very similar in chemical structure. Accordingly, there is considerable overlapping of pharmacologic properties, though relative potencies may be strikingly different in the various bioassays. Arginine and lysine vasopressins are antidiuretic principles in mammals, but arginine vasotocin performs this function in amphibians, reptiles, and birds. Antidiuresis is an adaptation to terrestrial life, and it is not surprising to find that the neurohypophysial principles have no effect on water conservation in fishes, though they may promote the loss of sodium through the kidneys. Oxytocin is known to regulate milk ejection, and probably facilitates parturition and sperm transport in the female reproductive tract of mammals; its function in male mammals, if it has any, is unknown. Arginine vasotocin acts in birds to stimulate contractions of the oviduct and to promote oviposition, effects that are essentially oxytocic. When administered to mammals, vasotocin has potent oxytocic and milk-ejecting actions, whereas the vasopressins do not. It appears that during evolution vasotocin was discarded in mammals, and a clear separation appeared between octapeptides promoting water conservation (vasopressins) and the octapeptide controlling uterine contractility and milk ejection (oxytocin). The functions of neurohypo-physial principles in aquatic vertebrates remain largely undetermined. It has been suggested, however, that these neurosecretions may be especially important in fishes in regulating the release of adenohypophysial hormones. The mere presence of these peptides in the pituitaries of the lowest vertebrates does not mean that they have to function as systemic hormones in these groups.

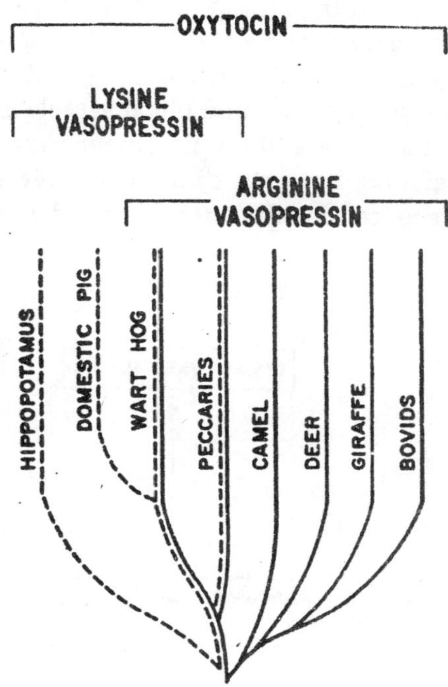

Fig. 5.4. Distribution of the neurohypophysial hormones among the artiodactyls superimposed on a scheme showing their probable evolution. Arginine vasopressin is indicated by solid lines; lysine vasopressin by broken lines.

FORMATION, STORAGE, RELEASE AND TRANSPORT

Oliver and Schafer demonstrated in 1895 that whole pituitary extracts, administered intravenously to the dog, induced vasoconstriction and elevated the blood pressure. Other workers promptly showed that this effect was attributable to the pars nervosa, and not to any other component of the pituitary. Since this part of the pituitary does not show cytologic characteristics of a gland, the source of the active principle became a problem. Some investigators postulated that the active material might be secreted by the adenohypophysis and passed to the pars nervosa for storage; others thought that it might be produced *in situ* by the pituicytes, even though these cells did not appear to be glandular. Kamm and his co-workers (1928) succeeded in separating two active fractions from neural lobe extracts, one being especially potent in elevating the blood pressure of mammals and the other acting principally to induce uterine contractions. Evidence contributed by many other workers made it clear that blood-borne hormones were present in neural lobe tissue, but it remained hard to believe that they could originate in such a nonglandular structure. The problem was finally clarified by the Scharrers and Bargman and their co-workers in the early 1950's. They established the concept that oxytocin and vasopressin are neurosecretory products which arise in the neurons of certain hypothalamic nuclei. The neurohormones are actually synthesized by these glandular neurons, moved along the axons of the hypothalamo-hypophysial tract, and discharged around blood vessels in the pars nervosa (a neurohemal organ). The pituicytes are neuroglia and are no longer regarded as the source of these secretions; there is still the possibility, however, that the pituicytes may be involved in either the release mechanism or the separation of the active peptides from the carrier substance.

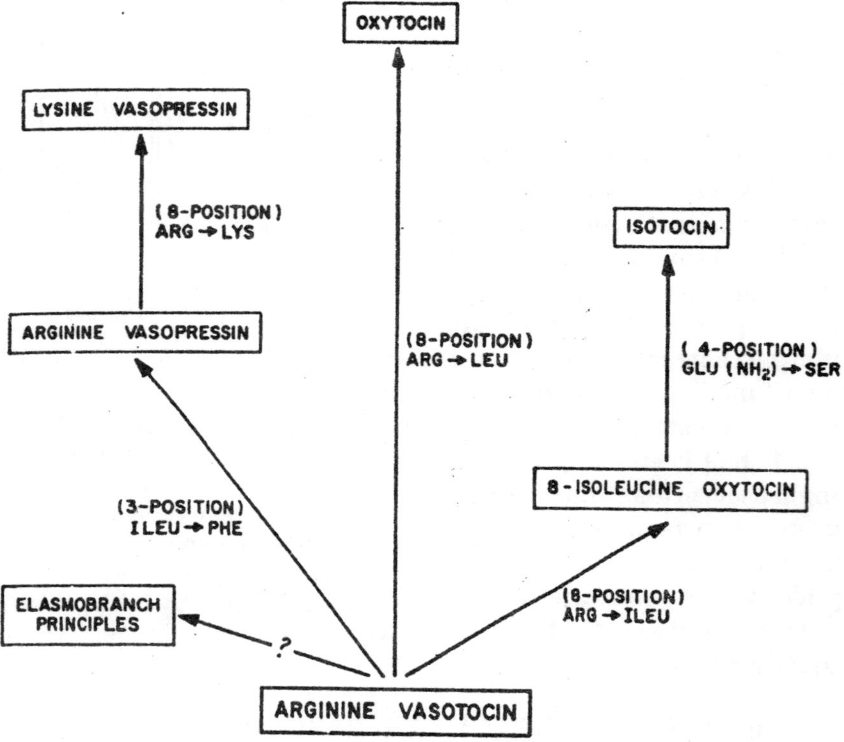

Fig. 5.5. Hypothetical scheme showing how vasopressins and oxytocins may have arisen from arginine vasotocin during the course of vertebrate evolution. Each step involves a single amino-acid substitution, and it is presumed that each change was the consequence of a gene mutation. The molecular positions where changes occurred are indicated in parentheses.

The paired supraoptic and paraventricular nuclei are the most important ones for the production of these neurohormones. These two pairs of nuclei in the dog contain about 100,000 neurons. Oxytocin and vasopressin are identifiable in these areas of the hypothalamus, and it must be assumed that under certain conditions the secretions can be freed directly into the hypothalamic circulation. The dog stores exceptionally large amounts of vasopressin in the hypothalamus. The administration of vasopressin and oxytocin to the snake *(Diadophis punctatus)* leads to a marked accumulation of neurosecretory material in the hypothalamic nuclei and pars nervosa.

In birds, in which a septum separates the two lobes of the pituitary, it is possible to remove the pars nervosa without disturbing the adenohypophysis. Ablation of the posterior lobe of the laying hen results in a prompt and permanent diuresis, but impairs neither ovulation nor oviposition. The consumption of water increases rapidly, and birds weighing 1,000 to 2,000 grams may drink as much as 1,000 grams of water daily. The diabetes insipidus in the operated birds may be controlled by the administration of neurohypophysial principles. The posterior lobe principles are known to facilitate oviposition in birds but, in the absence of the posterior lobe, enough oxytocin is apparently released from the hypothalamic neurons to take care of this need.

Totally hypophysectomized mammals seldom show any serious disturbances in water metabolism. In the hypophysectomized rat, neurosecretory material accumulates at the broken end of the pituitary stalk, the latter undergoing reorganization to form a normal but miniature neural lobe. On the other hand, if the hypothalamo-hypophysial tracts are severed near the supraoptic nuclei, the cell bodies of the latter deteriorate, and the organism is practically deprived of oxytocin and vasopressin. Under these conditions, a permanent diabetes insipidus results, the animals consuming large quantities of water and eliminating it through the kidneys.

The factors involved in the release of neurohypophysial secretions have not been completely elucidated. The release of anti-diuretic hormone (ADH; synonymous with vasopressin) may be altered by changes in osmotic pressure and volume of the blood, certain sensory reflexes, and actions occurring within the central nervous system itself. In snakes as in other amniotes, intraperitoneal injections of sodium chloride, or the withholding of water, cause neurosecretory material in the hypothalamic nuclei and pars nervosa to be rapidly depleted. The osmoreceptors in dogs are thought to be located in some area of the prosencephalon which is supplied by the internal carotid artery. If hypertonic solutions of sugar or salt are injected into the carotid artery of normal animals during water diuresis, a prompt fall in urine output occurs. Changes in blood volume and in electrolyte concentration of the plasma can bring about striking increases in ADH. Various emotions, as well as coitus and pain, reflex stimulation as in suckling and milking, changes in environmental temperatures, etc., are known to promote the release of such neurohormones.

Certain drugs, probably acting on hypothalamic nuclei, are known to reduce the flow of urine by increasing the release of ADH. Among these may be mentioned morphine, nicotine, ether, and various barbituates. Since the main nervous connection between the hypothalamus and pars nervosa is the hypothalamo-hypophysial tract, one wonders whether the neurons that elaborate neurosecre-tions can also conduct nervous impulses. There is some evidence that they can do both.

Various staining techniques have been employed in order to identify neurosecretory cells in different components of the neuro-hypophysial complex. It is probable that these techniques stain a protein "carrier substance" to which the peptide principles are adsorbed or attached, rather than the peptides themselves. Van Dyke *et al.* isolated a protein (30,000 mol. wt.) from neurohypophysial extracts which has vasopressor and oxytocic activities in the ratio of 1:1. These activities could be separated from the protein by treatment with acid. The van Dyke protein has more recently been named "neurophysin," and has been obtained from a number of mammalian species. The natural neurohormonal peptides may be combined with neurophysin *in vitro,* and subsequently dissociated from the protein. Neurophysin is so specific in its binding of neurohypophysial principles that it can be utilized in separating these principles from other peptides which are present in pituitary extracts. These studies suggest that the neurohormonal peptides are bound to neurophysin within the neurosecretory cells, and probably after they are released into the pars nervosa. It is not known whether these neurohormones exist as free peptides in the blood plasma or as peptide-protein complexes.

BIOLOGIC ACTIONS OF NEUROHYPOPHYSIAL PRINCIPLES

It appears that the principles of the neurohypophysis may not be indispensable for life. Although vasopressin has strong pharmacologic actions on the circulation, there is no proof that it performs a physiologic role. Oxytocin has been assigned a role in promoting uterine contractility in mammals during coitus and at the time of delivery, but individuals of some species seem to deliver young quite well after the supply has been eliminated, or reduced, by hypothalamic lesions. Most mammals could not rear their young without ejecting milk, but the individual could live without doing so. Even in the absence of the antidiuretic principle, mammalian organisms may rely upon other homeostatic mechanisms for the regulation of body water. To what extent desert animals conserve water through the release of ADH is not known. When the laboratory rat is deprived of water, it not only increases the output of ADH but also diminishes the consumption of food, thus reducing the loss of water through the feces.

Antidiuretic Effects

Even before the chemical identification of neurohypophysial principles, physiologists had demonstrated that they act upon the kidney to reduce the volume of urine. If a dog's kidney is isolated and perfused with blood from the trunk of another dog, it produces a large volume of dilute urine. The addition of small amounts of pituitary extract to the perfusate causes the kidney to excrete a smaller volume of more concentrated urine. Inclusion of the dog's head in the perfusion circuit causes the excised kidney to diminish its output of urine, but this effect of the head is abolished by removing the pituitary gland.

There can be no doubt that ADH (vasopressin) is of physiologic utility in the regulation of water balance in mammals. By acting upon the epithelial cells of the distal portion of the renal tubule, it encourages the reabsorption of water. Diabetes insipidus (excretion of a large volume of dilute urine) can be produced in mammals, such as the cat, dog, and monkey, by hypothalamic lesions that destroy the supraoptico-hypophysial tracts. Approximately 90 per cent of the fibers in these tracts must be severed before any profound disturbances occur in water metabolism. Such diabetic animals consume large quantities of water (polydipsia) and eliminate a large volume of urine (diuresis); the excessive water loss may be corrected by the administration of vasopressin. The essential effect of the peptide on the distal segment of the nephron is to increase the permeability of the epithelial surface to water.

Members of most mammalian species die when 20 per cent of the body water is lost, but the camel can survive after losing more than 40 per cent of its body water. Hibernating mammals, such as the hedgehog and certain desert rodents, have remarkable abilities to economize water. The Mongolian gerbil and similar desert species can survive for years without drinking water, living on a diet of dry grain without weight loss. The kangaroo rat *(Dipodomys)* obtains water through the oxidation of its food, but it does not store an excess of water in the tissues and has no especial resistance to dehydration. When excess water is administered by stomach tube, it is excreted with great difficulty, and the animals are likely to show symptoms of "water intoxication." The urine volume is about half that of the laboratory rat and is much more concentrated. The neural lobe of the kangaroo rat is relatively larger than that of the laboratory rat and contains more vasopressin per microgram of tissue. This suggests that vasopressin may have some special importance in desert rodents with respect to water metabolism.

Striking antidiuretic actions have been demonstrated in birds, reptiles, and certain amphibians. Most annurans, maintained in water and injected with neurohypophysial peptides, increase in weight. This is known as the "Brunn" or "water balance" effect. This weight increase results from a composite effect of the neurohormones; water loss through the kidneys is diminished, water uptake through the skin is increased, and water is absorbed from the urinary bladder and returned to the blood. Pieces of frog skin *in vitro* respond to ADH by accelerating the movement of water from the outside to inside, but only if the inside is hypertonic to the outside as it normally is *in vivo*. The urinary bladder of toads and frogs may also be used in *in vitro* experiments; it responds to ADH by increasing water absorption from the luminal surface. In many urodele amphibians that never leave the water (*e.g.*, *Necturus*), antidiuresis either cannot be produced by these peptides, or the required dose is so high that it is obviously unphysiologic. Posterior lobe principles have not been demonstrated to exert antidiuretic effects in fish, though they may be important in these vertebrates in promoting the renal loss of sodium. The administration of vasotocin to the teleost *Carassius auratus* causes sodium to be lost through the kidneys and promotoes the intake of sodium, perhaps by an effect on the gills.

Amphibian skin and bladder are useful tissues since they may be observed *in vitro* after the application of neurohypophysial principles. Evidence has been adduced that ADH acts upon amphibian membranes by dilating the pores. Isotopically labeled thiourea penetrates the toad's skin at a slower rate than water, presumably because of the larger molecules. ADH increases the rate of flux of the thiourea molecules. On the basis of the *pore theory* this is explained by assuming that the pores of the skin are normally just large enough to admit water molecules, but ADH dilates them sufficiently for thiourea molecules to pass through. It has been suggested that the pore theory may be applied to account for the actions of ADH on the kidney.

Oxytocic Effects

Two main functions have been ascribed to oxytocin in the promotion of uterine contractility: (*a*) to facilitate the ascent of spermatozoa in the female tract after intromission and (*b*) to expel the fetus from the female tract at parturition. Owing to the intermittent nature of these events it has been difficult to obtain quantitative information. Many workers have felt that in many species the spermatozoa reach the fallopian tubes too rapidly to be accounted for by the flagellate motility of the spermatozoa themselves. Moreover, nonmotile spermatozoa are known to ascend the female tract as quickly as motile ones. Oxytocin has been observed to increase the ascent of spermatozoa and various fluids introduced into the bovine uterus maintained *in vitro*.

Increased contractility of the uterus has been demonstrated in domestic animals after natural mating, tactile stimulation of the external genitalia, and mechanical stimulation of the uterus. In cows the content of oxytocin in external jugular blood was found to be increased after rectal palpation of the uterus and cervix. Milk ejection may accompany natural mating in several species, suggesting that there is an augmented release of oxytocin at this time.

Extracts containing oxytocin have been used clinically for many years to induce uterine contractions and facilitate delivery of the fetus. The neural lobes of rats and dogs contain very little oxytocin for several hours after labour, suggesting that the stored peptides have been used to support the process. The fact that parturition may occur normally in the absence of the pituitary gland does not prove that oxytocin is unimportant in parturition; sufficient quantities of neurohormone may continue to be produced by the hypothalamus. It is generally agreed that the sensitivity of the uterine muscle to oxytocin varies in accordance with the sex hormones that act upon it. Furthermore, there are differences in the reactivity of different regions of the same uterus. While the evidence is suggestive, no final conclusions can be made concerning the physiologic role of oxytocin in parturition.

There are indications that posterior lobe peptides may be important in the reproductive physiology of birds and reptiles. They effect oviposition in the hen by inducing contraction of the shell gland and of the vagina. Vasopressin stimulates contractions of the oviducts of the painted turtle, *Chrysemis picta.* Assay studies have shown that the oxytocic activity of the hen's neural lobe is practically depleted prior to oviposition.

Neural lobe secretions appear to influence reproductive behaviour in certain fishes. It has been reported that synthetic oxytocin induces the spawning reflex in *Fundulus.*

Galactogogic Effects

One of the best-established functions of oxytocin is its role in stimulating milk ejection. The milk ejection reflex manifests itself by a sudden rise in milk pressure within the glands and, like other reflexes, can be conditioned. The lactating female becomes conditioned to a variety of tactile, visual, and auditory stimuli associated with suckling or milking. In this manner the lactating mother unconsciously participates and makes it possible for the nursing young to obtain a full supply of milk. This reflex arc is a neurohormonal one, the efferent component being a neurohormone that is, in all probability, oxytocin. This principle is released from the neurohypophysis and causes contraction of the branching myoepithelial cells around the mammary alveoli.

The development of the mammary glands and milk secretion require a large number of hormones. There are indications that all of the anterior pituitary hormones may be involved in normal mammary functions. It has been proposed that an important effect of oxytocin is to prevent involution of the lactating mammary gland. Oxytocin, reflexively discharged, is thought to reach the anterior pituitary and cause it to release prolactin. The latter hormone then maintains the functional integrity of the mammary alveolar tissue.

The Hormones of Pregnancy and Lactation

The hormones that are most directly involved in pregnancy and lactation originate from the pituitary gland, the ovary, the placenta, and probably, in certain species, from the uterine endometrium. Like many other physiologic processes, these events involve a whole train of balanced forces and cannot be accounted for on the basis of hormones acting in isolation. Other endocrine glands, whose products are more directly concerned with systemic metabolism, exert indirect but important influences on pregnancy, parturition, and lactation.

THE EVOLUTION OF VIVIPARITY

With the evolutionary movement of organisms from aqueous to terrestrial habitats, many changes have occurred in the nature of the egg and in the structure and function of the genital system. Although widely separated taxonomic groups have exploited different devices for retaining the developing young on or within the body of a parent, there are suggestions that endocrine mechanisms have played an important role in making these adaptations possible. Evolutionary innovations are difficult to prove, and no unifying concept has emerged that accounts for all the facts. Although many specific exceptions can be found, the following evolutionary trends seem to be apparent: (1) a reduction in the number of eggs produced; (2) a closer association of the sexes and internal fertilization; (3) the addition of reserve food materials to the egg and compression of the larval stages into the embryonic period; (4) retention of embryos within the female tract and parental protection of the young; (5) reduction in size of the egg and the development of a placenta; and (6) endocrine regulation of the mammary glands for early nutrition of the newborn.

The most primitive method of sexual reproduction is that of discharging innumerable gametes into the surrounding water, fertilization occurring without any participation of the parents. Such eggs generally contain little yolk and are largely devoid of protective coverings. A closer association between the sexes became established, and in many fishes and amphibians the spermatozoa are discharged in close proximity to the eggs. In the salamander *Ambystoma punctatum,* the males discharge packets of sperms, which the females insert into their reproductive tracts. Thus fertilization is internal but there is no association of the sexes. With the production of fewer eggs, various kinds of mechanisms have arisen to assure the survival of the embryos. One finds among teleost fishes many kinds of parental care, nest building, and brooding. Male pipefishes and sea horses, for example, have specialized brood pouches on the ventral body wall. In other species, either the males or females take up the fertilized eggs and incubate them in their mouths. In the primitive ovoviviparous condition the eggs contained tremendous quantities of yolk and they were retained within the oviducts to provide greater protection and more uniform environments for development.

In most salamanders and a few reptiles, the zygotes only begin to cleave before leaving the female tract. The eggs of birds are in the early gastrula stage when they leave the oviduct, but reptilian eggs may be retained much longer. Almost all amphibians are oviparous and return to water for breeding, but there are a few exceptions. Adaptive viviparity has been described in a small toad, *Nectophrynoides occidentalis,* which breeds in arid mountainous environments. After a gestation period of nine months the young are born in a fully metamorphosed condition and do not require an aqueous habitat. Parturition is accomplished by inflation of the lung sacs and contraction of thoracic muscles, but delivery cannot occur unless the animal finds suitable mechanical support in the environment.

The oviducts secrete a great variety of materials for the protection and nutrition of the eggs. The aquatic eggs of fishes and amphibians are provided with jelly envelopes, membranes, and shells. In reptiles and birds, the albumenous covering of the egg contains large quantities of water, which are absorbed by the yolk before or after laying. In ovoviviparous selachians, with no intimate apposition of fetal and maternal vascular structures, the secretory products of the uterus provide an important source of nourishment. The viviparous selachians develop circulatory structures as an additional source of nourishment to supplement a reduced supply of yolk in the egg.

Corpora lutea have been identified in the ovaries of a number of vertebrates and they are not invariably associated with viviparity. They have been found in oviparous and ovoviviparous selachians, teleosts, amphibians, and reptiles, as well as in the oviparous monotremes and all higher mammals. They correlate with the retention of eggs in the oviducts in ovipara and with the retention of embryos in the uteri in the vivipara. The luteal structures of certain teleost fishes have been shown to release progesterone-like agents that control the growth of the ovipositor. The ovaries of some ovoviviparous snakes appear to contain progesterone, and similar materials are found in the blood plasma. It is clear that progesterone is not exclusively a mammalian hormone and this is another indication that there have been evolutionary changes in hormone emphasis rather than changes in the types of hormones.

The monotremes *(Ornithorhynchus, Echidna)* are the only mammals that lay eggs. The eggs are relatively small but contain enough nutriment to support development up to an advanced stage, though not to the level of self-sufficiency. In all other mammals, methods of uterine feeding have been perfected and the size of the egg is radically reduced. Mammalian eggs are typically fertilized in the oviducts and undergo early development while retained there. The cleaving eggs of many species are coated with albumenous secretions from the oviducts, but the nutritional requirements are met largely by the limited stores contained within the developing egg. During oviducal development, lasting from 4 to 10 days, the bulk of the embryo shrinks rapidly, indicating that there has not been any appreciable absorption of materials from the fluids of the oviduct. The embryotroph elaborated by the uterine glands is of great importance in many species for nourishment of the blastocyst before the placenta has been established. It is probable that in the Artiodactyla and Perissodactyla the uterine secretions are an important source of nourishment for the embryo and fetus, from the beginning to the end of pregnancy. In cases of delayed implantation, the uterine secretions apparently maintain the blastocysts for weeks or months. The first step toward viviparity in mammals has been the retention of embryos within the uterus, nothing more than a primitive yolk-sac placenta being established. The gestation period of the opossum *(Didelphis virginiana)* is only 12 or 13 days, the young being born as soon as the metamorphic changes have been completed. The luteal phase of the estrous cycle is prolonged and parturition correlates with the involution of the corpora lutea. Pregnancy may be terminated at any stage by removal of the ovaries, which suggests that in this species pregnancy does not involve any hormones other than those that regulate the estrous cycle. Extraovarian mechanisms operate in higher mammals and produce hormones that make it possible for embryos to be retained in the uterus for periods longer than the limits of the estrous cycle.

HORMONES IN PREGNANCY AND PARTURITION

The Hypophysis

We have seen that the anterior pituitary gonadotrophins are essential for the periodic release of eggs from the ovary and for the secretion of ovarian hormones that build up the type of uterus most suitable for the reception of the blastocysts. After the animal has become pregnant, the effects of hypophysectomy vary with the species and in accordance with the stage of gestation at which the operation is performed. In the rabbit, cat, and dog, hypophysectomy produces abortion no matter at what stage of pregnancy it is performed. In other species, such as the mouse, rat, guinea pig, and monkey, hypophysectomy at approximately midterm or later may not interrupt gestation or interfere appreciably with delivery. Lactation fails to occur or is of short duration. The gestation period of the dogfish *Mustelus canis* is 10 months, and hypophysectomy during the first 5 months does not interfere with embryonic development, the absorption of yolk, or the establishment of the yolk-sac placenta. The pituitary gland is necessary for the maintenance of normal gestation in viviparous snakes.

Rats may be hypophysectomized at midpregnancy and carried to term by the administration of estrogen and progesterone. Since the ovaries of the rat are indispensable throughout the course of pregnancy, it appears that the maintenance of pregnancy after hypophysectomy depends upon the continued function of the corpora lutea brought about by a luteotrophic principle (prolactin), which must be of extrahypophysial origin. The corpora lutea appear to be the main source of progesterone in the rat. Extracts possessing luteotrophic action have been prepared from the rat's placenta. There is circumstantial evidence indicating that in certain mammalian species the placenta functions as an adjunct to the anterior hypophysis and takes over the production of gonadotrophins, which are necessary for the maintenance of gestation.

Studies suggest that maintenance of the corpora lutea of the pregnant rabbit is a function of estrogenic hormones rather than a direct consequence of gonadotrophin action. The corpora lutea of hypophysectomized rabbits may be maintained either by the administration of estrogens or gonadotrophins. It may be that in this species the gonadotrophins do not act directly on the corpora lutea to cause the continued release of progesterone, but prolong luteal secretion by encouraging the production of estrogens. It is probable that the gonadal hormones *per se* have more important direct actions on the gonads than is commonly supposed. It is well known that estrogens have considerable ability to promote growth of the ovarian follicles of hypophysectomized rats. There are apparently many species variations in the relative importance of estrogens and pituitary gonadotrophins in maintaining luteal function.

The Ovary

Full differentiation of the endometrium requires both estrogen and progesterone, and removal of the ovaries while the developing zygotes are in the tubes invariably prevents implantation of the blastocysts and placentation. If pregnant females are ovariectomized while the blastocysts are floating free in the uterine lumen, the blastocysts generally die. However, if the armadillo is bilaterally ovariectomized at about the middle of the four-month period of delayed implantation, implantation occurs about 30 days later and is indistinguishable from normal implantation. Hence, in this species some non-ovarian tissue can assume the function of maintaining the uterus. As with the pituitary gland, the ovary is more essential for the maintenance of pregnancy in some species than in others. The ovaries are indispensable at practically all stages for the maintenance of pregnancy in the opossum, mouse, rat, rabbit, golden hamster, 13-lined ground squirrel, goat, and viviparous snakes. In women, monkeys, and mares ovariectomy during the early months of pregnancy usually does not cause abortion. In the guinea pig, cat, dog, and ewe, the ovaries may be dispensed with during the second half of gestation. The ovaries are not essential during the terminal stages of pregnancy in cows and pigs. Failure of implantation or abortion in castrate animals may be prevented by the administration of progesterone. There is ample evidence that the placenta performs endocrine functions during pregnancy, being capable of secreting both gonadotrophins and steroids of the ovarian type in certain species.

In a variety of mammals the quantity of estrogen excreted through the kidneys is increased strikingly during pregnancy. The total estrogen in the urine of pregnant women increases gradually after the first week, reaches a peak shortly before parturition and drops abruptly a few days after the birth of young. The increased quantities of estrogen in the urine are not the result of a diminished renal threshold, because after the second month of pregnancy such compounds are present in the blood in greater amounts than at any other time. Nonpregnant women excrete about 300 M.U. (mouse units) of estrogen per day, whereas during pregnancy the level rises to 20,000 M.U. per day. It is noteworthy that over 99 per cent of the urinary estrogen is in the form of estriol glucuronide and that such conjugated estrogens have little physiologic potency. The total amount of conjugated estrogen begins to fall shortly before parturition, whereas the amount of unconjugated, physiologically active estrogen undergoes a relative increase at this time. This would seem to suggest that estrogen performs some role during parturition that it does not exercise during the preceding months of pregnancy.

The nonpregnant mare excretes around 2,000 M.U. of estrogen daily, but during pregnancy it exceeds one million mouse units per day. Estradiol and estrone are found during the estrous cycle of the mare, but during pregnancy at least three additional steroids appear—equilin, equilenin, and dehydroequilenin, the first two being the most abundant. The abundance of estriol in pregnant women and of equilin and equilenin in pregnant mares supports the concept that these are estrogens of pregnancy and that they probably perform some role peculiar to gestation. Bilateral ovariectomy of pregnant women, monkeys, and mares does not abolish the high excretion of estrogens. The excretion rate of estrogens in pregnant monkeys is not altered by removing the ovaries and fetus, but it falls to non-pregnant levels after removal of the placenta. The development and regression of the fetal gonads of the horse correlate with the rise and fall of the maternal estrogens.

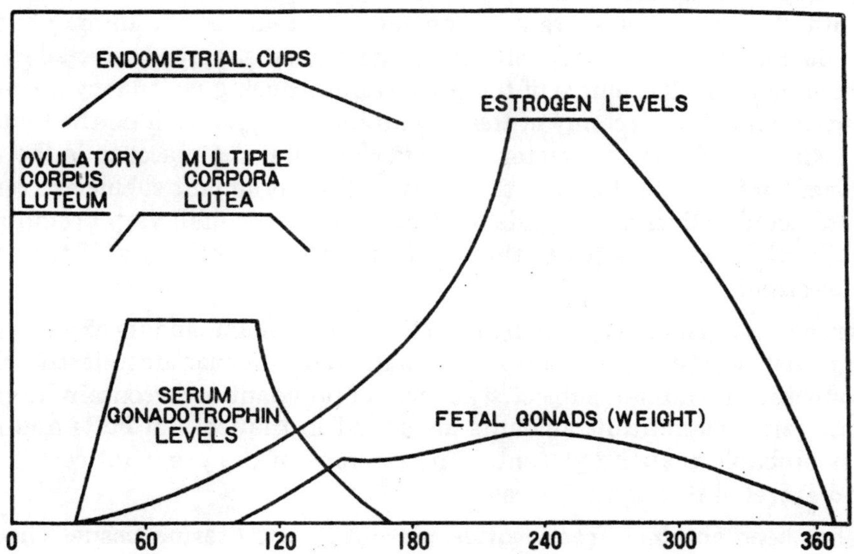

Fig. 6.1. Diagram showing the urinary excretion of chorionic gonadotrophin, total estrogen, and pregnanediol during human pregnancy.

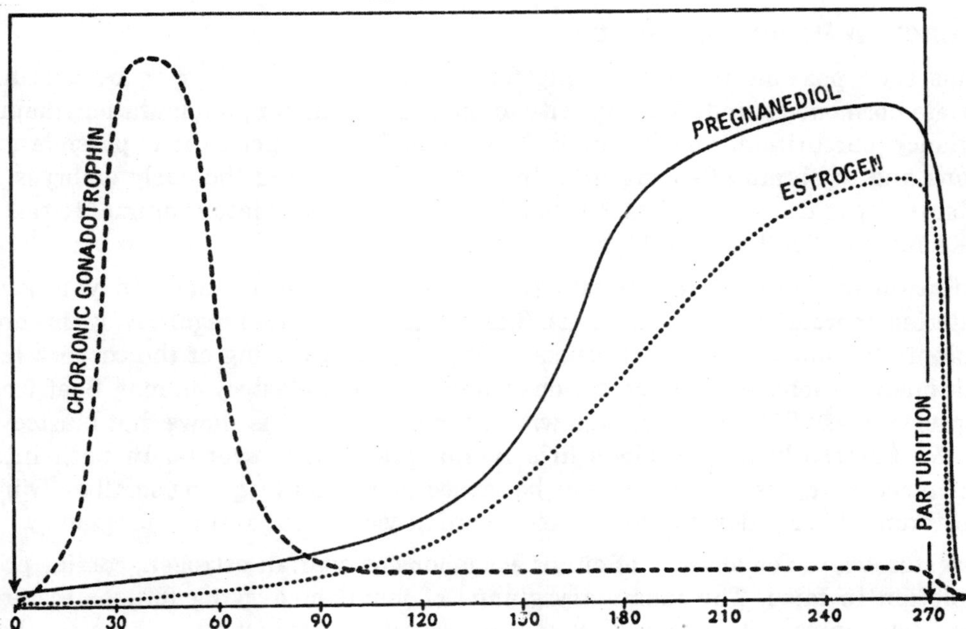

Fig. 6.2. Sequence of events during gestation in the mare. Note that the ovulatory corpus luteum wanes and accessory corpora lutea are formed when the serum gonadotrophin levels rise. Urinary estrogens reach a high peak at about 220 days and fall before parturition.

Small amounts of progesterone are thought to be secreted by the preovulatory follicle and the titers rise markedly during the course of pregnancy. Pregnanediol excretion in pregnant women rises progressively after the second or third month and falls precipitously after loss of the placenta. The indications are that the corpus luteum of the human ovary produces progesterone during early pregnancy, whereas an additional supply is secreted by the placenta after the second month. The source of the progesterone during pregnancy apparently varies with the species. Since ovariectomy of the ewe around midgestation neither causes abortion nor prevents the rise of blood progesterone, it is clear that the placenta is the major source of the hormone. On the other hand, if the ovaries of the pregnant rabbit are removed seven days before expected delivery, the levels of blood progesterone fall very promptly and this is followed by abortion. The ovaries of the rabbit seem to be the main, if not the exclusive, source of progesterone.

Progesterone disappears very rapidly from the blood stream, and its excretory metabolites seem to vary with the species. Sodium pregnanediol glucuronidate is the chief urinary catabolic product in the human subject. The feces of pregnant cows contain large amounts of androgen, but very insignificant amounts are found in the feces of bulls and nonpregnant animals. It is probable that the placental progesterone of the pregnant cow is converted to androgen and excreted through the feces.

Relaxin has been obtained from ovaries, placentae, and uterine tissue. The ovary seems to be the principal source of this hormone in the sow, whereas in the guinea pig and mouse the placenta may play an important role in its formation. Under suitable conditions all three tissues in the rabbit may produce relaxin.

Pregnancy in Prepuberal Mice

It has been possible to cause young female mice to rear offspring by administering appropriate hormones to induce receptivity to the male, ovulation, implantation, maintenance of pregnancy, parturition, and lactation. In mice and other species it is possible to induce ovulations and to obtain offspring from them by transplanting the early embryos into the uteri of sexually mature recipients, but to induce a sexually immature animal to rear its own young is a more difficult undertaking.

Ovulation and sexual receptivity to the male may be induced in young mice by administering appropriate gonadotrophins. Blastocysts are produced regularly by this procedure, but implantation does not ensue because of improper functioning of the corpora lutea and the consequent absence of a progestational uterus. Viable blastocysts may float free in the uteri for as long as 22 days post coitum without implanting. This shows that blastocysts may survive for relatively long periods in a nonprogestational uterus. In both intact and ovariectomized mice, the blastocysts may be caused to implant by giving small daily injections of progesterone. Larger doses of progesterone are necessary to continue pregnancy.

Progesterone, in the absence of either exogenous or ovarian estrogen, carries pregnancy to completion in mice. The mammary glands of females carrying fetuses to term have developed lobuloalveolar systems that contain secretory products. On the other hand, the mammary glands remain poorly developed in females receiving progesterone treatments in the absence of fetuses and placentae. These facts indicate that endogenous estrogen is being supplied from some source, most likely from the placentae or fetuses.

Parturition in these animals is effected by the administration of relaxin. Although progesterone maintains pregnancy and permits development of the mammary glands, relaxin is required to lengthen the interpubic ligaments and to promote uterine contractility. The pubic symphyses of females not carrying full-term young failed to respond fully to relaxin. This suggests that either the fetuses or the placentae add enough estrogen to act synergistically with relaxin in promoting relaxation of the pubic ligaments. It may be that in these mice relaxin initiates the release of oxytocin from the neurohypophysis and thus encourages uterine motility, progesterone failing to have this effect. Lactation occurs after delivery even in the castrated mothers as a consequence of progesterone and relaxin treatments.

Thus, it is possible through hormone manipulations to induce in young, sexually immature mice all of the events of gestation, parturition, and lactation.

Gonadotrophins Peculiar to Pregnancy

The placenta and, in certain species, the endometrium produce gonad-stimulating hormones that are similar in some respects to those produced by the anterior hypophysis but that differ from the latter both physiologically and chemically. The pregnancy gonadotrophins are present in a variety of mammals but are strikingly different in their physiologic properties.

Chorionic Gonadotrophin

This hormone is a glycoprotein and is characteristic of pregnancy in primates, although hormones similar in activity may be present in lower mammals. It is secreted by the chorionic villi of the placenta and appears in the blood and urine during early pregnancy. Human chorionic gonadotrophin (HCG) appears in the urine shortly after implantation and reaches a high peak about one month after the first missed menstrual period. At this time, around 100,000 R.U. (rat units) may be excreted daily. After this peak the blood and urinary titers of the hormone drop to low levels, which remain fairly constant until a few days after parturition. In the chimpanzee and monkey, chorionic gonadotrophin is produced for a brief period during early pregnancy and disappears thereafter.

HCG resembles luteinizing hormone (LH) from the anterior hypophysis in most of its actions, and in addition it seems to have the properties of a luteotrophin inasmuch as it prolongs the functional status of the corpus luteum. It converts the corpus luteum of the menstrual cycle into the corpus luteum of pregnancy, thereby prolonging the luteal production of hormones until the placenta becomes capable of secreting the high amounts of gonadal steroids required for the continuation of pregnancy. Although HCG has some ability to cause follicle stimulation in hypophysectomized rodents, its main action in the female is on the corpus luteum. Studies have shown that HCG undergoes certain qualitative changes from early to late pregnancy.

Although HCG is found normally only in pregnant women, it produces a great variety of gonadal actions in many other vertebrates. When given to the human male it causes the differentiation of Leydig cells and induces and maintains the production of testicular androgens. When administered to intact animals it produces follicular growth and ovulation, probably by acting synergistically with the circulating endogenous gonadotrophins of pituitary origin. It causes a release of spermatozoa when given to lower vertebrates such as amphibians. Thus

HCG may act directly in hypophysectomized animals or indirectly in the presence of pituitary gonadotrophins of intact subjects and affect the follicles and corpora lutea of the ovaries and the cells of Leydig or seminiferous tubules of the testis.

Chorionic gonadotrophin may be found in certain pathologic states associated with pregnancy and in others that have no connection with pregnancy. In such conditions as hydatidiform mole and chorioepithelioma of the female, high titers of HCG may be produced. Some neoplastic diseases.of the testis may likewise result in the secretion of this substance.

Pregnancy Tests

Since chorionic gonadotrophin is secreted by placental tissue, its presence in the blood or urine forms the basis of many tests for pregnancy. The Aschheim-Zondek test depends on the capacity of pregnancy urine to induce corpora lutea or "blood points" in the ovaries of mice and rats within 96 hours after treatment. The production of vaginal estrus in the immature rat, 72 to 96 hours after injecting the sample of urine, is a sensitive pregnancy test. In the Friedman test, pregnancy urine causes the formation of corpora lutea within 24 hours after being injected into immature or isolated mature rabbits. The amphibian tests are quicker and perhaps more reliable than those done on mammals. Female frogs and toads ovulate within six to eight hours after pregnancy urine has been injected into the dorsal lymph sacs. The Galli-Mainini test depends on the prompt evacuation of sperms from the testes of frogs. If the injected urine contains HCG, sperms can be identified in the cloacal fluid within three hours after the injection. The African frog *(Xenopus laevis)* is about ten times more sensitive to pregnancy urine than the female rat.

A number of immunologic pregnancy tests are now in common use. Most of these are based on the agglutination-inhibition reaction utilizing either erythrocytes or latex particles sensitized with HCG. These methods are accurate and can be completed within an hour or less.

Equine Gonadotrophin

The blood serum of pregnant mares contains a gonadotrophin (PMS) that is apparently constant in composition and is a sugar-containing protein. Its biologic properties are similar to a mixture of pituitary FSH and LH, the predominant effect depending upon dosage. PMS is, however, quite different from human chorionic gonadotrophin and from the hypophysial gonadotrophins. All attempts to split the PMS complex into FSH and LH components have failed so far." Unlike HCG and pituitary FSH and LH, PMS remains in the blood and lymph and is practically absent from the urine. Even when PMS is injected into other animals it remains in the blood for long periods, not being so rapidly metabolized as the other gonadotrophins. If small doses of PMS are given subcutaneously to hypophysectomized rats, growth of the ovarian follicles ensues. When the subcutaneous injection is followed by intravenous injections, ovulation or luteinization occurs. Since the hormone has high FSH activity, cystic follicles are often produced, especially if large doses are given repeatedly.

PMS appears in the blood of the mare on about the fortieth day of pregnancy and remains high until about day 120; then it drops and is absent after day 180. The corpus luteum of ovulation has begun to wane by the time that PMS appears in the blood. Under the influence of PMS the mare's ovaries form large vesicular follicles. Some of these follicles ovulate and form corpora lutea, whereas others undergo luteinization without ovulating. In this manner a crop of accessory corpora lutea is normally formed during pregnancy. The accessory corpora persist until about the 180th day of gestation, and from this time until the end of pregnancy the ovaries contain neither corpora lutea nor large follicles. It is apparent that the ovaries of the mare do not provide a source of progesterone throughout the whole course of gestation. The placenta apparently secretes both estrogen and progesterone during the second half of pregnancy. It appears significant that the formation and persistence of the accessory corpora lutea coincide with the period of high levels of PMS in the blood. The ovaries of the nilgai and African elephant also contain accessory corpora lutea during pregnancy. The gestation period of the elephant is about two years, and accessory corpora are found only between the sixth and ninth months. It is inferred that these animals produce a pregnancy gonadotrophin similar to that of the mare, but it remains to be determined.

The gonads of equine fetuses also respond to the circulating hormones of the mother. Under the influence of serum gonadotrophin and estrogen, the fetal ovaries become larger than those of the mother. The fetal testes also are larger than the testes of newborn males.

Whereas human chorionic gonadotrophin is produced by the placenta, specifically by the Langhans cells of the fetal chorion, equine gonadotrophin appears to be formed in the endometrial cups and hence is uterine in origin. The following evidence indicates that the endometrial cups produce PMS: (*a*) the hormone can be extracted from the cup areas but not from other parts of the endometrium; (*b*) the hormone appears in the blood coincident with the development of endometrial cups; and (*c*) when the endometrial cups first appear, they contain higher concentrations of PMS than are present in the blood.

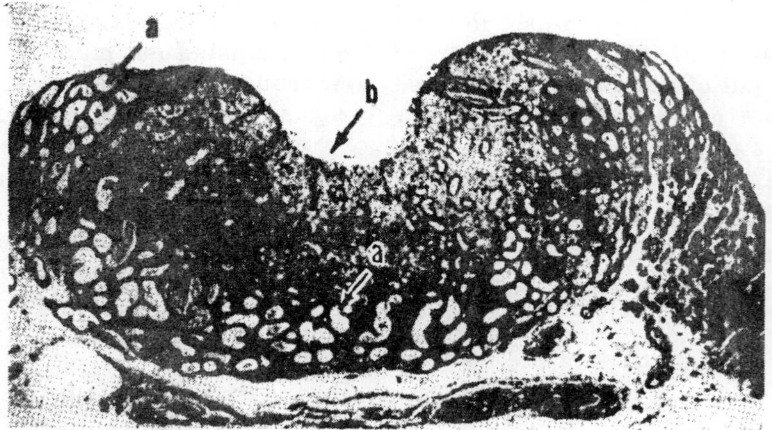

Fig. 6.3. Section through an endometrial cup of a pregnant mare. Note the dilated uterine glands *(a)*. The coagulum has been removed from the central region of the cup *(b)*.

MECHANISMS OF UTERINE ACCOMMODATION

During gestation a new individual develops inside a hollow, muscular organ, the uterus, and it is apparent that the latter must undergo profound modifications in order to accommodate and support the products of conception. Provision must be made for the nutritional, respiratory, and excretory requirements of the retained embryo. As pregnancy progresses there must be gradual enlargement of the uterus to permit growth of the fetus, but at the same time its muscular walls must remain quiescent enough to prevent premature expulsion. At parturition the myometrium must be activated in order to deliver the new individual to the exterior, and in mammals the mammary glands must be called into action for the postpartum nutrition of the young.

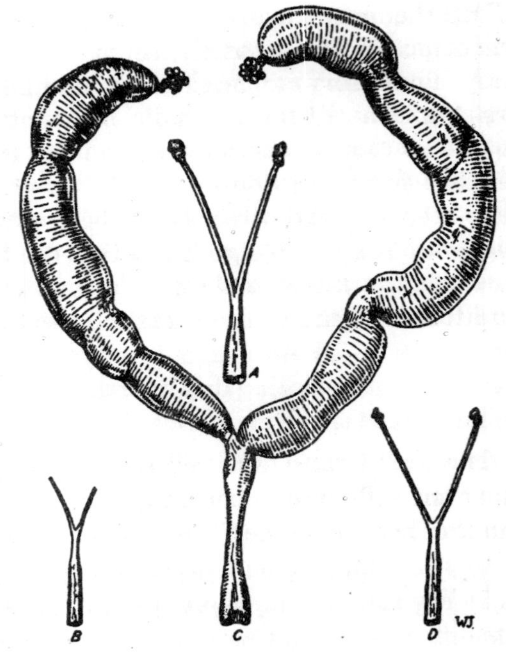

Fig. 6.4. Female genitalia of the rat, showing size changes during different natural and experimental conditions. A, Tract removed from a normal, adult, virginal animal during diestrus. B, Tract from an animal which had been oophorectomized at thirty days of age and autopsied six months later. C, Tract removed on the nineteenth day of normal pregnancy. Observe the large corpora lutea of the ovaries and the marked distention of the uteri produced by the products of conception. Compare with A. D, Tract from an animal which had been hypophysectomized for six months. (All the tracts were dissected out *in toto* and drawn to scale.)

The uterus passes through three stages in its adjustments to the products of conception: uterine preparation, growth, and stretching. During the period of *preparation,* before the blastocysts implant, estrogen augments the blood supply to the uterus and causes some increase in the number of cells. With the release of increasing amounts of progesterone there is a decided wave of hyperplasia involving all tissues—particularly the smooth muscle cells. Thus, by the time of implantation the number of cells has increased tremendously. Since the resulting cells are smaller than those from which they arose by mitosis, the marked hyperplasia produces no significant increase in the size and weight of the uterus. The period of uterine *growth* begins immediately after implantation. As the products of conception enlarge sufficiently to constitute an effective physical stimulus, the uterus adapts itself through hypertrophy of the cells already present. The tremendous degree of hypertrophy occurring in the smooth muscle cells is sufficient to account for most of the increase in uterine size during middle and late pregnancy. During the period of uterine *stretching,* the products of conception grow at an accelerating rate, whereas uterine growth diminishes. The period of uterine stretching lasts for only a day and a half in the hamster, but it continues for better than three weeks in the guinea pig.

Distention of the uterine lumen is recognized as an important factor conditioning enlargement of this organ. Clinicians have been aware for many years that in advanced ectopic pregnancy the empty uterus remains much smaller than in normal intrauterine pregnancies. Implantations often occur unilaterally in laboratory rodents having a duplex type of uterus, the equivalent uterus remaining empty or sterile. This condition may be produced experimentally by ligating or sectioning the oviduct on one side. In all these instances it has been obvious that the sterile uterus does not enlarge to the same degree as the gravid uterus, even though both are exposed to the same circulating hormones. Moreover, it has not been possible through the administration of hormones to modify the size and structure of a nongravid uterus until it is comparable with a gravid one. These observations, as well as many others, indicate that the tension produced by the growing products of conception constitutes a stimulus that enables the uterus to become structurally adapted to its contents. Tension produced by introducing paraffin pellets or rolled rubber dam into the lumen of the uterus exerts a growth-promoting action comparable to that resulting naturally from the products of conception. Pathologic and surgical conditions that enforce the accumulation of uterine secretions within the lumen produce a similar effect. From the foregoing statements one would be justified in conjecturing that the mature uterus is capable of undergoing two types of normal growth: (1) that which occurs periodically during the sexual cycle mainly as a result of chemical stimulation by the ovarian hormones, and (2) that which takes place during pregnancy under the influence of a complex of hormones together with the added physical stimulus of tension produced by the growing products of conception.

Spacing and Migration of Blastocysts

Little is known about the mechanisms that operate in polytocous mammals to regulate the spacing of embryos in the uterus. In species such as the rabbit, the transplantation sites are approximately the same distance apart. Moreover, in mammals having bicornate uteri, the blastocysts may shift from one horn to the other in order to balance the two sides. In the pig, for example, the left ovary produces more eggs than the right one, but the right and left horns of the uterus accommodate approximately the same number of fetuses. In monotocous species, the fetus may be found in the uterine horn opposite to the ovary that gave rise to the egg.

The spacing and implantation of blastocysts have been carefully studied in the rabbit. The three principal mechanisms are muscle activity, adhesion, and invasion. From three to five days post coitum the blastocysts are moved along the uterus in a random manner and increase rapidly in diameter after this period. It is thought that progesterone conditions the uterus so that muscular contractions arise from both ends and wherever it happens to be stimulated by a blastocyst that has attained a certain size. The contractions emanating from each blastocyst spread in both directions but they lose propulsive strength with increasing distance from the source. In this manner each blastocyst may repel others above or below it, but remain unaffected by the contractions it has produced. Since the repulsions are mutual and become weaker with distance, the blastocysts become separated from each other by about the same distances. At seven days post coitum the blastocysts have become so enlarged that they cannot be moved up or down the uterus. The antimesometrial muscle over each blastocyst loses tone and forms a pocket in which the blastocyst is held. The adjacent circular muscle behaves as an incomplete sphincter and holds the blastocyst in the pocket. The adhesion of the rabbit's blastocyst to the uterus is probably alkali induced. The trophoblast penetrates the uterine lining by both displacement and destruction of cells. Adhesion and invasion occur where the epithelium has an underlying capillary and may result from local elevations of pH attendant upon the transfer of carbon dioxide or similar materials to the maternal circulation.

When a silk suture is placed in the antimesometrial wall of one horn of the rat's uterus before mating, blastocysts do not implant in the sutured horn, though they do implant normally in the unsutured horn. Insertion of the suture before estrus does not induce deciduoma, nor does it interfere with fertilization and tubal transport of the ova. The foreign body may produce an unfavourable intrauterine environment which causes death of the blastocysts, or, by interfering with tone and motility of the uterus, it may allow the blastocysts to escape into the vagina before they have had a chance to implant.

Parturition

The mechanisms involved in the onset of labour are very complex and remain poorly understood. It is certain that we cannot think of parturition as being triggered by any single substance but must regard it as the consequence of many synchronized events that have occurred during the course of gestation. The uterus is relatively quiescent during gestation, but as labor approaches there are signs of increasing myometrial irritability and the development of more efficient patterns of contraction. During the end of gestation actomyosin, the contractile protein of muscle, increases in quantity and improves in quality, and this facilitates the forceful muscle contractions that are required to expel the fetus. There is increased sensitivity of the uterine muscles to hormones and various kinds of mechanical stimuli.

When the fetuses are surgically removed, the placentae and extraembryonic membranes being left *in situ,* gestation continues for the characteristic period. The placentae and empty membranes are carried to term and delivered at the normal time. This shows that hormones or toxic materials released by the fetus do not give the signal for parturition. The length of the gestation period is genetically determined, although it can be modified by many internal or external environmental factors. In certain genetic strains of cattle, the fetus continues to grow *in utero* during a greatly prolonged gestation. The fetuses become so large that normal delivery is impossible, and they must be removed by caesarean section. On the other hand, some strains may carry abnormally small fetuses for prolonged periods. These considerations indicate that fetal size itself does not determine when labor will begin.

The hormones generated from the placenta and ovaries are known to play key roles in determining the onset of labour. In most mammalian species progesterone exerts a pregnancy-stabilizing effect, and labour cannot occur until its influence is effectively diminished. In some species progesterone is produced by the ovary throughout pregnancy but, in others, it arises from both the ovary and placenta. Estrogens promote rhythmic contractility of the uterus. It is probably significant that estrogen increases in amount and effective form as the end of gestation draws near. Oxytocin from the neurohypophysis is also known to have a powerful effect on uterine contractility, and there is circumstantial evidence that it may be involved in the labor mechanism. The fact that totally hypo-physectomized animals may deliver young normally does not necessarily prove that oxytocin is not involved in the process; the hypothalamic nuclei that form the hormone may add effective amounts of it to the circulation in the absence of the posterior pituitary. If oxytocin is involved, there is the problem of how it is released at exactly the right time. Relaxin is definitely a hormone of pregnancy and is known to be secreted by the placenta as well as by the ovary.

Without the proper hormonal balance and timing, labour would be abnormal and injure the fetus and mother. For example, it is well known that labour may be precipitated by the administration of large doses of oxytocin. However, if the cervical canal has not been softened and the pubic ligaments relaxed by the action of relaxin and other hormones, the violent uterine contractions would probably kill the fetus and rupture the uterus instead of expelling it through the vagina. Whatever the exact mechanism of labor may be, it is certain that all of the events are nicely synchronized and that no single factor can account for the process. The hormones and other factors that tend to stabilize pregnancy are gradually overcome by forces that act in an opposite direction and tend to end it.

The Theory of Progesterone Blockage

Estrogens encourage the muscles of the uterus to contract, whereas progestogens inhibit uterine contractions and thus prevent premature expulsion of the fetus. As long as progesterone is dominant in the uterus, the myometrium is "blocked" and cannot deliver the fetus. Since progesterone does not affect the actomyosin content of muscle, it must act at a higher level of organization. Progesterone has been found to reduce the excitability of uterine muscle, and no excitation wave spreads from the point of stimulation. Many types of experiments prove that the myometrium is unable to respond effectively to stimulants as long as it is under the influence of progesterone. The concept has been advanced that it is the *progesterone block* that maintains pregnancy, and withdrawal of the block is responsible for the onset of parturition.

Some insight into the mechanism of labour has been gained by studying a rare anomaly of human pregnancy. On very rare occasions a septum divides the human uterus into two horns, and twins, having separate placentae, may develop in the separate chambers. The fetus in one horn of the uterus may be born prematurely, whereas the other may not be born until as much as two months later. Seemingly the onset of labour centers around some local effect of the placenta in these cases. It has been proposed that the placenta discharges its progesterone directly into the uterine tissues rather than into the general circulation. If this is true, the highest levels of progesterone would be found at the implantation site, with a diminishing concentration gradient extending into the myometrium from this area. This would explain why little or no progesterone from the placenta of one twin reaches the myometrium of the opposite horn. If placental progesterone is carried by the blood stream, the failure of one placenta should not cause delivery, since the producing placenta would provide enough progesterone to block myometrial contractions in both horns of the uterus. Experiments on laboratory animals indicate that the concentration of progesterone in the uterus during late pregnancy declines with distance from the placenta.

The postion of the placenta in the uterus appears to be a very important matter. If the placenta is attached at or below the midline of the uterus one would expect, according to the progesterone block concept, that the fundus would recover first from progesterone inhibition after the placenta ceases producing this hormone. Thus myometrial contractions would start at the superior end of the uterus and push the fetus through the cervix. If, on the other hand, the placental attachment is at the upper end of the uterus, the activity gradient would start from the cervical end of the uterus and push the fetus in the wrong direction. These theoretical assumptions strongly imply that a local factor, presumably the progesterone block, performs a decisive role in delivery, but they do not rule out the possibility that substances other than progesterone may be important in conditioning myometrial activity.

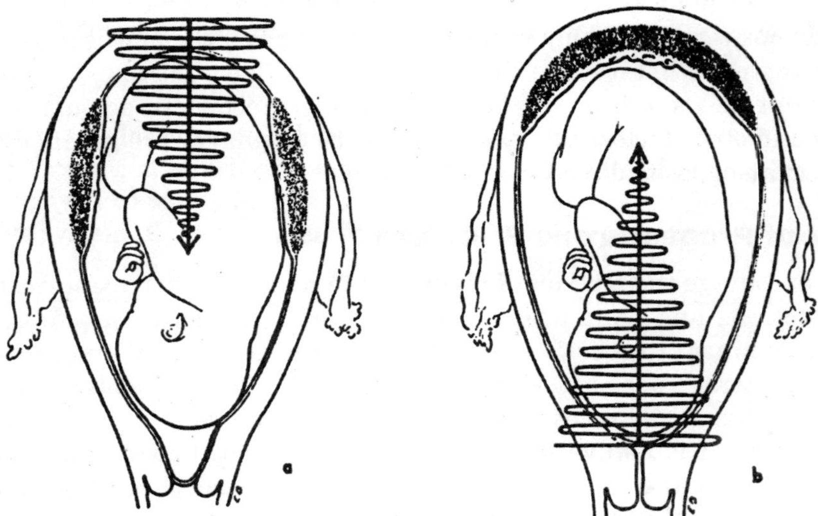

Fig. 6.5. Hypothetical illustration of a progesterone gradient in the parturient myometrium determined by the position of the placenta. The normally located placenta *(a)* releases progesterone well down in the uterus. At parturition the corpus of the uterus recovers first from the progesterone block, and uterine contractions push the fetus toward the vagina. When the placental progesterone is concentrated in the upper end of the uterus *(b)*, the lower end of the uterus recovers first from the progesterone block and uterine contractions push the fetus in the wrong direction.

THE MAMMARY GLAND AND LACTATION

Anatomy

The mammary glands are regarded as homologous with sweat glands, since they originate as integumentary ingrowths. The manner of embryonic origin is the same in both sexes. Each mammary gland of the human female is composed of 16 to 25 lobes that radiate from the nipple. Each lobe of the gland is drained by a lactiferous duct that extends toward the areola, the pigmented area around the nipple. At this level each lactiferous duct dilates into a sinus, again constricts, and opens at the tip of the nipple. The lactiferous ducts branch repeatedly, producing an extensive arborization within the mammary lobe. The resting gland consists mostly of an extensive duct system; however, a few end buds and alveoli may be proliferated from the ducts in the nonpregnant woman. Each lobe is subdivided by connective tissue into lobules of various sizes. At puberty a rapid and extensive deposition of fat occurs in the breasts.

The immature glands consist of a few short ducts radiating from the nipple. The glands of the male do not differentiate much beyond this infantile condition, but a conspicuous growth and branching of the duct system occurs in the prepuberal female under the influence of increasing titers of ovarian hormone. In the postpuberal virginal female slight fluctuations in the mammae may be correlated with ovarian changes during the reproductive cycle. An extensive and characteristic differentiation of the glands occurs during pregnancy. The duct system becomes extensively arborized, and the terminal twigs end in secretory alveoli. The alveolar lining constitutes a secretory surface from which milk arises. At parturition the secretion of milk is intensified, and the gland gradually involutes until lactation ceases.

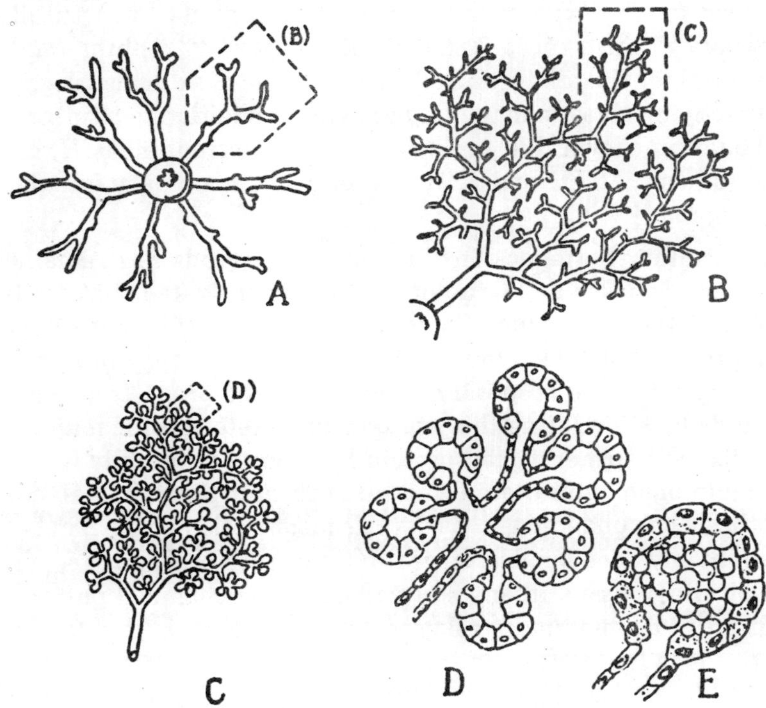

Fig. 6.6. Diagrams illustrating the differentiation of the mammary glands. *A*, The gland of prepuberal animals consists of relatively simple ducts radiating from the nipple. *B*, Small segment of A enlarged to show the condition of the adult, virginal gland. Under the influence of estrogens the duct system becomes extensively branched. *C*, Small area of *B* enlarged again to show the condition of the gland during pregnancy. Note the great development of terminal alveoli. *D*, Diagram illustrating the cell structure of a few terminal alveoli. *E*, The accumulation of milk globules within an alveolus.

Development of the Mammary Glands (Mammogenesis)

Complete functioning of the mammary glands, like pregnancy and parturition, is an exceedingly complex phenomenon involving the interplay of many hormones as well as nervous factors. It is another example of hormones working in concert rather than in isolation. Mammary gland physiology logically falls into two categories; development of the glands to a functional state, and the formation and evacuation of milk.

Early studies involving the administration of ovarian hormones to intact or ovariectomized animals pointed to the conclusion that estrogen is particularly concerned with growth of the duct system and that progesterone, acting in concert with estrogen, is required for full alveolar growth. In species such as the mouse, rat, and monkey, large doses of progesterone, without estrogen priming, have been shown to evoke the development of both the duct and alveolar systems. There are many species variations in the manner in which the mammary rudiment responds to the ovarian steroids. Without discussing the many species differences, it is sufficient to state that there is general agreement that estrogen and progesterone function synergistically and both are needed for the experimental production of glands structurally comparable to those of midpregnancy. Since the anterior pituitary gonadotrophins *(viz.,* FSH, LH, and prolactin) are essential for the production of ovarian hormones during pregnancy, it is obvious that they are indirectly involved in mammary growth. In addition to pituitary effects mediated by the ovary, it has been shown that prolactin and somatotrophin (STH) can act directly on mammary tissue and that ACTH and TSH can influence mammary functions through their respective target organs.

The action of pituitary hormones on the mammary glands has remained a live field of experimentation since Stricker and Grueter (1928) demonstrated that crude extracts of the anterior hypophysis have lactogenic effects. It was soon found that estrogen, alone or in combination with progesterone, fails to induce mammary development in hypophysectomized animals. This finding led to the proposal that estrogen and progesterone acted indirectly on the mammary glands by stimulating the pituitary to secrete specific mammogenic hormones in addition to its six well-authenticated protein hormones. Originally it was suggested that one pituitary "mammogen" stimulated duct growth and the other alveolar growth. The mammogen hypothesis has undergone considerable modification at the hands of its originators, but it must remain highly questionable, since it has not been possible to separate these special substances from pituitary tissue in any state approaching chemical purity. Most workers in the field feel that pituitary-mammary relationships can be accounted for on the basis of the established anterior lobe hormones and that it is not necessary to postulate the existence of other hypophysial factors. With the availability of highly purified anterior pituitary hormones, many aspects of the problem are beginning to be clarified.

Studies on the rat have been systematically pursued by a number of investigators, and, while the results may not be directly applicable to all species, the differences are likely to be in the relative importance of the various hormones rather than in the kinds of hormones required. It appears probable that at least five of the anterior lobe hormones, in addition to ovarian steroids, are involved in the pro-duction of a fully developed mammary gland such as is found during late pregnancy. Some of the most pertinent facts may be summarized as follows:

1. When estrogen is administered to castrated male or female rats, it produces mainly growth of the duct system. If larger doses are given for prolonged periods, some alveolar development may appear. Physiologic doses of estrogen plus progesterone produce full mammary growth (pregnancy type) when administered to castrates. Estrogen alone, given to intact females, produces both ductal and lobuloalveolar development. In the latter instance, it is probable that estrogen encourages the hypophysial release of prolactin and this hormone stimulates the secretion of progesterone by the corpora lutea.

2. In hypophysectomized rats estrogens and progestogens, alone or in combination, fail to induce mammary development. FSH and LH, administered to hypophysectomized subjects, do not stimulate the mammary glands. This indicates that mammary development requires hormones in addition to FSH, LH, estrogens, and progesterone.
3. In immature rats, lacking both pituitaries and gonads, a combination of estrogen and somatotrophin (STH) is necessary to produce growth of the arborescent system of milk ducts. If the adrenal glands, in addition to the pituitaries and ovaries, are removed, ductal growth requires estrogen plus STH plus adrenal corticoids.
4. In order to obtain full lobuloalveolar growth, comparable to that of late pregnancy, prolactin and progesterone are needed in addition to estrogen, STH, and adrenal corticoids. In the presence of the adrenal glands, ACTH may be used instead of the corticoids. Thus, a quintet of hormones is needed to build up the gland to the prolactational stage.
5. Milk secretion (lactogenesis) may be induced in the prolactational gland by withdrawing or diminishing the estrogen and progesterone and continuing the prolactin and adrenal corticoids (or ACTH). Although STH and TSH are not necessary for lactogenesis in the rat, both hormones probably contribute to the normalcy of the process in the intact animal.

Many types of experiments, particularly those involving hypophysectomized animals, suggest that the placenta performs an important role in mammary development during the second half of pregnancy. Ablation of the pituitary gland at midpregnancy does not prevent full mammary growth in the rat and certain other species. There is a tendency to think of milk secretion as beginning at or shortly after parturition but, on the contrary, the transformation from a prolactational to a lactational gland in the rat is very gradual, and there is considerable evidence of secretion during the second half of pregnancy. There is ample evidence that the placenta secretes estrogen, progesterone, and a potent prolactin-like hormone. The latter hormone, like its pituitary counterpart, affects two target organs during pregnancy, namely, the corpora lutea of the ovaries and the mammary glands. It may be supposed that these two targets compete for the prolactin present in the system; but, as the corpora lutea wane during the end of pregnancy, the mammary apparatus gains priority. This ascendancy is retained by the mammary glands after parturition, even though a new crop of corpora lutea is formed as a result of the postpartum estrus.

Milk Secretion (Lactogenesis)

The anterior hypophysis is necessary for the initiation and maintenance of milk secretion. The discovery that crude anterior pituitary extracts are lactogenic .was followed by the isolation and purification of prolactin. It was found that minute amounts of purified prolactin, injected into an appropriate teat canal of the rabbit, produced lactation in localized sectors of the mammary gland, neighboring untreated sectors remaining unaffected. Since the rabbits were not hypophysectomized, the possibility that other pituitary hormones participated in the response was not ruled out.

The importance of the adrenal cortex in the initiation of lactation in a variety of mammals has been clearly established. Although animals that are adrenalectomized during pregnancy may deliver normal litters, lactation is so meager that they cannot raise their young. Lactation may be induced in the pseudopregnant rat by prolactin plus adrenal cortical hormone, but purified prolactin alone is ineffective. That lactation can be induced in hypophysectomized rats and guinea pigs by giving either prolactin plus adrenal cortical hormone or prolactin plus ACTH has been shown.

After parturition there is a striking increase in milk secretion. The exact mechanisms that operate at this time have not been fully agreed upon. There is evidence that estrogen and progesterone act synergistically to inhibit lactation. At parturition there is a fall in the circulating titers of ovarian and placental steroids, and this operates in some manner to turn the full force of the pituitary and adrenal cortex upon the mammary apparatus. Low levels of estrogen in the blood stimulate the pituitary to increase its output of prolactin, whereas high levels of estrogen tend to inhibit lactation. It is not known whether this inhibitory action of estrogen is effected through the pituitary or at the level of the mammary gland, or both. It may be that at parturition the fall in the relative ratio of progesterone to estrogen allows the latter hormone to exert its positive effect in promoting the release of prolactin from the anterior hypophysis. Whatever the exact mechanism may be, it is reasonably certain that prolactin and adrenal corticoids are the most essential hormones for the initiation of lactation.

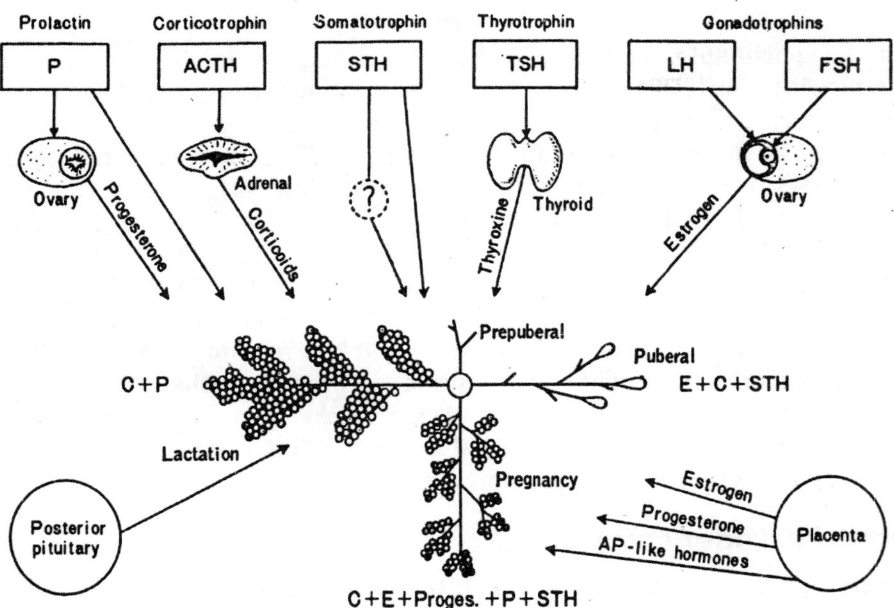

Fig. 6.7. A simplified diagram showing the action of hormones on mammary growth and lactation. In the diagram of the gland: upper—rudimentary gland; right—prepuberal to puberal gland; lower—prolactational gland of pregnancy; left —lactating gland.

The hormones that are involved in mammogenesis and lactogenesis in the rat may be summarized. FSH and LH function synergistically to promote the secretion of estrogen by the ovary. ACTH stimulates the adrenal cortex to secrete corticoids. Growth of the duct system is produced by the action of estrogen, STH, and adrenal corticoids. Prolactin stimulates the corpora lutea to secrete progesterone. Full lobuloalveolar (prolactational) development requires a combination of prolactin, STH, estrogen, progesterone, and corticoids. Milk secretion by the developed gland ensues when the influence of estrogen and progesterone is diminished and prolactin and adrenal corticoids attain supremacy. It is probable that lactogenesis is facilitated by STH and TSH, although neither is absolutely necessary in the rat.

Maintenance of Lactation (Galactopoiesis)

After delivery of the young, milk yield rises rapidly and then declines slowly until the young are weaned. The hormonal mechanisms involved in galactopoiesis are very similar to those that initiated milk production. Hypophysectomy at any period during lactation terminates the process. The continued production of prolactin is probably essential throughout the period of lactation, and there can be little doubt that ACTH, STH, and TSH are likewise of importance. When milk production begins, pressure develops within the glands and, if it is not relieved by suckling or milking, it rises to the point where milk secretion is retarded. The mammary glands of mice and rats involute quickly after removal of the litters, but the administration of prolactin to such animals tends to maintain the alveolar epithelium and retard involution. When the young are removed from rats on the fourth day of lactation, injections of oxytocin markedly delay mammary involution. Ablation of the neural lobe of the hypophysis of lactating rats abolishes the milk-ejection reflex, and the young die from starvation unless injections of oxytocin are given the mother. When such operated mothers become pregnant a second time, parturition and lactation are normal, indicating that regeneration of the neural lobe occurs in this species. After removing the litters from postpartum rats, milk production can be maintained for as long as 75 days by injecting a combination of prolactin, oxytocin, and cortisol.

Studies have shown that the stimulus of sucking, either with or without the removal of milk, temporarily lowers the content of prolactin in the anterior pituitaries of rats, guinea pigs, and rabbits. There are also indications that regular applications of the suckling stimulus maintain prolactin secretion at a high level. The rat normally lactates for about three weeks, since the young are weaned at this time, but experiments show that the mammary glands are capable of responding for a much longer time. The lactation period has been prolonged for 70 days by providing rats with fresh litters every 10 days. The lobuloalveolar system shows little or no evidence of involution during this extended period, although there is a decline in milk yield as judged by litter weight gains. It is generally agreed that stimulation of the nipple during the act of suckling reflexly prolongs the release of prolactin and other galactopoietic hormones from the animal's hypophysis.

Section of the spinal cord immediately craniad of the first lumbar vertebra in lactating rats completely paralyzes that portion of the body that bears the last three pairs of nipples. If the cranial nipples are covered so as to force the young to suck the denervated nipples, the young rapidly lose weight and succumb from inanition. When other lactating litters are introduced, they die in the same manner. If two of the innervated glands are exposed to suckling litters, lactation continues in all the glands, irrespective of the nerve supply. It may be inferred that this surgical procedure has destroyed nervous pathways that are essential for the stimulus of suckling to prolong the release of galactopoietic hormones from the pituitary gland.

Involution of the mammary glands of mice is delayed when the nipples are irritated by the local application of spirits of turpentine. If this substance is applied to selected nipples, some remaining untreated, lactation from all the glands is prolonged in the absence of nursing litters. Turpentine applied to the skin of the back produces no such effect; such nonirritating materials as water applied to the nipples do not delay mammary involution.

Milk Ejection

For clear thinking about lactation it is necessary to distinguish between milk secretion and milk removal, since the two processes appear to be regulated by different neuroendocrine mechanisms. The former, which we have already discussed, centers around the release of prolactin and other hormones from the anterior hypophysis. Milk removal, on the other hand, involves the reflexive release of oxytocin from the neurohypophysis. Oxytocin is conveyed by the circulation and causes milk evacuation by contracting the myoepithelial cells that surround the mammary alveoli. The act of suckling or milking reflexly stimulates both milk secretion and milk ejection. If a young animal is permitted to nurse an anesthetized mother, only the milk stored in the cistern and larger ducts can be withdrawn without any active participation of the mother. The greater part of the milk in the gland can be obtained only if there is activation of a neurohormonal reflex and a consequent contraction of the mammary alveoli. Within some 30 to 90 seconds after applica-tion of the suckling or milking stimulus, intramammary duct pressure suddenly rises and milk begins to flow freely. This results from the onset of the milk-ejection reflex and, in popular language, is called the "let-down" or the "draught."

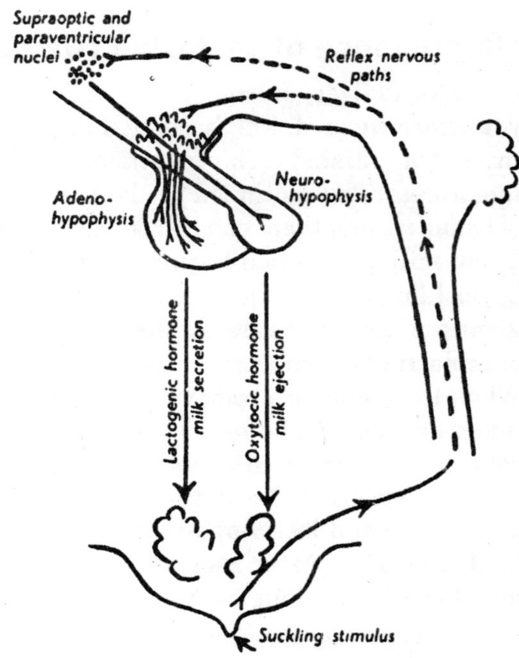

Fig. 6.8. Diagram illustrating probable neurohormonal reflexes involved in milk secretion and milk ejection.

The afferent component of the reflex arc is nervous, whereas the efferent limb is hormonal. The sensory stimuli associated with suckling excite receptors in the mammary glands and impulses are conveyed over spinal and brain stem tracts to the diencephalon and supraoptic nuclei of the hypothalamus. Through the mediation of the supraoptico-hypophysial tracts, the neural division of the hypophysis discharges oxytocin and probably some vasopressin. Oxytocin in the blood forms the efferent component of the arc, and the myoepithelial basket cells around the mammary alveoli constitute the effector tissue. This or a similar mechanism operates in a great variety of laboratory and domestic animals, and there is much evidence that it functions in lactating women as well. The milk-ejection reflex may be excited by various kinds of conditioned stimuli, and it is also subject to inhibition by emotionally stressful situations.

Many factors other than mechanical stimulation of the mammary glands can induce the milk-ejection reflex. The importance of milking cows at a set time every day is well recognized. The sound of buckets, the presence of calves, washing the udder, the sight of food, and so on may all constitute stimuli to which the animals become conditioned. Manipulation of the external genitalia and other events associated with mating, rather than with nursing or milking, may be sufficient to cause the discharge of milk, presumably through the reflexive release of oxytocin.

It had been known for many years that crude extracts of the posterior hypophysis facilitate milk evacuation, but the situation has been clarified since purified and synthetic preparations became available. Although oxytocin is five to six times more effective than vasopressin in eliciting milk ejection, the latter hormone possesses some intrinsic capacity to do so. Milk ejection occurs after the administration of oxytocin to lactating subjects even without mechanical stimulation of the mammary glands, as well as after the glands have been denervated. Suckling does not lead to milk ejection after severance of the pituitary stalk or after certain areas of the hypothalamus are destroyed by electrolytic lesions. Electrical stimulation of the supraoptico-hypophysial tract of goats and rabbits causes a copious discharge of milk. Jugular vein blood drawn from goats thus treated has the capacity of inducing milk ejection when administered to lactating test animals. These and many other types of experiments indicate that the reflexive release of milk is mediated through the hypothalamus and neurohypophysis.

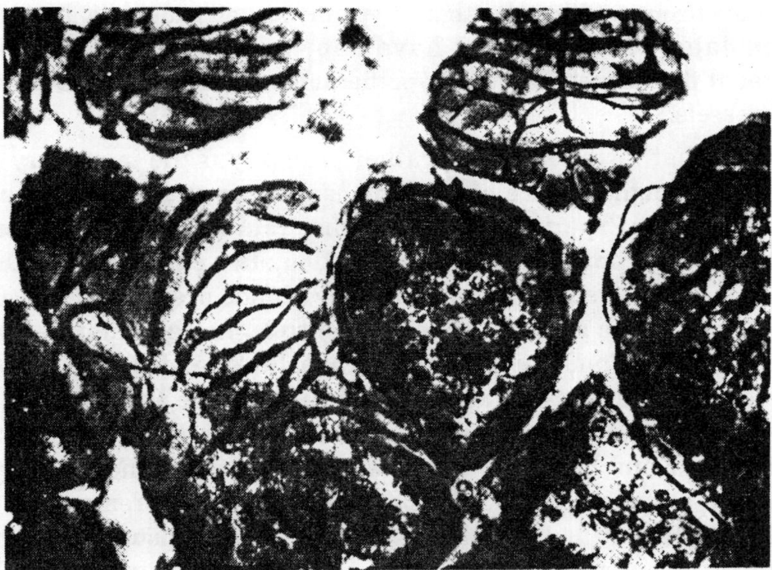

Fig. 6.9. Section of the mammary gland of the goat, showing the myoepithelial cells surrounding the alveoli.

The Myoepithelial Cells

For many years it had been assumed, on rather inadequate evidence, that smooth muscle tissue in connection with the alveoli served the function of squeezing milk from the gland. Histologic studies, however, failed to reveal the presence of this tissue in association with the alveoli, except in very meager amounts. Earlier histologists had identified myoepithelial cells or "basket cells" intimately surrounding the alveoli, but their distribution and structural features remained vague until they were studied after silver impregnation. There is now no reasonable doubt that these cells are the effector contractile tissue of the mammary gland.

Direct microscopic studies of the living gland have shown that the myoepithelial cells are capable of contracting. Direct mechanical or electrical stimulation causes them to contract and evacuate the alveolar contents. They may be caused to contract by the local application of oxytocin or vasopressin. Various drugs, such as histamine, acetylcholine, pilocarpine, and 5-hydroxytryptamine, also contract the myoepithelial cells after local application. Epinephrine and norepinephrine, on the other hand, have no effect on these cells, though they do not prevent oxytocin from exerting its usual effect.

Inhibition of Milk Ejection

Clinicians are aware that worry, fear, embarrassment, sadness, and other strong emotions at the time of nursing may inhibit the flow of milk in women. In human beings as well as in many other mammals, milk ejection may be blocked by the administration of epinephrine. It is probable that the epinephrine blockade is due to vasoconstriction within the mammary gland, since epinephrine does not inhibit the action of oxytocin on mammary alveoli *in vitro*. In the majority of cases, emotional stress seems to block the ejection of milk by preventing the nervous stimulus from reaching the neuro-hypophysis and causing a discharge of oxytocin. Experiments on laboratory animals have shown that it may also result from a sympatheticoadrenal discharge from the hypothalamus, which produces constriction of the mammary blood vessels.

It is well known that lactation ceases rather promptly if the mammary glands become engorged with milk through absence of suckling or milking or through disturbances in the milk ejection mechanism. Experiments on lactating rats have provided some interesting information on the mechanism of mammary involution after engorgement. On the day after parturition, the teat ducts of some of the mammary glands are ligated subcutaneously, the unoperated teats of the same animal serving as controls. As the animals continue nursing their litters, the operated glands became conspicuously distended by 8 hours and reach a maximum at 24 hours. If the ligatures are loosened at this stage, milk ejection can be elicited by the intravenous injection of oxytocin. If the ducts remain tied for 48 hours, the glands lose their pinkish colour and histologic examination reveals that the capillary meshwork of the gland is collapsed and devoid of blood cells. If the sutures are removed at this stage and oxytocin administered intravenously, milk ejection no longer ensues. However, contraction of the myoepithelial cells occurs when minute amounts of oxytocin are topically applied to the surface of the gland. These observations indicate that the failure to eject milk is due to capillary reduction rather than to myoepithelial incompetence. It is not known whether such constriction of the capillary bed occurs naturally in cases of engorgement, although there are suggestions that it does.

7 Biochemical Communication

The regulation of growth and metabolism of complex multicellular organisms depends heavily upon chemical messages sent between cells. This includes secretion of hormones into the circulatory system, chemical transfer of information through communicating cell junctions, and passage of signals between neurons in the brain. This chapter deals with these matters and also with communication between organisms, *i.e.,* with the biochemistry of ecological relationships. Embryonic development and differentiation of tissues also require communication between cells as does the functioning of the immune system.

A. THE HORMONES

The term hormone has traditionally been applied to substances synthesized in and secreted by one tissue and which act to influence distant target organs or tissues. However, many peptide hormones also act as neurotransmitters, passing across very short gaps between cells. In addition, many chemical messengers, including the peptide growth factors, act more locally. Looking at lower invertebrates as simple as *Hydra,* we find peptides resembling our own hormones and neurotransmitters. These are secreted by neuroendocrine cells of *Hydra* and diffuse throughout the body. In higher animals hormones regulate the concentrations of nutrients such as glucose and of ions such as Ca^{2+} and phosphate in the blood. They control the volume and osmotic pressure of body fluids, as well as digestion, growth, reproduction, and responses to stress.

1. Receptors, Feedback Loops, and Cascades

Every hormone must have one or more receptors, most of which are proteins. These may be found embedded in the outer surface of the plasma membrane, in the cytoplasm, or in the cell nucleus. Binding of a hormone to its receptor often elicits both a rapid response and a slower one. For example, we have seen that glucagon, adrenaline, and vasopressin bind to cell surface receptors and promote the synthesis of cyclic AMP. The cAMP induces rapid chemical modifications of many proteins. Some of these may diffuse into the nucleus and affect transcription of genes, a slower response. Insulin also exerts both rapid and slower responses.

Receptor Types

Many different kinds of protein can serve as hormone receptors. The most abundant are the G protein-coupled 7-helix receptors such as that of a β adrenergic receptor pictured Glucagon, adrenaline, ACTH, and gastrin are a few of the hormones that bind to receptors of this type. Similar receptors respond to light and over 1000 different 7-helix receptors respond to smell and taste. The reality of the dissociation and reassociation of the α and subunits in response to binding of a hormone has been demonstrated in living cells by the use of fluorescence resonance energy transfer (FRET). Not all receptors activate G proteins. One large group of membrane-associated receptors have single transmembrane helices but require dimerization to be effective.

The bacterial chemoreceptor has a very small ligand-binding domain and a larger internal domain that activates a histidine kinase. Many growth-factor receptors, including the insulin receptor, have internal domains with protein tyrosine kinase activity.

In contrast, steroid hormones, thyroxine, and retinoids bind to internal receptors. In 1968, Gorski *et. al.,* and Jensen *et. al.,* proposed independently that steroid hormone receptors in the cytoplasm bind incoming steroid molecules and after an "activation" step carry the hormone into the nucleus, where the hormone-receptor complex would bind at many sites in the chromatin inducing transcription of selected genes. Doubt has been cast on the assumption that the steroid hormone receptors must bind hormone initially in the cytoplasm. However, the role of steroid receptors in regulating transcription is well established.

Feedback Loops

Maintenance of a steady state within an organism depends upon numerous negative-feedback loops. Hormones assist in adjusting reaction rates to maintain a steady state when conditions are changed. For example, blood glucose rises after a meal. This increase is sensed in the pancreatic beta cells, which release insulin. The released insulin promotes uptake of glucose by cells and its conversion into glycogen and lipid stores. When the glucose level falls, inhibitory mechanisms that decrease insulin release are allowed to operate.

Similar regulatory loops can be traced for nearly all hormones. Sometimes they involve several stages and involve sensing devices in the central nervous system. In such cases neural impulses stimulate the *hypothalamus* of the brain to release *neurohormones,* which travel to the anterior lobe of the pituitary gland. The pituitary, in turn, releases hormones such as *corticotropin* (adrenocorticotropic hormone, *ACTH*), which stimulate the adrenal cortex to release its hormones. The latter exerts feedback inhibition upon the hypothalamus to decrease the secretion of ACTH by the pituitary. Steroids also participate in feedback loops to the hypothalamus. Using ^{3}H-labeled hormones or fluorescent analogs, it has been possible to locate specific brain cells sensitive to a given hormone by autoradiography.

A characteristic of hormonal effects is that they are seldom unique, and are often balanced by counter-acting effects of other hormones. For example, both glucagon and adrenaline promote the release of glucose from liver glycogen into the bloodstream. The glucocorticoids stimulate the rate of production of glucose from other body constituents. Growth hormone tends to increase glucose levels by inhibiting utilization of sugar by tissues. On the other hand, insulin acts to promote uptake of glucose by tissues and a more efficient utilization. The thyroid hormone increases the overall rate of metabolism of cells and also tends to promote a decrease in blood glucose.

Signaling Cascades

It is a well known fact that hormones frequently elicit the synthesis of second messengers such as cAMP, inositol phosphates, or diacyglycerol. This not only provides amplification of the initial hormonal signal but also allows a single hormone to control a "domain" of many metabolic

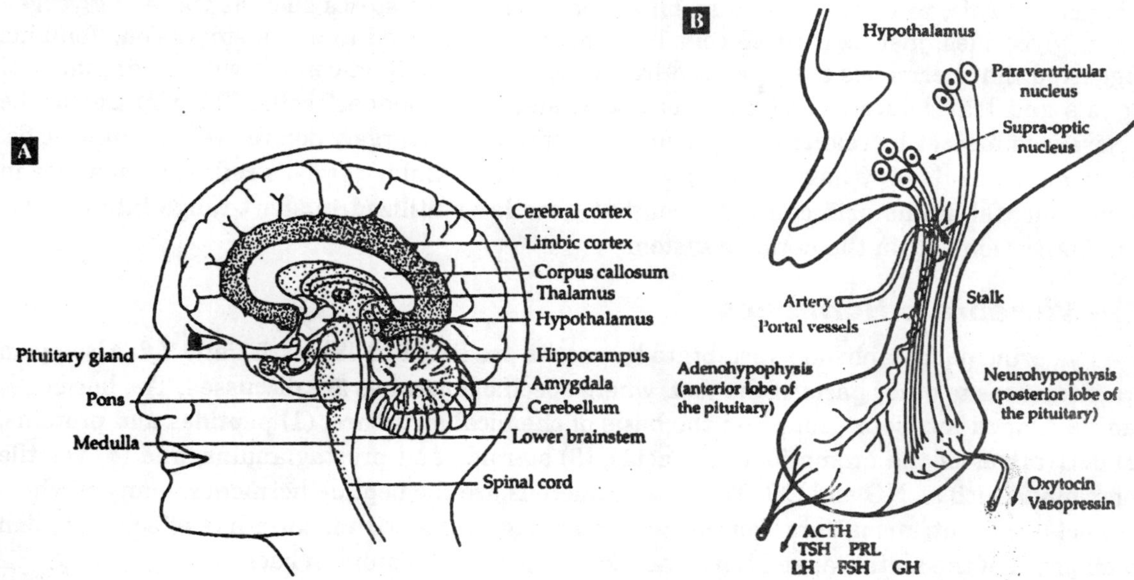

Fig. 7.1. (A) Median sagittal section of the human brain. From Maya Pines. (B) Drawing illustrating the synthesis of peptide hormones in the hypothalamus and transport via portal blood vessels into the anterior lobe of the pituitary gland or via nerve tracts into the posterior lobe.

processes. Each of these processes, in turn, can influence others. Many processes are affected by several different hormones and by more than one second messenger. Since we are far from knowing how many hormones exist and how many second messengers are released, the network of regulatory interactions within cells may be one of overwhelming complexity. An abbreviated version of a mitogen-activated kinase (MAP kinase) cascade. These cascades are not only initiated but also are propagated by a series of phosphorylation reactions catalyzed by more than 1000 protein kinases encoded by the human genome. Together with more than 500 protein phosphatases, which are often joined together as a bifunctional protein they form a complex branching network of interactions. These help to control responses not only to hormones but also to varying metabolite concentrations and physical stimuli.

Second Messengers as Hormones

The same compounds that serve as intracellular second messengers sometimes act as hormones. Tomkins suggested that cAMP and some other small molecules serve as "symbols" indicating a metabolic need. For example, in bacteria ppGpp serves as a symbol of nitrogen or amino acid deficiency. In cells ranging from those of bacteria to animals, cAMP is a symbol for carbon-source starvation. In E. *coli* cAMP levels increase during carbon-source starvation and stimulate the initiation of transcription of many bacterial operons. In *Dictyostelhim discoideum* cAMP is released by cells, when substrate depletion occurs. In this instance, the cyclic nucleotide acts as a hormone transmitting a signal to other cells.

Whereas cAMP is sometimes used by lower organisms as a hormone, its metabolic lability makes it unsuitable for higher animals. Thus, in our bodies the hormones glucagon and adrenaline carry a message to cell surfaces, where binding to receptors stimulates cAMP production. This, in turn, leads to mobilization of metabolic stores such as those of glycogen and triglycerides, just as if these cells had also been subjected to acute starvation. Tomkins suggested that hormones are produced by "sensor" cells in direct contact with environmental signals and travel to and activate more sequestered "responder" cells. The picture can be generalized further by realizing that neurotransmitters are largely derivatives of amino acids. These amino acids may have originally served as intracellular symbols reflecting changes in environmental amino acid concentration but were later utilized in short-range intercellular communication within the nervous system.

The Vertebrate Hormones

The principal established vertebrate hormones are listed in Tables 7.1 and 7.2. Also given are references to other parts of the text, where specific hormones are discussed. The hormones can be divided into four groups on the basis of chemical structure: (1) peptides and proteins, (2) derivatives of the aromatic amino acids, (3) steroids and prostaglandins, and (4) volatile compounds such as NO and CO. The most numerous are the peptide hormones, many of which also act as neurotransmitters. Peptide hormones, *e.g.,* those with insulin-like effects, function in all phyla of the metazoa, and hormone-like molecules are found in bacteria.

2. Hormones of the Pituitary Gland (Hypophysis) and Hypothalamus

Connected to the brain by a stalk (Fig. 7.1), the pituitary gland releases at least ten peptide or protein hormones that regulate the activity of other *endocrine* (hormone-producing) glands in distant parts of the body. The pituitary is composed of several distinct parts: the anterior lobe (*adenohypophysis*), a thin intermediate portion (*pars intermedia*), and a posterior lobe (*neurohypophysis*). Each has its own characteristic endocrine functions.

The anterior lobe of the pituitary secretes a series of ten or more peptide hormones ranging in size from the ~20-residue β-melanotropin to the ~200-residue growth hormone (somatotropin). Several of these contain a common heptapeptide unit, which is marked in green in the following structure:

Ac • Ser • Tyr • Ser • Met • Glu • His • Phe • Arg • Trp • Gly • Lys • Pro • $ValNH_2$

Structure of a-melanotropin from pig, beef, and horse

Not only this heptapeptide but also the entire amino acid sequence of *α-melanotropin* is found within the sequence of *corticotropin* (Fig. 7.2), which has an additional 29 amino acids at the C-terminal end. The same heptapeptide was also found in the *lipotropins.* The explanation is, in part, that several of these hormones arise from a single 31-kDa precursor protein called *prepro-opiomelanocortin.* It contains an N-terminal signal sequence that is removed shortly after synthesis, as well as pairs of adjacent basic residues (Arg-Arg, Arg-Lys, Lys-Arg and Lys-Lys) at a number of places (Fig. 7.2). After removal of the signal sequence, further cleavage is thought to occur within the secretory vesicles by proteases, which cut either on the carboxyl side of these basic pairs or between them. The same precursor is made in both anterior and intermediate lobes and is rapidly cut to ACTH, β-lipotropin, and an N-terminal part. In the intermediate lobe the ACTH is then cleaved at the Lys-Lys and Arg-Arg pairs to form αMSH and another peptide called corticotropinlike intermediate lobe peptide (CLIP). Beta lipotropin is degraded rapidly in the intermediate lobe and more slowly in the anterior lobe to γ-lipotropin

Table 7.1. Peptide and Protein Hormones of Vertebrates

Source and name of hormone	*No. residues*	*Principal site action*
A. Pituitary gland (hypophysis)		
1. Adenohypophysis (anterior portion)		
Corticotropin (ACTH)[a]	39	Adrenal cortex, adipose tissue
β-Melanotropin (β melanocyte-stimulating hormone, β-MSH)[a]	18-22	Skin
β-Lipotropin (β-LPH)[a]	91	Precursor of β-MSH
γ-Lipotropin (γ-LPH)[a]	58	Precursor of β-MSH
β Endorphin[a]	31	Brain
Somatotropin (growth hormone, GH)	~200	All tissues
Prolactin (mammotropin)	~200	
Thyrotropin (thyroid-stimulating hormone, TSH)[b]		Thyroid Ch 25, B2
Follitropin (follicle-stimulating hormone, FH)[b]		Ovaries, testes
Lutropin (luteinizing hormone, ICSH or LH)[b]		Ovaries, testes
2. Pars intermedia (intermediate portion)		
α-Melanotropin (α-melanocyte-stimulating hormone, α-MSH)[a]	13	Skin
3. Neurohypophysis (posterior portion)		
Oxytocin (ocytocin)	9	Uterus, mammary glands
Vasopressin (antidiuretic hormone)	9	
B. Pancreas		
Insulin	51	All cells
Glucagon	29	Liver, adipose tissue
C. Ovary (corpus luteum)		
Relaxin	–	Pelvic ligaments
D. Thyroid		
Calcitonin (thyroclacitonin)	32	Bones, kidney
E. Parathyroid		
Parathyrin (parathyroid hormone)	84	Bones, kidney
F. Kidney		
Erythropoietin		Bone marrow
Renin (an enzyme)		Adrenal cortex

[a] Arise by cleavage of pro-opiomelanocortin.
[b] Related two-subunit (αβ) proteins with a common β subunit for these three hormones, for human chorionic gonadotropin (hCGH), and for mitogen-regulated protein (proliferin).

and the opioid peptide β-endorphin. Precursor proteins have been identified for many other peptide hormones, even those with very short chains. Proteolytic cleavages and other processing reactions occur within the secretory pathways of organisms from yeast to humans.

Many pituitary hormones have a pyroglutamate (5-oxoproline) residue at the N terminus. This presumably arises by attack of the terminal $-NH_2$ group on the amide carbon of an N-terminal glutamine side chain with displacement of NH_3. The C terminus is often an amide of the carboxyl group with ammonia, which usually arises from a peptide chain containing one additional glycine residue at the C terminus. The processing of peptide hormones doesn't end with their synthesis. They are usually degraded quickly or are converted into derivatives with weaker hormonal activity.

Pituitary Growth Hormone and Related Hormones

The pituitary growth hormone (*somatotropin*) and *prolactin* are 22– to 23-kDa proteins, which share homology also with human *placental lactogen,* a lactogenic hormone secreted by the placenta, and with a growth factor called *mitogen-regulated protein* (or proliferin). The polypeptide chain of the 191-residue porcine and human growth hormone folds into an antiparallel four helix bundle but with two long irregular connecting strands. The high degree of homology among somatotropins of many other species indicates a near identity of three-dimensional structures. However, biological function is species-specific. Humans and monkeys respond only to growth hormone from primates. The interaction of an aspartate side chain at position 171 with arginine 43 of the receptor protein may account for some of this specificity. The receptor is a member of a large superfamily of receptor proteins with single transmembrane helices and extracellular domains similar to that of tissue factor and also, in some respects, to immunoglobulin domains. Receptors for growth hormones bind in specific ways to two molecules of receptor protein.

Human growth hormone produced in bacteria is used to help very short children to grow. Bovine growth hormone produced in bacteria is used to increase milk production from cows. However, this use may be damaging to the cows. Some humans produce too much growth hormone, often as a result of tumors in the pituitary. The resulting condition of *acromegaly* causes excessive bone growth and many other problems. Growth hormone has a broad range of other effects, *e.g.*, mimicking the action of insulin.

Lactogenic hormones also have 4-helix bundle structures, and the prolactin receptor structure resembles that of growth hormone as well as those of a large cytokine family. Prolactin affects placental development during pregnancy. However, during the latter half of pregnancy the placenta has a dominant endocrine effect, synthesizing both progesterone and the placental lactogen.

The Pituitary Glycoprotein Hormones

The thyroid-stimulating hormone *thyrotropin* (TSH), together with *folitropin* (FH) and *lutropin* (LH; Table 7.1), form a family of related ~28-kDa dimeric glyco-proteins in which each subunit has a three-loop structure stabilized by a characteristic "cystine knot." Also included in the family is the placental *chorionic gonadotropin,* which is found only in human beings and a few other species. LH has a central role in promoting both spermatogenesis and ovulation by stimulating synthesis of steroid hormones in the testes and ovary, respectively. Human chorionic gonadot-ropin (hCG) is also essential for maintenance of pregnancy and acts by stimulating the ovaries to secrete required steroid hormones.

All of these glycoprotein hormones are αβ dimers, and within a single species the subunits of TSH, FH, LH, and CG are identical. However, the β subunits are all different. In the human there are at least six genes or pseudogenes for the hCG β chain in a cluster that also contains a single LH β chain gene. The hormones undergo glycosylation and sulfation in the Golgi before secretion. The hormones bind to 7-helix receptors, which are coupled to formation of cAMP or inositol trisphosphate. Mutations in LH may cause male infertility, while mutations in the corresponding receptor may cause male precocious puberty.

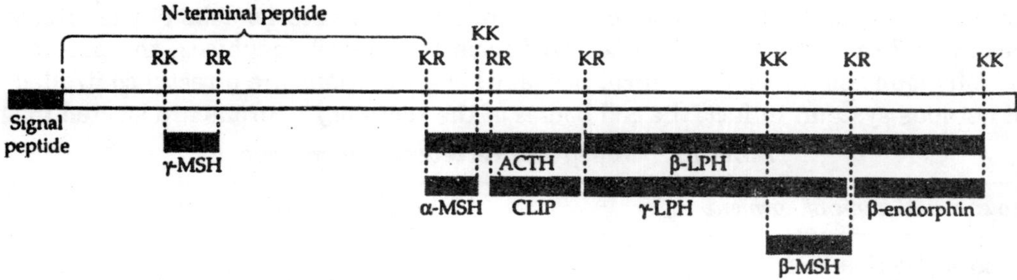

Fig. 7.2. Schematic structure of preproopiomelanocortin and of some of the products of its processing by protolytic enzymes. The abbreviations K and R are for lysine and arginine.

Hypothalamic Releasing Hormones

As was mentioned previously, the anterior lobe of the pituitary releases its hormones in response to at least nine neurohormones known as *releasing hormones* or releasing factors. They are secreted in minute quantities by the hypothalamus into a special portal vein that carries them directly to the pituitary where they exert their effects (Fig. 7.1B). As is indicated in Fig. 7.4 and in Table 7.3, several releasing factors are small peptides, but others are quite large. *Thyrotropin-releasing hormone* (THR, thyroliberin) is a tripeptide; but human *growth hormone-releasing hormone* (somatoliberin) is a 44-residue peptide. Both are synthesized as larger proteins, which are cleaved and processed to form the mature C-amidated hormones. *Corticotropin-releasing hormone* (CRH; CRF; corticoliberin), the 41-residue ACTH-releasing factor, is also cut from the much larger prepro-CRF. Release of both LH and FSH is stimulated by a single *gonadotropin-releasing hormone* (GnRH). The releasing factors bind to 7-helix G-protein coupled receptor. Both the releasing factor and gonadotropin are released into the appropriate parts of the bloodstream in a pulsatile fashion emphasizing the neural origin of their release.

The hypothalamus also synthesizes *release-inhibiting factors*. One of these, *somatostatin* (Table 7.3), inhibits release of somatotropin, thus counteracting the effect of the growth hormone releasing hormone. Somatostatin acts both in the pituitary and also in the pancreas, where it inhibits the release of both insulin and glucagon. The result is a lowering of blood glucose. This suggested a new approach to the treatment of diabetes. However, because of the many other effects of somatostatin and its rapid degradation it has not been useful clinically. Nevertheless, hundreds of analogs of somatostatin have been synthesized, some of which may be of practical value. The biological activity of somatostatin resides largely in the sequence FYKT at positions 6 – 10, a sequence that is thought to form a beta turn (Fig. 7.3). Much of the rest of the molecule can be left off and the disulfide bridge moved up as far as positions

6 – 11 with retention of high potency. Human somatostatin is synthesized initially as a 116-residue precursor.

A 56-residue peptide, which is formed from the 10-kDa precursor to GnRH, inhibits secretion of prolactin. Inhibition of FSH release is accomplished by feedback inhibition. Hormones known as *inhibins* are produced in the gonads and act to inhibit release of FSH from the pituitary.

Vasopressin and Oxytocin

In contrast to the large peptide hormones made in the anterior lobe of the pituitary are *vasopressin* and *oxytocin,* which are secreted from the neurohypophysis, the posterior lobe. The neurohypophysis consists of neural tissue, whose secretions are directly controlled by the central nervous system. In fact, the cell bodies of the secretory neurons are located in specific nuclei of the hypothalamus (Fig. 7.1B). About 4000 vasopressin-secretory neurons and a similar number of oxytocin neurons are present in the neurohypophysis of the rat. Vasopressin is a

Table 7.2. Nonpeptide Vertebrate Hormones

Type, source, and name of hormone	*Principal site of action*
A. Amino acid derivatives	
1. Thyroid	
Thyroxine and triiodothyronine	Most cells
2. Adrenal medulla	
Adrenaline, noradrenaline (epinephrine, norepinephrine)	Most cells
3. Pineal gland	
Melatonin	Melanophores
4. Nerves and other cells	
Serotonin (5-hydroxytryptamine)	Arterioles, central nervous system
B. Steroids and prostaglandins	
1. Testes	
Testosterone	Most cells
2. Ovaries	
Estrogen (estradiol-17β)	Most cells
3. Corpus luteum	
Progesterone	Uterus, mammary glands
4. Adrenal cortex	
Corticosterone, cortisol	Most cells
Aldosterone	Kidney
5. Various tissues	
Prostaglandins	Smooth muscle
C. Volatile hormones	
1. Nitric oxide, NO	Endothelium, brain
2. Carbon monoxide, CO	Brain

major regulator of blood volume and pressure, and its secretion is influenced by stress. It increases the water permeability of the kidney collecting duct cells by inducing translocation of aquaporin proteins from intracellular storage vesicles into the apical plasma membrane. Vasopressin binds to G-protein-coupled receptors. A defect in the type 2 vasopressin receptor leads to the condition of nephrogenic *diabetes insipidus* in which the body fails to concentrate the urine. Oxytocin acts on smooth muscles of the uterus during childbirth and triggers the release of milk from the mammary glands. The latter response is partially controlled by the suckling of the infant, which induces the nervous system to release oxytocin into the bloodstream.

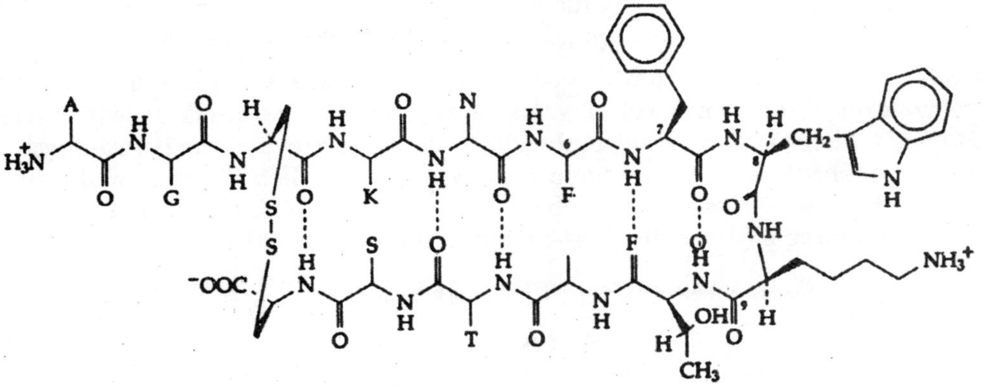

Fig. 7.3. Possible secondary structure of somatostatin with a beta turn at residues 7-10 and a disulfide bond between positions 3 and 14. The true conformation would have the plane of the beta sheet puckered and twisted.

Table 7.3. Releasing and Inhibiting Hormones from the Hypothalamus

Name	*Number of amino acid residues*	*Sequence*[a]
Thyrotropin-releasing hormone (thyroliberin, TRH)	3	*p*EHP–NH_2
Gonadotropin-releasing hormone (GnRH, LH- and FSH- releasing hormone)	10	*p*EHWSYGLRPG–NH_2
GH-releasing hormone (somatoliberin)	44	
Corticotropin-releasing hormone (corticoliberin, CRH)	41	
MSH-releasing factor (melanoliberin)[b]	5	CYIQNC (S–S between C1 and C6)
Somatostatin (GH-release-inhibiting hormone)	14	AGCKNFFWKTFTSC (S–S)
MSH-release-inhibiting factor[b]	3	PLG
Prolactin-releasing factor		
Dopamine (prolactin-release-inhibiting factor)		

[a] Standard one-letter abbreviations are used. pE is pyroglutamyl (5-oxoprolyl) and –NH_2 at the right indicates a C-terminal *carboxamide*.

[b] Ring and tail fragments of oxytocin.

Hormones related to oxytocin and vasopressin occur in most vertebrates, the compound *vasotocin* shown in Fig. 7.4 being the most common. Substitution of phenylalanine for isoleucine at position 3 gives arginine vasopressin, the vasopressin found in our bodies. Structure of oxytocin and related hormones are also shown in Fig. 7.4. Like somatostatin, vaso-pressin and oxytocin may also form antiparallel pleated sheet structures with p turns. The structural requirements for hormone activity have been studied intensively. Both the macrocyclic hexapeptide ring and the tripeptide side chains are necessary for maximal activity.

The gene for arginine vasopressin is that of a 166-residue precursor protein carrying a 19-residue signal sequence at the N terminus. This sequence is followed by that of vasopressin, then after a GKR linker by the 95-residue *neurophysin II*. Finally, after one additional arginine there is a 39-residue glycopeptide. Oxytocin originates in a parallel way from its own precursor.

The 93- to 95-residue neurophysins act as carriers for vasopressin and ocytocin, forming specific complexes with them. Neurophysins contain 14 cysteine residues, which form seven disulfide bonds. There is a striking similarity in sequence between the neuro-physins, snake venom toxins, a wheat germ lectin (agglutinin), a ragweed pollen allergen, and a small plant protein called hevein. On the basis of the alignment of cysteine residues, Drenth proposed that all of these proteins have a disulfide-linked core whose structure.

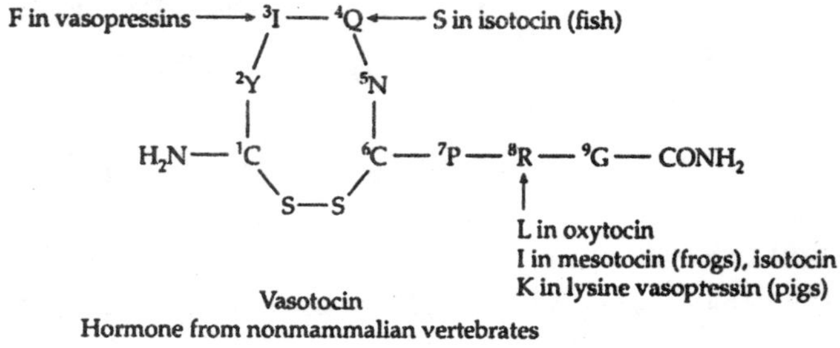

F in vasopressins → ^{3}I — ^{4}Q ← S in isotocin (fish)

^{2}Y ^{5}N

H_2N — ^{1}C ^{6}C — ^{7}P — ^{8}R — ^{9}G — $CONH_2$

S—S

L in oxytocin
I in mesotocin (frogs), isotocin
K in lysine vasopressin (pigs)

Vasotocin
Hormone from nonmammalian vertebrates

Fig. 7.4. Structure of the nonmammalian hormone vasotocin and of related hormones including ocytocin and vasopressin.

Melanocortins

The melanocortin peptides, which are derived from pro-opiomelanocortin as indicated in Fig. 7.2, are formed in varying amounts in the pituitary, in two brain nuclei, and in some peripheral tissues. In some animals α-MSH arises primarily in an intermediate part of the pituitary. The hormone has a direct effect on the melanocytes (Box 8-F) causing darkening of the skin. In addition, the various melanocortin peptides (ACTH, α-, αβ-, and γ-MSH) bind to five different types of receptors. These have been linked to the control of energy homeostasis, appetite, and obesity in both mice and humans. The sequence His-Phe-Arg-Trp (of the green-shaded sequence in Fig. 7.2) is essential for binding. In keratinocytes α-MSH may form a 1:1 complex with tetrahydrobiopterin, the coenzyme for tyrosine hydroxylase, and a regulator of tyrosinase, an essential enzyme for melanin formation.

3. Pancreatic and Related Hormones

The functions of the 51-residue insulin begin early in life. Mammalian preimplantation blastocysts already show a response to insulin. The glucose transporter GLUT1 is present at the earliest stages; synthesis of GLUT2 and GLUT3 begins at the eight-cell stage. However, the insulin-regulated transporter GLUT4 is not present in the blastocyst. A newly discovered insulin-regulated GLUT8 may function during preimplantation development. Although the secretion of insulin is of primary importance to the regulation of the glucose concentration in mammals, it is still not clearly understood. The beta cells have insulin receptors and other components of the insulin signaling system such as the insulin receptor substrates IRS-1 and IRS-2 and phosphatidylinositol 3-kinase (PI3-K). The sensing of glucose by the beta cells is also not yet well understood. This lack of knowledge has made it difficult to improve the treatment of diabetes. A new approach is to engineer non-beta cells to secrete a steady supply of insulin. Such a possibility has been demonstrated in mice using gut K-cells.

The *insulinlike growth factors* (IGF-I and IGF-II) are produced in many different tissues and promote growth of other cells. *Relaxin,* which is produced in the corpus luteum of ovaries during pregnancy, is responsible for inducing widening of the birth canal during the late stages of pregnancy and inhibits contraction of uterine muscle, perhaps by decreasing the activity of the kinase that phosphorylates the 20-kDa light chains of myosin. Relaxin is found throughout the animal kingdom, even in the protozoan *Tetrahymena.* Its structure is apparently identical in pigs, whales, and in a primitive tunicate. In human males relaxin is apparently produced in the prostate, where it may function as a sperm motility factor. Relaxin, IGF-I, and IGF-II are all structurally homologous to proinsulin and contain the characteristic 3-disulfide structure of insulin. The IGF-I receptor structure resembles that of the insulin receptor (Fig. 7.11) and also that of the epidermal growth factor (EGF) receptor.

Glucagon belongs to a family that also includes the gastrointestinal hormones *secretin, gastrointestinal inhibitory peptide (GIP), vasoactive intestinal peptide* (VIP), and *glicentin* (Table 7.4). A complex processing pathway converts 14- to 16-kDa preproglucagons into the active hormone. Proglucagon is processed to glucagon in the pancreas, but in the endocrine L cells of intestinal mucosa it yields glicentin, a polypeptide containing the entire glucagon sequence, and other products. Glucagon receptors generate both cAMP and Ca^{2+} as second messengers.

The 27-residue secretin stimulates secretion of bicarbonate into the pancreatic juice and inhibits gastric secretion of acid. The 28-residue VIP is found throughout the gastrointestinal tract of mammals and birds as well as in the brain and the lungs. It is a potent vasodilator and may be the major relaxant of pulmonary smooth muscle. It has been reported totally absent from lungs of asthma patients. The gastrointestinal inhibitory peptide (GIP) is larger than VIP but also has a close homology with glucagon.

The 36-residue *pancreatic polypeptide* is a hormone of uncertain functions. The crystalline polypeptide has at the N terminus an 8-residue collagenlike helix that lies parallel to a C-terminal α helix. The overall shape resembles that of both insulin and glucagon. This PP-fold includes also neuropeptide Y, which is considered in the next section, and neuropeptide YY.

4. Gastrointestinal and Brain Peptides

The largest endocrine gland in the body is the gastointestinal tract, which produces a profusion of peptide hormones, many of which are also found in the brain. Indeed, a majority of the known vertebrate peptide hormones occur in the brain. For example, glucagon has been found in the brainstem and hypothalamus. Many of these peptides, or closely related ones, are also found in lower invertebrates. For example, the 10-residue *Hydra head activator* is also present in mammalian brain. The concentrations of these peptides in the brain is very low (10^{-12} 10^{-15}M).

Gastrin is produced in the lower portion of the stomach and regulates the secretion of acid as well as growth of the gastrointestinal mucosa. It may also function as an Fe^{3+} carrier. The shorter gastrin 17 as. well as the longer 34-residue gastrin 34 are both active as is a synthetic pentapeptide with the hormone's C-terminal sequence. The family of *pancreozymin-cholecystokinins* (CCK) are 8- to 58-residue peptides produced in the upper intestinal tract. They have a 4-residue amidated C-terminal sequence in common with gastrin. This tetrapeptide has some biological activity, but eight residues are required for full activity as is conversion of the tyrosine at position seven from the C terminus to an O-sulfate ester. Both gastrin and CCK molecules are partially converted to sulfate esters. Most regions of the brain contain CCK peptides in amounts exceeding those of other neurotransmitters. These arise (in pigs) from a 114-residue preproCCK. The sulfated insect neuropeptide, *leukosulfakinin* (Table 7.4), is homologous to gastrin and CCK. Another hormone *gastrotropin* is produced by cells of the intestinal mucosa in the distal ileum and stimulates gastric secretion.

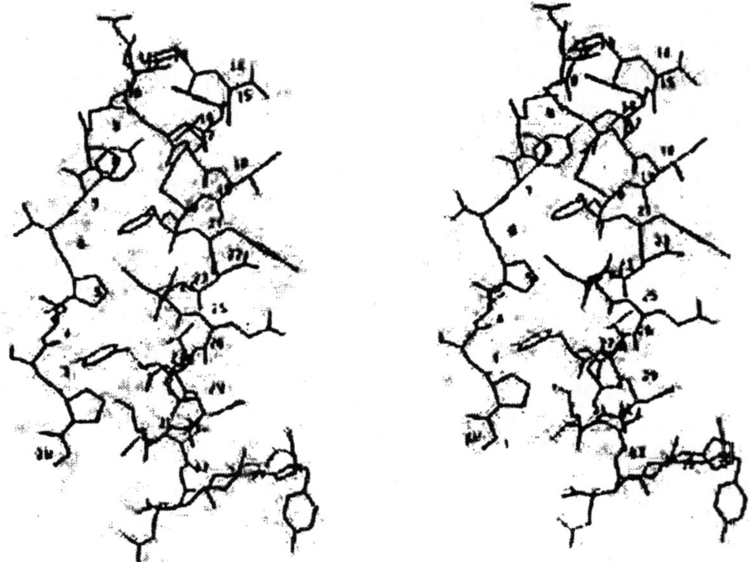

Fig. 7.5. Structure of the avian pancreatic polypeptide, a small globular protein. From Blundell *et. al.*

Motilin, a 22-residue intestinal neuropeptide, stimulates motor activity of the gastrointestinal tract. *Bombesin* was first isolated from frog skin but probably also functions in both the intestinal tract and the brain. It has a powerful hypothermic effect. A mammalian homolog of bombesin, the 27-residue *gastrin-releasing peptide* (GRP), is found throughout the gastrointestinal and pulmonary tracts as well as the central nervous system. Bombesin-like material, possibly GRP, is produced by some cancers and may serve as an autocrine growth factor. The 29-residue *galanin* was originally isolated from procine intestine but is found throughout the central nervous system. It may function as a neurotransmitter or modulator. The 15-residue *guanylin* is an important regulator of epithelial transport in the intestine and probably in other tissues. The active hormone, which is cut from the 99-residue proguanylin, contains two disulfide bonds.

Guanylin receptors activate guanylate cyclase with production of cyclic GMP, which functions in the regulation of intestinal fluid and electrolyte absorption. The 18- or 19-residue heat stable enterotoxins of some strains of *E. coli* bind to and activate the guanylin receptors. The resulting overproduction of cGMP causes severe diarrhea.

Neuropeptides Y (NPY) and YY are 36-residue amidated peptides that are members of the pancreatic polypeptide (PP) family. NPY is produced both in the peripheral nervous system and in the brain, where it is one of the most abundant neuropeptides. Another member of the PP family is *semi-nalplasmin,* a regulator of calcium ion transport in bovine sperm. NPY is best known for its stimulation of appetite. It also inhibits anxiety and increases memory retention. It has a vasoconstrictive effect on blood vessels, participating in cardiovascular regulation. Peptide YY is formed in endocrine cells of the intestine, while NPY is formed in neurons of the parasympathetic system. Both participate in regulation of fluid and electrolyte secretion. Both are found in other vertebrate species.

NPY is one of the most important of several regulators of feeding behavior of animals. PYY_{3-6}, another member of the neuropeptide Y family, suppresses appetite by antagonizing the action of NPY. A large variety of hormonal effects seem to be involved in control of appetite. There are both short-term and long-term mechanisms. For example, when introduced into the gut of rats prior to feeding, CCK and various other gastrointestinal peptides decrease the amount of food eaten.

Much attention has been focused on the 146-residue cytokine *leptin,* a hormone produced by adipose tissue. Leptin, which is sometimes described as the antiobesity hormone, was recognized by mutations of the *obese* gene (OB) or of the OB receptor in genetically obese mice. When food is scarce, the fat cells shrink and decrease their secretion of leptin. The decrease is sensed by receptors in the hypothalamus, which signal for increased NPY secretion and decreased secretion of αMSH. NPY increases appetite, while αMSH has an opposing role of blocking feelings of hunger. Nevertheless, there are doubts that leptin's primary role is control of obesity.

The 13-residue *neurotensin* was first isolated from the hypothalamus but is more abundant in cells of the ileum. It induces gut contraction, lowers blood pressure, and has a variety of other effects. *Substance P* (SP; Table 7.4) has been regarded as a possible neurotransmitter for some time but is also found in the digestive tract. It is the most abundant of a family of five neurokinins (or tachykinins). Others include neurokinin A (substance K), neurokinin B, neuropeptide K, and neuropeptide γ. They have a common C-terminal sequence $FXGLM\text{-}NH_2$. Substance P is thought to be involved in the perception of pain, and mice lacking a substance P receptor appear to have reduced sensitivity to pain. Substance P as well as the related substance K are derived from two large precursor proteins, which appear to arise as a result of alternative modes of splicing of mRNA.

5. Other Mammalian Peptide Hormones

The action of the 32-residue thyroid hormone *calcitonin* has been described in Box 22-C. This calcium-regulatory hormone is produced in the thyroid C cells from a precursor having an extra 82 residues at the N terminus and 16 residues at the C terminus. The same gene gives rise in neural tissues to a neuropeptide, possibly a neurotransmitter, called *calcitonin gene-related polypeptide* (CGRPP). The 84-residue *parathyrin* (parathyroid hormone) is present in secretion granules as a 90-residue prohormone containing six extra residues at the N terminus. The primary biosynthetic product *preproparathyrin* contains an additional 25 residues. An N-terminal 34-residue fragment of the hormone, when injected subcutaneously daily, causes an increase in bone density in persons with osteoporosis. The hormone acts via a G-protein-coupled receptor in bone and kidney. A calcium ion receptor, which binds Ca^{2+} cooperatively, acts as a sensor that regulates release of the parathyroid hormone to regulate the serum Ca^{2+} concentration. A 141-residue *parathyroid hormone-related protein* has an N-terminal sequence homologous with that of parathyroid hormone, eight of the first 13 residues being identical. It is secreted by a variety of cells and serves as a growth factor.

Endogenous Opioid Peptides

Extensive processing is also involved in formation of analgesic opioid peptides, which are present naturally in the brain. The formation of β-endorphin in the hypothalamus from prepro-opiomelanocortin has already been mentioned. Prior to the discovery of P-endorphin, the pentapeptides *Met-enkephalin* and *Leu-enkephalin* were discovered and were found to compete with opiate drugs for receptors in the brain. The larger β-endorphin, which contains the Met-enkephalin sequence at its N terminus, is a far more potent opiate antagonist than are the enkephalins. Since the Met-enkephalin sequence within P-endorphin is not flanked by basic residues, it apparently is normally not released. Two other recently discovered brain peptides are *endomorphin*-1 (YPWF-NH_2) and *endomorphin*-2 (YPFF-NH_2). They are also potent agonists for the opioid receptors, especially the μ receptor.

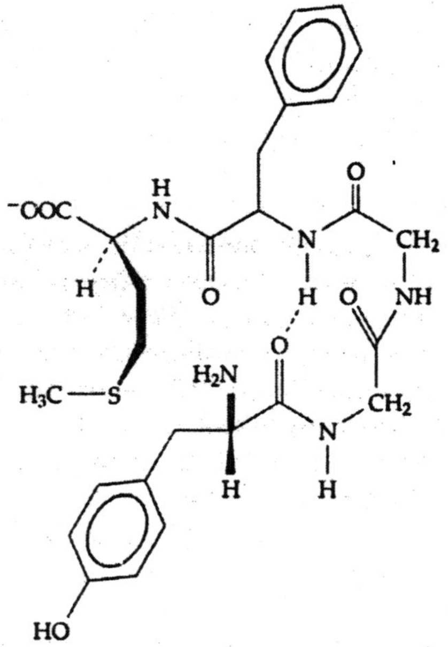

Tyr-Gly-Gly-Phe-Met
(Met-enkephalin)

Table 7.4. Some Pancreatic and Gastrointestinal Hormones and Neurohormones and their Sequences

Name and source	*No. of residues*	*Sequence*[a]
Glucagon	29	HSQGTFTSDYSKYLDSRRAQDFVQWLMNT
Secretin (pancreas)	27	HSDTFTSELSRLRDSARLQRLLQGLV-NH_2
Vasoactive intestinal peptide	28	HSDAVFTDNYTRLRKQMARKKYLNSILN-NH_2
Gastrointestinal inhibitory peptide (GIP)	43	YAEGTFISDYSIAMDKIRQQDFVNWLLAQ-Q
Glicentin	100	A proglucagon containing the entire glucagons Sequence in residues 64-92
Pancreatic polypeptide	36	
Neuropeptides Y (NPY) and YY	36	
Gastrin (stomach)		
Gastrin-17	17	pEGPWLEEEEEAYGWMDF–NH_2[b]
Cholecystokinin, CCK or pancreozymin (gallbladder, pancreas), many forms exist		
CCK 58	58	
CCK8	8	DYMGWMDF-NH_2
Motilin (porcine)	23	PVPIFTYGELQRMOEKERNKGQ
Bombesin	14	*p*EQRLGNOWAVGHLM–NH_2
Gastrin-releasing peptide	27	
Galanin	29	
Guanylin	15	PNTCEICAYAACTGC (S–S bridges)
Neurotensin	13	pELYENKPRRPYIL
Substance P	11	RPKPQQFFGLM-NH_2
Physaelemin (frog skin)	11	pEADPNKFYGLM
Neurophysins	93-95	
β-Endorphin	31	YGGFMTSEKSQTPLVTLFKNAIIKNAHKKGQ
Dynorphin	17	YGGFLRRIRPKLKWDNQ
Met-enkephalin	5	YGGFM
Leu-eukephalin	5	YGGFL
Angiotensin II	8	DRVYIHPF
Bradykinin (BK)	9	RPPGFSPFR
Lys-bradykinin (kallidin)	10	KRPPGFSPFR
Sleep peptide	9	WAGGDASGE
Atrial natriuretic hormone	28	
Chemotactic factors		
for neutrophils	3	*f*-MLF
for phagocytes	4	TKPR
Speract	10	GFDLNGGGVG

[a]Standard one-letter abbreviations; pE, 5-oxoprolyl; *f*, formyl;-NH_2, C-terminal carboxamide.
[b]Y-12 may be sulfated.

Both Met-enkephalin and Leu-enkephalin have their own pro- and prepro forms. Bovine *prepro-enkephalin A* is a 268-residue protein containing a 20-residue signal sequence and four sequences of Met-enkephalin and one of Leu-enkephalin, each flanked by pairs of basic residues. There are also Met-enkephalin-Arg-Gly-Leu (YGGFMRGL) and Met-enkephalin-Arg-Phe sequences. Not all of these are cut out cleanly, and other peptides such as Met-enkephalin-Arg-Arg-Val-NH_2 are also found in brain. *Proenkephalin B* contains three copies of Leu-enkephalin contained within longer peptides. One of these, *β-dynorphin* (Table 7.4), is also a potent opioid compound. The enkephalins are thought to act as neurotransmitters, which are rapidly degraded after their release by two or three membrane-bound peptidases. Attempts are being made to design inhibitors that might inactivate these enzymes allowing buildup of enkephalin concentrations with a resultant analgesic effect.

ATP, ADP, and Adenosine

Usually regarded as a strictly intracellular compound, ATP is also released into extracellular space. There the ATP, as well as ADP and adenosine, have a variety of local hormonal functions. ATP receptors are found in many tissues and are present in some nerve synapses. ATP is one of the substances that induces sensations of pain. It may affect secretion of saliva, signal a full urinary bladder, induce a feeling of warmth, and have functions in the immune system, in platelet clotting, and as a neurotransmitter. Adenosine has been recognized for many years as an extracellular signaling molecule, a local hormone that can arise by breakdown of ATP or by secretion from cells. At least four types of receptor are present in the human body. Adenosine is thought to modulate neural responses in many tissues. It may be involved in sleep, in regulation of serotonin transport, and in control of appetite. Extracellular ADP appears to have a role in controlling bone osteoclasts.

Kinins

These hormones are small peptides that induce contraction of smooth muscles, lower blood pressure, and increase vascular permeability. They also have a function in contact-activated blood coagulation. The most important human kinins are the nonapeptide *bradykinin* and the related decapeptide *lysine-bradykinin* (Table 7.4). Other forms such as Met-Lys-bradykinin and Ile-Ser-bradykinin (T-kinin) are also known. The precursors to the kinins, the ***kininogens,*** are cleaved by the protease *kallikrein* (Fig. 7.17) or by kallikreinlike enzymes to form the kinins. Kinins are suspected of being important producers of pain in inflammatory conditions such as arthritis.

Endothelins

Endothelial cells of blood vessels produce *endothelins* that cause vascular smooth muscle contraction and a rise in blood pressure. Three human genes code for the closely related endothelins-1, -2, and -3. A 203-residue preproendothelin-1 is processed to form the 39-residue prohormone called *big endothelin*-1. Some of this peptide is secreted and circulates in plasma, where it may have various hormonal functions. Cleavage of the prohormone by a cellular metalloprotease yields endothelin-1, a 21-residue peptide held in a looped configuration by two disulfide bridges. It is homologous to a group of neurotoxins that includes the α-scorpion toxins and ω-conotoxin. These toxins act on voltage-dependent ion channels. Endothelin-2 is produced largely in the kidneys and intestine, while endothelin-3 is found in high concentrations in the brain. Type A endothelin receptors are 7-helix G-protein-coupled proteins, which activate phospholipase C with generation of inositol 1,4,5-trisphosphate and diacylglycerol. The Ins-P_3 causes release of Ca^{2+}, while diacylglycerol mediates mitogenic responses.

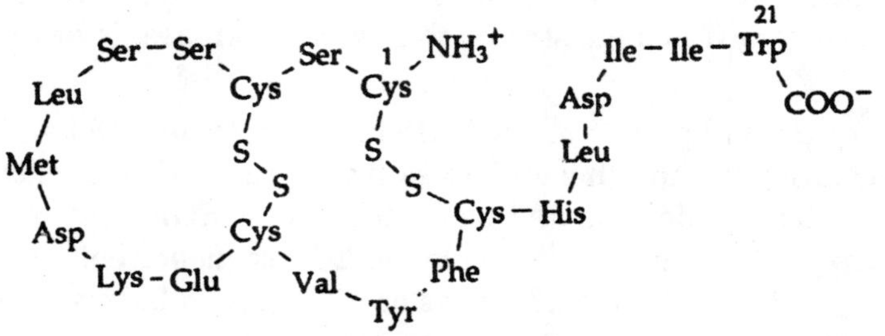

21
Ser – Ser Ser 1 NH_3^+ Ile – Ile – Trp
Leu Cys Cys Asp COO^-
Met S S Leu
S S
Asp Cys Cys – His
Lys - Glu Val Phe
Tyr

Endothelin-1

Opposing the effects of the endothelins, which act slowly, is a fast-acting endothelium-derived *relaxing factor,* which has been identified as nitric oxide, NO. Also affecting blood pressure is the potent vasorelaxant *atrial natriuretic factor.* This 28-residue peptide, which is discussed in Box 22-D, is produced by the cardiac atria and stimulates the excretion of Na^+ and of water by the kidneys. It also promotes hydrolysis of lipids within human adipocytes.

Peptides as Attractants

Small peptides as well as larger polypeptides serve to attract cells within the human body and other multicelled organisms. Both unicellular and multicellular organisms also use peptides as pheromones. The human immune system depends upon hormonelike *chemotactic factors.* Neutrophils are attracted by such peptides as formyl-Met-Leu-Phe, which have a bacterial origin, while the basic tetrapeptide Thr-Lys-Pro-Arg activates the phagocytic polymorphonuclear leukocytes and macrophages. Larger 8- to 10-kDa proteins known as *chemokines* (chemotactic cytokines) attract leuko-cytes to sites of inflammation (Fig. 7.6). Some proteins serve as pheromones. Examples range from the 40-residue mating pheromones of protozoa of the genus *Euplotes* to the 17-kDa sex pheromone of the female hamster. The decapeptide *speract* (Table 7.4) is produced by sea urchin eggs and stimulates the respiration of spermatozoa. Similar factors probably function in fertilization of human ova.

6. Protein Growth Factors and Cytokines

The pituitary growth hormone is only one of a large family of protein growth factors that are secreted by cells and which promote the growth of other cells. Many of the growth factors are also described as *cytokines,* local protein hormones that conduct cell-to-cell communication to regulate growth, development, and differentiation. Among the first growth factors to be recognized were the *insulinlike growth factors* (IGF or somatomedins), mitogenic peptides isolated from plasma. They share some of the metabolic effects of insulin but are less active. On the other hand, they are much more active than insulin in promoting cell growth and proliferation of cells. The abundant IGF-I (somatomedin C), a 70-residue single-chain basic peptide with a sequence and three-dimensional structure homologous to that of proinsulin, is considered a major mediator of the action of the pituitary growth hormone (GH, somatotropin). Studies in cell culture suggest that GH may induce differentiation of cells, and that IGF-I may then cause a rapid proliferation of the newly differentiated cells. The homologous 67-residue IGF-II may have a similar function in fetal development. The cell surface receptor for IGF-I is similar to the insulin receptor, but IGF-II receptor is structurally different. It is a monomeric 250-kDa protein; and although it is a substrate for a tyrosine kinase, it has no kinase activity of its own.

The 53-residue *epidermal growth factor* (EGF or *urogastrone)* is found in human urine and in very high concentration in the submaxillary salivary glands of male mice. Like the pancreas these glands contain both endocrine and exocrine tissues. EGF is synthesized in mice as a 1217-residue precursor, which contains not only the EGF sequence but also seven other related sequences. Related growth factors include transforming growth factor-α (TGF-α), neuregulins, betacelulin, and epiregulin, all of which promote growth of epithelial cells and are involved in wound healing. The EGF and related peptide chains are each crosslinked by three disulfide bridges. The three-dimensional structure of EGF, deduced from NMR measurements, contains largely β structure and loops and is organized into two domains in a "mitten shape." The receptor for EGF is a 1186-residue transmembrane glycoprotein. The extracellular glycosylated N-terminal region of the receptor contains the EGF-binding site. It also contains two cysteine-rich repeat sequences homologous to one of those in the insulin receptor A chain (Fig. 7.11). The cytoplasmic C-terminal part of the EGF receptor contains a 250-residue tyrosine-specific protein kinase sequence. Following dimeri-zation the EGF receptor phosphorylates tyrosine residues in various proteins including itself (auto-phosphorylation). The receptor is also phosphorylated on Thr 654 and other residues through the action of the Ca^{2+}- and phospholipid-dependent diacylglycerol-activated *protein kinase C.* Serines 1002,1046, and 1047 may also become phosphorylated, perhaps resulting in desensitization of the receptor.

Binding of EGF to its receptor produces within minutes an increased transcription rate for the prolactin gene and other nearby genes. The urinary form of EGF, urogastrone, is an inhibitor of ulcer formation. It is found in relatively large amounts in the urine of pregnant women (who tend not to develop ulcers).

Platelet-derived Growth Factor (PDGF) is released from the α-granules of blood platelets during clot formation and is thought to stimulate the growth and mitosis in fibroblasts that is necessary for wound healing. It consists of two chains, A and B. The 31-kDa precursor of the A chain is encoded by the cellular oncogene *c-sis*. The *PDGF receptor* is another transmembrane glycoprotein with a C-terminal tyrosine kinase domain. However, its construction differs from that of the insulin or EGF receptors. The external part of the single-chain receptor appears to contain five immunoglobulinlike domains. Binding of PDGF to the receptor causes responses within minutes. These include activation of the tyrosine kinase, hydrolysis of phosphatidylinositides, increases in the levels of cAMP and of Ca^{2+}, and increased transcription of a group of genes. The last include the proto-oncogenes *c-myc* and *c-fos,* which encode proteins that regulate transcription. The PDGF receptor is part of a recognized autocrine stimulatory loop in cells infected with a virus carrying the *v-sis* oncogene. The oncogene product resembles PDGF and binds to the PDGF receptors of the cell producing the *v-sis* product. In this way the cancer cell stimulates its own growth.

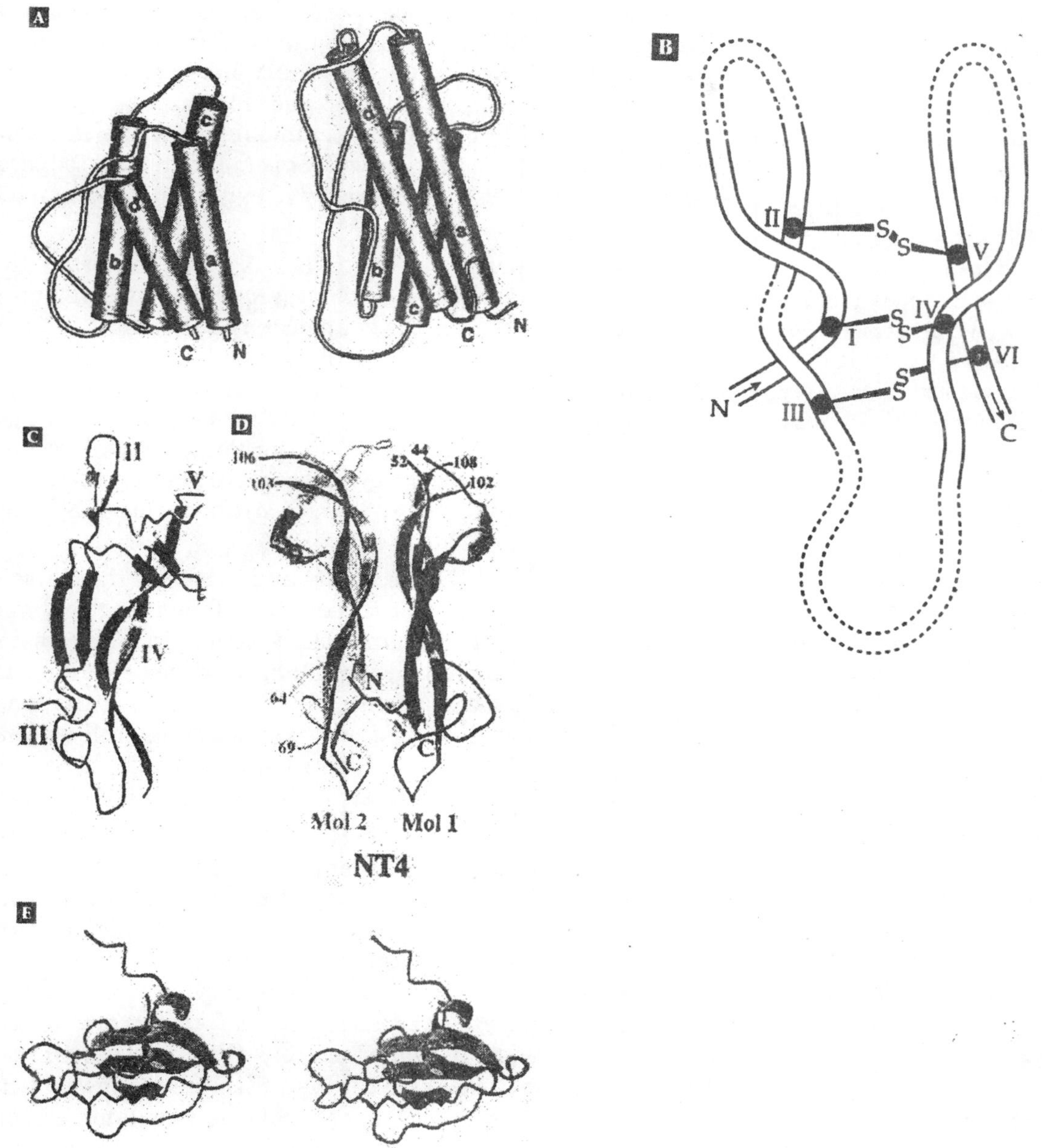

Fig. 7.6. Structures of some cytokines and other growth factor proteins. (A) Schematic drawing of representative structures of "short" and "long" helical cytokines. Note the difference in the topology of the connection between helices A and B in the two models. The short cytokines (left) include interleukins-2 and -4. Human growth hormone has a long cytokine structure. Both groups include various colony-stimulating factors. (B) Schematic representation of the disulfide knot topology of cystine knot cytokines. The cysteine residues are numbered I to VI in order of occurrence. (C) Ribbon drawing of a monomer of nerve growth factor (NGF). (D) Ribbon drawing of a dimer of the closely related neurotrophin 4 (NT4). (C) and (D) are from Robinson *et. al.*, (E) Stereoscopic ribbon drawing of the chemokine *eotaxin-3*. From Ye *et. al.*

Transformation of kidney fibroblasts into cancer-like cells can be induced by the concerted action of PDGF, an analog of EGF, and the *transforming growth factor* (TGF-β). The latter is one of 30 or more related growth factors that have numerous functions in normal tissues. Platelets produce a relatively large amount of the 25-kDa TGF-β and it too may be involved in wound healing. TGF-α is a smaller protein with a structure resembling that of EGF. While TGF-β inhibits epidermal cell growth, TGF-α stimulates growth. It is found in elevated levels in the skin lesions of *psoriasis* and may be the cause of the excessive epithelial growth in that disease.

There are at least nine *fibroblast growth factors* (FGFs). Originally found in brain, they act on many cells including the endothelial cells that line blood vessels. Basic FGF and acidic FGFs have homologous sequences and are also related to the lymphokine interleukin-1. *Vascular endothelial cell growth factor* (VEGF), which is similar to PDGF, is essential for maintenance of the endothelium. The FGFs and VEGF as well as TFG are potent *angiogenic factors* needed for growth of blood vessels. These proteins are important not only to normal blood vessels but also to invasive tumors that must develop blood vessels in order to grow. Excessive production of angiogenic factors may also be a factor in eye diseases including the retinal deterioration caused by diabetes. Another protein, *angiogenin,* is a ribonuclease, which is discussed.

There are four closely related transmembrane FGF receptors and subforms that arise by alternative mRNA splicing. The receptor structures include three external immunoglobulinlike domains and an internal tyrosine kinase domain at the C terminus. Mutations in FGF receptors are associated with a variety of skeletal defects and other hereditary problems. For example, the Gly380Arg substitution in the transmembrane segment of FGF receptor 3 is the major cause of *human dwarfism* (achondroplasia). The fibroblast growth factors, as well as other proteins such as the IGFs, HGF, and TGF-β, bind not only to their receptors but also to heparan and heparin. This binding appears to be a major factor in controlling the availability of the growth factors.

The *nerve growth factor* (NGF) was identified over 40 years ago by Rita Levi-Montalcini on the basis of its activity in promoting the profuse out-growth of neurites from embryonic neurons. The 118-residue monomer consists largely of three (β-hairpin loops, which are held together by three disulfide bridges that form a "cystine knot." The C15-C80 disulfide passes through a ring formed by the C58-C108 and C68-C110 disulfide bridges (Fig. 7.6 B, C). A similar folding pattern and disulfide core are found in TGF-β2 and also in several other *neurotrophins,* growth factors involved in the development and survival of neurons (Fig. 7.6 D). NGF may also have a more general function in promoting tissue repair. Like EGF nerve growth factor is most abundant in the submaxillary glands of male mice. Larger oligomers containing bound Zn^{2+} are present in mouse submaxillary glands. Two different receptor proteins, one of which is a tyrosine kinase, are present on many cell surfaces. The glial cells, which lie between the neurons, have their own growth factors.

Bone, formation and resorption are influenced by several protein factors. For example, IGF-I stimulates formation of bone, but EGF promotes breakdown. Additional *bone-derived growth factors and morphogenetic factors* also have been described. A *cartilage-inducing* factor has been identified as TGF-β.

A group of glycoproteins function as hematopoietic growth regulators in the development of blood cells. The 166-residue cytokine *erythropoietin* is the primary regulator of red blood cell formation in mammals. At least four glycoprotein *colony-stimulating factors* (CSF) promote proliferation of granulocytes and macrophages. The lymphocyte-produced *lymphokines* include the *interleukins* and other proteins. Two species of interleukin-1 (IL-1) serve as mediators of inflammation. They induce proliferation of T lymphocytes and fibroblasts, bone resorption, release of acute phase proteins (Section C), breakdown of cartilage, and fever.

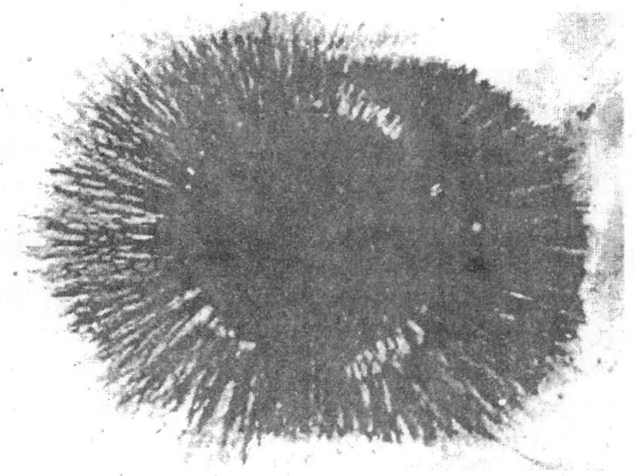

Fig. 7.7. Effect of one ng of nerve growth factor in promoting the production of neurites in a chick embryonic sensory ganglion. From Frazier *et. al.*

Interleukin-2 (T-cell growth factor; Fig. 7.6 A) is secreted by some activated T-lymphocytes. This 133-residue largely helical protein is involved in generation of cytotoxic T-cells, stimulation of interferon release, and of release of a B-cell growth factor. Considerable excitement has accompanied the possibility of activating lymphocytes with IL-2 produced from cloned genes in bacteria to increase their ability to kill cancer cells. However, IL-2 is toxic, and this is limiting its use.

IL-3 is one of the colony-stimulating factors, which stimulates the growth of many types of blood cells. Other lymphokines include one derived from T helper cells, which activates resting T lymphocytes thus amplifying an immune response. Others are *α-interferon* and the neurotrophic factor (autocrine motility factor) *neuroleukin*. It acts in monomeric form, but as a dimer it seems to be identical to the enzyme phosphoglucose isomerase. While most hormones regulating growth and differentiation seem to be large peptides or proteins, *bursin*, which induces differentiation of lymphocytes, is the amidated tripeptide Lys-His-Gly-NH_2. The corresponding differentiation hormone for T lymphocytes is the 49-residue thymopoietin, a hormone of the thymus gland.

Tumor necrosis factor (TNF, also called cachetin) is a 157-residue hormone secreted by macrophages. It is a mediator of inflammatory responses including fever, shock, and *cachexia,* the wasting of the body during chronic diseases including cancer. TNF was isolated as the causative agent of cachexia and also as a factor produced in acute bacterial infections, which sometimes caused death of tumor cells and spontaneous recovery from cancer. In the latter case, it is the lipopolysaccharide (Fig. 7.30) and other bacterial endotoxins that induce the release of TNF by macrophages. Its extreme toxicity has prevented immediate harnessing of the tumor-killing potential of TNF. One function of TNF is regulation of transcription factor NF-κB (Fig. 7.40) in neutrophils and macrophages, a key part of the inflammatory response. TNF also mediates programmed cell death (apoptosis) and has been linked to obesity-induced insulin resistance. The cell surface TNF receptors have a variety of modular structures consisting of various disulfide-linked subdomains.

This long list of vertebrate peptide growth and regulatory hormones is not complete. The biological actions of these hormones are also complex. Growth factors usually have pleiotropic effects, which may involve many tissues as well as many regulatory systems. Are there any simplifying generalizations? Loret *et. al.*, point out that some growth factors such as IFG-1 and EGF are ubiquitous, affecting virtually all tissues. Others, such as PDGF and thrombin

(Fig. 7.17), are more localized in their effects. Some, such as the lymphokines, are more specialized. For one group of hormone receptors the effects are mediated by tyrosine kinases and internalization of the receptors. Another group of receptors activate G proteins and, in turn, adenylate cyclase or phospholipase C. The regulatory domains of the various receptors overlap, a property that allows different tissues to respond differently to hormonal stimuli. The result is the network of interactions that makes the body so sensitive and responsive.

7. Nonpeptide Mammalian Hormones

The free radical molecule nitric oxide, commonly abbreviated as either • NO or simply NO, is formed by hydroxylation of guanidine groups of arginine. First recognized as the *endothelium-derived relaxing factor,* •NO has received increasing attention because of its involvement in a broad range of physiological processes. These include regulation of blood pressure through effects on smooth muscles of the vascular endothelium, regulation of several aspects of the innate immune system, and neurotransmitter functions both in the brain and in the peripheral nervous system. Roles for NO in bacteria, other microorganisms, and plants have also been discovered. These often involve regulation of transcription. As mentioned in many of the effects of NO are a result of activation of soluble guanylate cyclase. In the endothelium other hormones, such as the endothelins, atrial natriuretic factor, and bradykinin, cooperate in the regulation of NO synthase. Neuronal NO synthase functions in the brain in olfaction and in formation of memory. In the peripheral system it mediates penile erection and plays a variety of roles in the enteric nervous system. Neuronal NO synthase is often localized to synaptic regions by binding to tissue-specific proteins. NO may also regulate cellular respiration by inhibition of cytochrome *c* oxidase.

In high enough concentrations NO is toxic. It is formed of phagocytic cells and utilized in the killing of ingested pathogens. It also contributes to the inflammatory response of tissues. Even the firefly's flash is triggered by a pulse of NO. The dangerous *stonefish,* whose sting causes death within six hours, apparently utilizes NO to kill its victim. A 148-kDa lethal protein (stonustoxin) in its venom induces rapid formation of NO, which causes a fatal drop in blood pressure.

Like carbon monoxide, NO binds tightly to many metal centers within a cell. This has added greatly to the problem of understanding the mechanisms of its action. NO also reacts rapidly with thiol groups of proteins and of small molecules such as glutathione. Because of its importance in the regulation of blood pressure, reactions of NO with hemoglobin and the related myoglobin have been studied intensively. NO binds to hemoglobin 1000 times more tightly than either O_2 or CO, preferentially occupying the hemes of the α subunits. Because there is so much hemoglobin in red blood cells, at most one NO per hemoglobin molecule can react. This allows as much as one NO to be carried to tissues along with three O_2 molecules. If the NO could be released in the capillaries, it would activate guanylate cyclase. The resulting cGMP would induce relaxation of smooth muscles and reduce blood pressure. However, tight bonding of NO to deoxyhemoglobin would prevent this release. A plausible possibility (with experimental support) is that NO is not bound to Fe but to the SH group of the conserved cysteine 93 of a β subunit of hemoglobin as SNO (S-nitrosothiol) hemoglobin. The NO may bind initially to the iron atom of an α subunit, but then be transferred to the nearby β Cys 93 to form the SNO-Hb. NO may then move from SNO-Hb to thiol groups in the tissues. Recent evidence suggests that the transfer occurs first to an SH group in the anion exchange *AE*1. An alternative explanation, which does not involve SNO-Hb, is that hemoglobin Fe-NO is converted to *nitrite* via oxidation of the iron to form a methemoglobin subunit (Eq. 7.1), and that it is nitrite which serves as the endothelial relaxing agent.

$$\underset{\substack{\text{Oxyhemoglobin} \\ \text{or} \\ \text{Mb-Fe(II)} \bullet O_2 \\ \text{Oxymyoglobin}}}{\text{Hb-Fe(II)} \bullet O_2} \xrightarrow{NO \quad NO_3^-} \underset{\substack{\text{Methemoglobin} \\ \text{or} \\ \text{Mb-Fe(III)}^+ \\ \text{Metmyoglobin}}}{\text{Hb-Fe(III)}} \quad ...(7.1)$$

The function of the monomeric myoglobin has often been assumed to be participation in facilitated diffusion of O_2. Although this is an important function under some circumstances, an additional role for myoglobin may be to scavenge NO, via Eq. 7.1, preventing its buildup to dangerous levels. The metmyoglobin produced can be reduced by methemoglobin reductase. A different situation is met by the parasitic nematode *Ascaris,* whose hemoglobin binds O_2 so tightly that it can't serve as an O_2 carrier. It may serve as an *NO-activated deoxygenase,* again using Eq. 7.1 to remove O_2, which can be toxic to the nematode. Free myoglobin can also react with NO to form heme-NO and hemenitroxyl complexes.

Like NO, CO also binds tightly to heme iron and is able to activate guanylate cyclase. CO is formed in the human body by the action of heme oxygenases. Synthesis of heme oxygenase-1 (HO-1) in smooth muscle is induced by a low oxygen tension (hypoxia). The resulting elevated level of CO not only may produce increased vasodilation, but also may inhibit synthesis of vascular smooth muscle cells. Heme oxygenase-2 (HO-2) is found in the brain, where it is colocalized with soluble guanylate cyclase. Some other organisms have a more active CO metabolism. The CO oxidation system of *Rhodospirillum rubrum* is activated by a CO-sensing heme protein, which acts as a transcriptional regulator. The CO binds to the heme iron, apparently inducing a conformational change that allows the protein to bind to its target DNA sequence.

Hormonal Lipids

We have already considered a number of hormones that are not water-soluble but may have to be transported by carrier proteins to their sites of action. These include retinoic acid (Box 22-A), metabolites of vitamin D, and the platelet-activating factor. The last functions in the brain as well as in blood. Hormonal lipids also include the prostaglandins. (Fig. 7.7), leukotrienes, and lipoxins. These are products of the eicosenoid cascade or network, which is activated by receptors linked to phospholipase C (Fig. 7.9). Ceramide formed by hydrolysis of sphingomyelin initiates additional responses. Sphingolipids may also be important mediators of apoptosis.

The sleep-inducing oleamide modulates signaling by serotonin-dependent and Gaba-dependent neurons and blocks gap junction signaling in brain glial cells. Oleamide is one of a family of fatty acid amides found in human plasma. One of these, found also in the brain, is *anandamide* (arachidonoyl-ethanolamide). It is an endogenous activator of the brain cannabinoid receptors. The 22-carbon *erucamide* (cis-13-docosenamide) stimulates growth of blood vessels. The fatty acid amides are apparently synthesized from corresponding acylglycines (Eq. 7.2) by the action of the peptidylglycine a-amidating enzyme using the mechanism of Eq. 7.11.

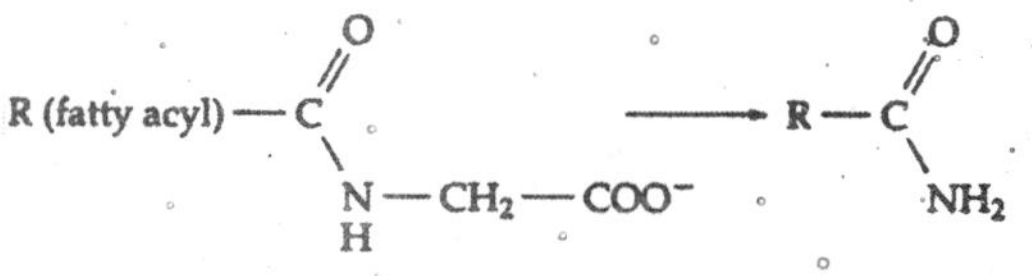

$$R\ (\text{fatty acyl})-C(=O)-NH-CH_2-COO^- \longrightarrow R-C(=O)-NH_2$$

The fatty acid amides are destroyed by an integral membrane protein, a *fatty acid amide hydrolase.*

8. Nonvertebrate Hormones and Pheromones

The chemical signals that are passed between bacteria and other microbial cells often resemble hormones of vertebrates. Thus, some bacteria secrete peptide mating pheromones. The sequence of an octapeptide of this type from *Streptococcus faecalis* is given in Table 7.5. Many bacteria utilize "quorum signaling." They do not secrete signaling molecules until they sense that there are enough of them to be effective if they act in unison. Then they all secrete an inducer. The best-known example is the induction of bioluminescence of *Vibrio fischeri* (Eq. 7.49). Long-chain fatty acyl derivatives of L-homoserine lactone act as secreted inducers. A lactonase hydrolytically inactivates the inducer to avoid excessive accumulation. Depending upon the types of bacteria and the specific response a variety of different fatty acyl groups may be present in the inducer. Other responses include the formation of bacterial film (biofilms) on a surface and release of virulence factors that induce attack on a host.

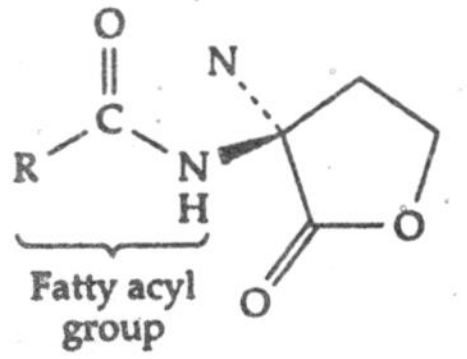

Fatty acyl group

N-Acylhomoserine lactones

Sexual conjugation in yeast is also induced by pheromones (mating factors). Yeast cells of mating type **a** synthesize the 12-residue mating factor **a** which contains a C-terminal cysteine methyl ester S-alkylated with a frans, *trans-tarnesyl* group (Table 7.5). Cells of type **a** synthesize a 13-residue factor **a**. Cells are attracted to the pheromone produced by cells of the opposite type. The *tremerogens,* sex hormones of certain basidiomycetes, have related structures (Table 7.5).

Peptides are not the only fungal hormones. The water mold *Blastodadiella* releases a *zoospore maintenance factor,* a cyclic phosphate derivative of 5'-phosphoribosyl-5 aminoimidazole-carboxamide. It is similar to the succinocarboxamide, which is an intermediate in *de novo* synthesis of purines (Fig. 7.15). Substitution of homocysteine for L-aspartate in step *g* of that sequence could generate a precursor to the zoospore maintenance factor.

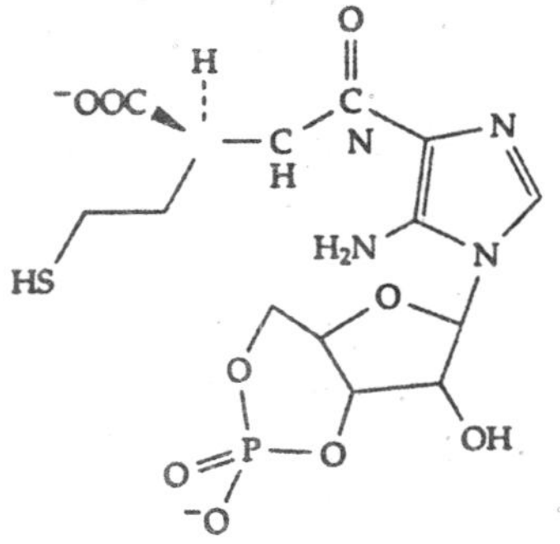

Proposed structure of zoospore maintenance factor of *Blastocladiella.*

Male sperm cells of the alga *Chlamydomonas allensworthii* (Fig. 7.11) are attracted to female gametes by a pentosylated isoprenoid quinone.

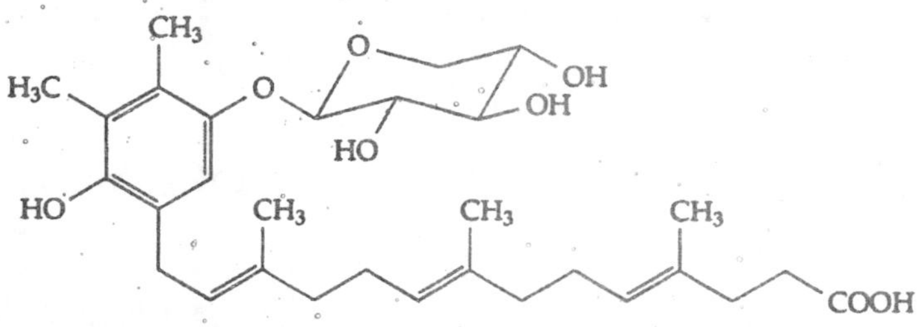

Compound 1

Neurohormones of Invertebrate Animals

The Cnidaria (coelenterates) have the simplest known nervous system. Simple amidated tetrapeptides, some of which are also found in molluscs, are among their neurotransmitters. EGRFamide and L-3- phenyllactyl-LRNamide are found in some sea anemones. When a hydra (Fig. 7.13) is cut into two pieces, one containing a head and one a foot, each piece reforms the missing end. The decapeptide *head activator* (Table 7.5) diffuses upward from the foot end and induces formation of a head. Similarly, a hormone produced by head cells may induce growth of a new foot end. Glutathione, flowing out from the hydra's prey after wounding by a nematocyst, is the feeding attractant for *Hydra vulgaris*. Termination of the response is dependent upon nitric oxide in this primitive invertebrate.

Certain large anemones enter into a symbiotic relationship with fishes, which recognize chemical signals from the anemones and are also chemically protected from the anemones' stings. One of the fish attractants, *amphikuemin,* is effective at a concentration of 10^{-10} M.

A variety of peptide neurohormones are produced by molluscs. Among these are the sea snail *Aplysia,* which is studied because of its simple nervous system and giant neurons. Proteolytic processing of precursors within single cells often yields neurohormones specific to those cells. The structures of two *small cardioactive peptides* secreted from single neurons are shown in Table 7.5. A 4.4-kDa egg-laying hormone is formed in at least eight processing steps. Among the peptides identified in the freshwater snail *Lymnaea* are many that are characteristic of mammalian pituitary gland, pancreas, brain, and intestinal tract. These include TRH, ACTH, αMSH, arginine vasopressin, ocytocin, calcitonin, gastrin, gastrointestinal peptide, glucogen, insulin, Met-enkephalin, pancreatic polypeptide, secretin, somatostatin, substance P, and vasoactive intestinal peptide. Also present are FMRF amide and arginine vasotocin.

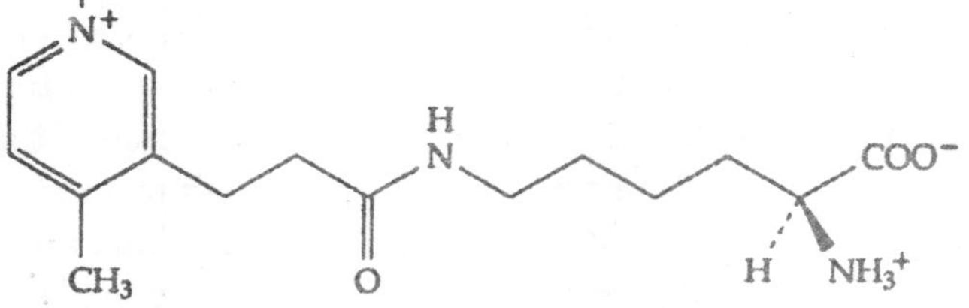

Amphikuemin, a powerful fish attractant

Table 7.5. Some Microbial and Invertebrate Peptide Hormones

Source name	*Number of residues*	*Sequence*[a]
Streptococcus faecalis sex hormone	8	FLVMFLSG
Yeast mating factor a	12	YIKGV(L)FWDPAC–OCH_3 (C: farneshl[b])
Tremerogen A-10	10	EHDPSAPGNGYC–OCH_3 (C: farnesyl[b])
Hydra head activator	10	pEPPGGSKVIF
Antho-RF amide (sea anemone)	4	pEGRF-NH_2
Small cardiovascular peptides (*Aphysia*)		
SCP-A	11	ARPGYLAFPRM-NH_2
SCP-B	9	MNYLAFPRM-NH_2
FMRFamide (coelenterates, molluscs)	4	FMRF-NH_2
(octopus)	7	YGGFMRF-NH_2
Shrimp blanching hormone	8	pELNFSPGW-NH_2
Fidler crab pigment-dispersing hormone	18	NSELINSILGLPKVMNDA-NH_2
Proctolin (cockroaches)	5	RYLPT
Myotropic neuropeptide (cockroaches)	11	EQFEDYGHMARF-NH_2 (F→sulfate ester in leukosulfakinin)
Adipokinetic hormone (locust)	10	PELNFTPNWGT-NH_2
Crustacean cardioactive peptide	9	EPFCNAFTGC-NH_2 (C–S–S–C)

[a] One-letter abbreviations, pE, 5-oxoproline; –NH_2, C-terminal carboxamide.
[b] Alkylated on S.

The cardioacceleratory peptide FMRFamide (Table 7.5), which was discovered in 1977, was the first in a large series of related neuropeptides that are found in organisms ranging from the nematode *Caenorhobditis elegans* to vertebrate animals. At least 18 genes in *C. elegans* encode 53 distinct FMRFamide-related peptides. Disruption of one of these genes causes hyperactivity, uncoordination, and other behavioral difficulties in the nematodes. Several groups of FMRFamide-related peptides have been found in *Drosophilia* and other insects. One group has the C-terminal FLRFamide and another HMRFamide. Among the latter are sulfate esters such as the cockroach neuropeptide shown in Table 7.5. The sequence of the sea urchin sperm chemoattractant *speract* (sperm attractant peptide-1; SAP-1) is shown in Table 7.4. This is one of a family of egg-associated peptides that stimulate sperm metabolism and mobility. The DNA sequence that codes for speract predicts formation of a 296-residue protein that contains four speract sequences plus six related decapeptide sequences, each separated by a single lysine residue. Many other SAP peptides, some containing the unusual amino acid *o*-bromo-L-phenylalanine, are formed.

Hormones of Insects and Crustaceans

Peptide neurohormones of insect brains include the pentapeptide *proctolin* (Table 7.5), which was first isolated from the cockroach and has since been found in crustaceans and in mammalian brain. It has been traced to specific insect neurons. A nonapeptide neurohormone from the shore crab does not resemble any other known vertebrate or invertebrate hormone.

The prawn *Pandalus borealis* changes its body color by means of movable pigment granules. The neurosecretory octapeptide *blanching hormone* (Table 7.5) controls the process. This is a member of a larger family of peptide hormones with related functions such as the 18-residue *pigment-dispersing hormone* of the fiddler crab and similar hormones in insects.

One of the insect neurohormones, the *activation hormone,* controls the secretion of the corpora allata, paired glands that synthesize the *juvenile hormone* (Fig. 7.4) in insect larvae. While the structure of the juvenile hormone varies somewhat with species, it is usually a polyprenyl ester. A specific binding protein provides the hormone with protection from degradative enzymes. However, in the tobacco hornworm an esterase, able to hydrolyze the protein-bound juvenile hormone, is produced at the start of pupal differentiation. The exact mechanism of action of juvenile hormones has been difficult to determine. However, it affects polyamine synthesis.

The *mandibular organ-inhibiting hormone* of the crab *Cancer pagurus* produces two neurohormones that inhibit the secretion of methyl farnesoate, which is thought to function as the juvenile hormone in crustaceans.

A related role of the insect corpora allata is to store and release the *prothoracicotropic hormone,* a peptide neurohormone formed in a single neurosecretory cell of the brain. The steroid hormone *ecdysone* (Fig. 7.12) is secreted by the insect's prothoracic gland. Also known as the *molting hormone,* ecdysone is required for the periodic replacement of the exoskeleton of the larvae. It induces molting in crayfish and other arthropods and appears to be needed by such members of lower phyla as schistosomes and nematodes. It also controls the biting behavior of mosquitos. In addition to α-ecdysone, the 20- and 26-hydroxyecdysones and 20, 26-dihydroxyecdysone have been identified in insects. It has been suggested that different ecdysones may function at different stages of insect development.

Ecdysone stimulates the synthesis of RNA in tissues. Visual demonstration of the effect is provided by its action on polytene chromosomes of fly larvae (Fig. 7.14). Fifteen minutes after the application of ecdysone, a puff is induced on one band of the chromosome; a second puff forms at a later time while a preexisting-puff diminishes. Thus, like steroid hormones in mammals, ecdysone appears to have a direct controlling effect on transcription. The cuticle-shedding process (ecdysis) is initiated by the brain peptide *eclosian.* However, the brain may be responding to the *ecdysis-triggering hormone,* a peptide that is secreted by a series of epitracheal glands located in various segments of the body.

Adipokinetic hormones control metabolism of insects during long-distance flight. In the migratory locust these hormones consist of a pair of related octapeptides and a decapeptide (Table 7.5). The hormones stimulate triacylglycerol lipase in the insects, fat bodies, induce release of carbohydrates from body stores, and affect many other aspects of metabolism. Insects also have hormones of the insulin family, proteins consisting of disulfide-linked A and B chains as in insulin. The silkworm *Bombyx mori* has 38 genes for the insulinlike *bombyxins,* which are synthesized in the brain.

Insects produce many different types of sex attractant pheromones. By 1995 more than 300 structures had been determined for pheromones from > 1600 insect species.

9. Plant Hormones

Plants possess a kind of circulatory system by which fluids are transported from the roots upward in the xylem and downward from the leaves through the phloem. Many compounds are carried between cells in this manner, while others are transported across cell membranes and against concentration gradients by active transport. A number of compounds that move between cells in either of these two manners have been classified as hormones. The major plant hormones consist of five compounds or groups of compounds: *auxins* (p. 1446), *gibberellins* (Eq. 7.5), *cytokinins* (Fig. 7.33), *abscisic acid* (Fig. 7.4), and *ethylene* (Fig. 7.16). A number of other plant regulators, some involved in defensive reactions, are sometimes also described as hormone. These include the *brassinosterols* (Fig. 7.9) and related compounds, *jasmonic acid* (Eq. 7.18), *salicylic acid*, bacterially produced *lipooligosaccharides* such as the NOD factors (Box 20-E), and polypeptides such as *systemin.* The sugars glucose and sucrose have hormonelike functions as does light, which controls many plant functions.

Plant hormones have multiple and overlapping functions, which are exerted predominantly by repression of gene expression. This makes it difficult to discuss their functions briefly. Most studied are the auxins, of which the principal member is *indole-3-acetic acid*. This compound, has been implicated as a controlling agent for cell division and cell elongation. In this capacity auxin influences a great variety of plant processes. Produced principally by growing shoots, auxin diffuses down the stem aided by special *efflux carriers* and inhibits the growth of lateral buds. However, the hormone stimulates the growth of stems, thus establishing the apical dominance of the tip of a plant. Other hormones also have an influence. Auxin is well established as the controlling agent in phototropism, the tendency of a plant to bend toward the sun.

A sensitive test for auxin, which is dependent on the bending of the coleoptile of *Avena saliva* (the common oat) in response to the hormone, allows detection of as little as three pmol of auxin. Using this assay, it was shown that auxin is transported laterally away from the illuminated side of plants, causing the darker side to elongate more rapidly. Both membrane-associated and soluble binding sites, which may represent natural auxin receptors, have been identified, and auxin response elements have been located in DNA. The membrane-bound receptors may regulate an ATP-dependent proton pump, while the soluble receptors may act to regulate gene transcription.

The gibberellins together with brassinosterols are active in helping to determine the *form* of plants. There are 66 different known gibberellins, which are synthesized (Eq. 7.5) in mature leaves and are transported downward. Their very active effect in stimulating RNA synthesis in dwarf varieties of vegetables suggests that gibberellins also serve as gene activators to promote RNA synthesis. A possible function in the *geotropic* response of plant roots is suggested by the presence of higher concentrations of glibberellins in the upper half of horizontal roots than in the lower half. On the other hand, auxin has long been known to have a higher concentration on the lower side of the root, and it has been assumed to inhibit elongation (in contrast to stimulation in stems). Because they are structurally and in physical properties somewhat similar to sterols, gibberellins have been assumed to act by a steroid hormonelike mechanism. The brassinosterols, which are true sterols, may also be expected to bind to soluble receptors and to regulate transcription.

The cytokinins are isopentenyladenosine derivatives (Fig. 7.33), which may be hydroxylated or substituted in the 2 position by a methylthio group. Cytokinins are synthesized in roots and translocated upward to other parts of the plant. They may originate in part from degradation of cytokinin-containing tRNA, but there is evidence that they may also be synthesized independently. The N^6-isopentenyladenine in tRNAs is generated by transfer of the isopentenyl group from Δ^2-isopentenyl pyrophosphate. The role of cytokinins may be at the level of gene transcription, but it has been difficult to identify signaling sequences. Their hormonal influence on plants appears to be independent of their function in tRNA. The most striking effect of cytokinins in solution is on differentiation of plant cells.

The terpenoid abscisic acid (Fig. 7.4) is synthesized by degradation of a carotenoid precursor. It is formed by plants in response to *stress* from low temperature, high salinity, or drought. Abscisic acid appears to block the growth-promoting effects of hormones such as the gibberellins and cytokinins. It is sometimes regarded as a general gene represser, which prepares plants for dormancy. Synthesis of abscisic acid occurs in response to the short day and long night pattern of the fall. It often opposes the action of gibberellins. The signaling pathway for abscisic acid apparently involves release of Ca^{2+} ions and formation of cyclic ADP-ribose (cADPR; p. 564). Among other effects this induces the closing of stomata in leaf surfaces.

Ethylene not only hastens the ripening of fruit but also tends to promote senescence in all parts of plants. Its signaling mechanisms are the best-known for any plant hormone. Ethylene is metabolized slowly in plants by oxidation to ethylene oxide. The latter is hydrolyzed to form ethylene glycol, which is metabolized further to CO_2 (Eq. 7.3).

$$CH_2{-}CH_2 \xrightarrow{O_2} \underset{\text{(epoxide, O bridging)}}{CH_2{-}CH_2} \xrightarrow{H_2O} HO{-}CH_2{-}CH_2{-}OH \longrightarrow \longrightarrow CO_2 \qquad ...(7.3)$$

A postulated flowering hormone, *florigen,* has not been isolated. Flowering seems to be controlled by a variety of different hormonal effects. Jasmonic acid (jasmonate) and salicylic acid act as plant defense signals. Salicylic acid activates a large number of transcription factors, which induce resistance to a variety of pathogenic organisms, a response referred to as *systemic acquired resistance.* Among other compounds synthesized as part of the systemic acquired resistance of plants are proteins known as *systemins.* Initially discovered in tomatoes, systemins have been discovered in more than 100 other species of plants. Jasmonic acid emitted from tomato plants acts as a pheromone that attracts wasps to attack caterpillars that feed on the tomato plants. In fact, wounding by herbivores may stimulate emission of a variety of volatile compounds that may attract predators to the attacking herbivores.

Many other compounds influence plant growth. Among them are the vitamins, thiamine, pyridoxine, and nicotinic acid, which are synthesized in the leaves and transported downward to the roots. Since they promote growth of roots, they are sometimes referred to as *root growth hormones.* However, they are nutrients universally needed by cells. Various compounds secreted by other organisms can either stimulate or inhibit growth of a given plant. Some are powerful toxins. Others, such as the previously mentioned NOD factors, and evidently also the riboflavin degradation product *lumichrome*, are beneficial. These *plant growth regulators* may be produced by other plants, by microorganisms, or by fungi. Much use is made in agriculture of synthetic growth regulators.

10. Secretion of Hormones

The effects of binding of hormones to cell surface receptors have already been emphasized. Equally important are the mechanisms that control the secretion of hormones. Both hormones and neurotransmitters are secreted by exocytosis of vesicles. Cells have two pathways for secretion. The *constitutive pathway* is utilized for continuous secretion of membrane constituents, enzymes, growth factors, viral proteins, and components of the extracellular matrix. This pathway carries small vesicles that originate in the trans-Golgi network. The *regulated pathway* is utilized for secretion of hormones and neurotransmitters in response to chemical, electrical, or other stimuli.

Many neurotransmitters are packaged into *small synaptic vesicles* ~50 nm in diameter. These may originate from large endosomes rather than from the Golgi. They are usually recycled and refilled repeatedly. Secretion of hormones, and of some neurotransmitters, occurs via *large dense-core vesicles* of ~100 nm diameter. These originate from the TGN and are not recycled. They are prominent in chromaffin cells and other cells that secrete large amounts of a signaling molecule. Secretion of hormones and that of neurotransmitters have several common features. Indeed, hormones of the hypothalamus, neurohypophysis, and the adrenal medulla are secreted by specialized neurons. However, while hormones are often carried in the bloodstream, neurotransmitters are most often secreted into the very small volume of a single synapse. The exocytosis must occur very rapidly from a small number of SSVs.

A common feature, and also a puzzle, of vesicular signaling is the nearly universal response to calcium ions. Exocytosis is usually triggered by a rise in the concentration of Ca^{2+}, and most receptor signaling also leads to an increase in cytosolic Ca^{2+}. The puzzle lies in the ability of cells to use a common mechanism for so many specific purposes. There are also many other factors that can control exocytosis. Recent evidence suggests that NO may play a role.

B. NEUROCHEMISTRY

The nervous system, a network of neurons in active communication, reaches its ultimate development in the 1.5 kg human brain. Many invertebrates, such as leeches, crayfish, insects, and snails, have brains containing no more than 10^4 to 10^5 neurons, but the human brain contains ~10^{11}. Each of these neurons interconnects through *synapses* with hundreds or thousands of other neurons. The number of connections is estimated to be as many as 60,000 with each Purkinje cell of the human cerebellum. There may be many more than 10^{14} synapses in the human brain.

In addition to neurons, the brain contains 5 – 10 times as many *glial* cells of several types. The neuroglia occupy 40% of the volume of brain and spinal cord in the human. Some glial cells seem to bridge the space between neurons and bloodcarrying capillaries. Others synthesize myelin. Some are very irregular in shape.

1. The Anatomy and Functions of Neurons

Although neurons have many shapes and forms, a common pattern is evident. At one end of the elongated cell (Fig. 7.8) is a series of *dendrites,* thin fibers often less than 1 μm in diameter. The ends of the dendrites form synapses with other neurons and act as receivers of incoming messages. Additional messages come into synapses on the *cell body,* while the *axon* serves as the output end of the cell. The axon, a long fiber of diameter 1 – 20 μm, is also branched. As a consequence, the nervous system contains both highly branching and highly converging pathways. Many of the axons are wrapped in a myelin sheath (Fig. 7.9).

The ends of the fine nerve fibers are thickened to form the *synaptic knobs,* which make synaptic contacts with dendrites on cell bodies of other neurons. In most instances the arrival of a nerve signal at the *presynaptic* end of a neuron causes the release of a transmitter substance (neurohormone). The transmitter passes across the 10 – 50 nm (typically 20 nm) *synaptic cleft* between the two cells and induces a change in the electrical potential of the *postsynaptic* membrane of the next neuron (Fig. 7.10). Excitatory transmitters usually cause *depolarization* of the membrane. By this we mean that the membrane potential, which in a resting neuron is –50 to –70 mv (Chapter 8), falls to nearly zero often as a consequence of an increased permeability to Na^+ and a resultant inflow of sodium ions. The resulting *postsynaptic potential*

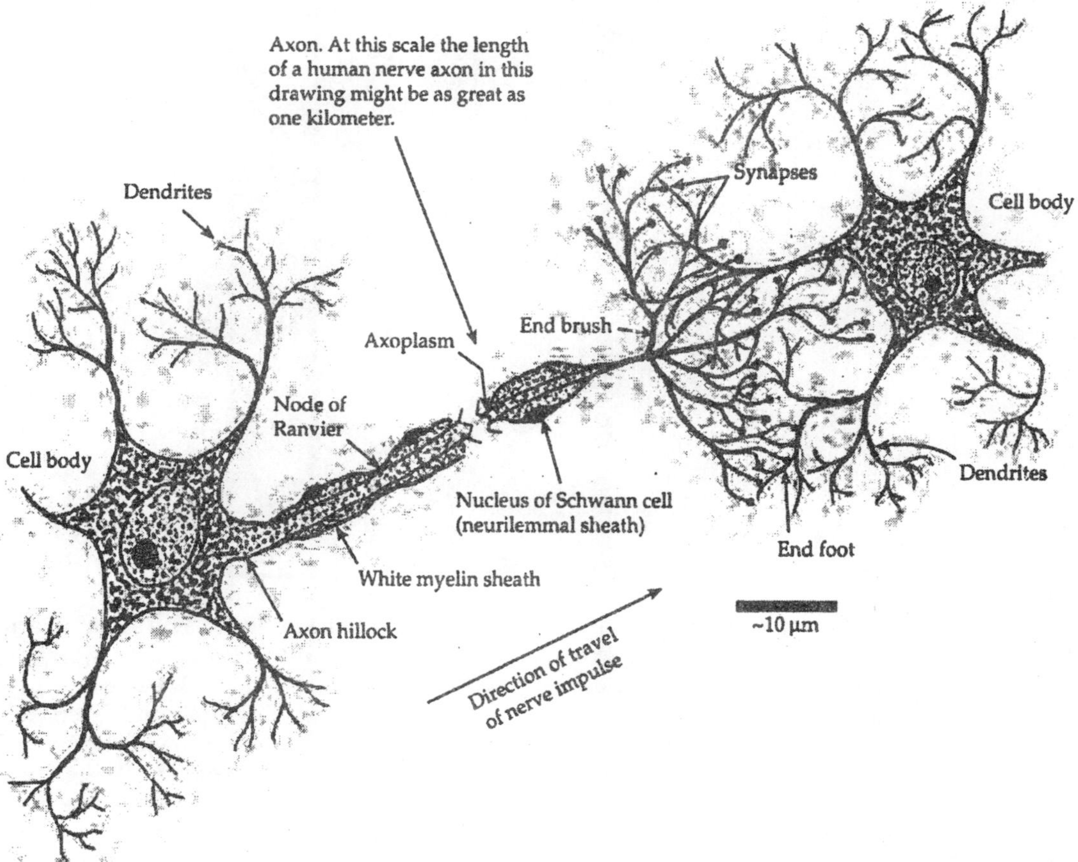

Fig. 7.8. Schematic drawing of a neuron (after Brand and Westfall.

(really a drop in the potential difference) is propagated to the cell body and axon and under appropriate circumstances may initiate an *action potential.* This is a narrow spikelike region of depolarization that travels down the axon at a constant velocity and with undiminished intensity (Fig. 7.11).

A characteristic of many neurons is an *all-or-none response* or firing. An action potential passes down the axon only if there is sufficient depolarization. In general, a stimulus must reach a neuron through *more than one synapse* before the neuron will fire. Furthermore, neurons are often *inhibitory,* releasing transmitters that counter the excitatory synapses and tend to prevent firing. Inhibition is important in damping out small excitations; thus sharpening the response of the nervous system toward strong stimuli. Another characteristic of basic importance to the operation of the brain is that neurons fire at longer or shorter intervals depending upon the strength and duration of the stimulus. The stronger the stimulus to a given neuron, the more rapid the train of spikes that passes down the axon. Thus, the brain functions to a large extent in decoding trains of impulses. The frequency of the impulses from neurons varies from a few per second to a maximum of about 200 s^{-1} in most nerves (up to 1600 s^{-1} in the Renshaw cells of the spinal cord). The maximum frequency is dictated by the refractory period of ~1 ms (Section B, 3).

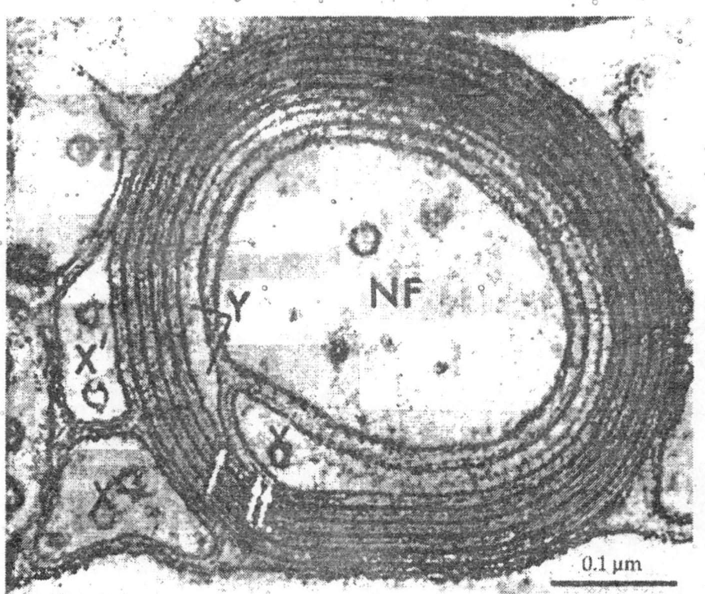

Fig. 7.9. Micrograph of a section through an axon of a neuron from rat brain. The structure of the myelin sheath can be seen clearly. The growing lip of cytoplasm (X) from a neuroglial cell is advancing around the axon process (NF) and insinuating itself into the space between the plasma membrane of the axon and the membrane that limits the thin layer of cytoplasm (Y) left behind by the growing lip during its previous turn. This cytoplasmic layer disappears as the inner leaflets of its plasma membrane fuse to form the major dense line of the myelin sheath. This process is occurring at the point indicated by the single arrow. The outer leaflet of the plasma membrane surrounding the lip fuses with its own outer leaflet laid down on the previous turn. The two outer leaflets thus give rise to the less dense intermediate line of the sheath (double arrow). The cell body from which the investing cytoplasmic sheet originated cannot be seen in this micrograph, but cytoplasm within the lateral margins of the sheet does appear (X).

Although the concepts of neuronal function outlined in the preceding paragraphs have been accepted for many years, more recent discoveries require that they be modified somewhat. Dendrites seem to be able to transmit information as well as to receive it. Furthermore, while information is certainly transmitted long distances by spike action potentials, shorter neurons and dendrites may communicate extensively by exchange of chemicals through low resistance gap junctions, also called *electric synapses*. Small changes in membrane potential transmitted through these junctions may alter the behavior of adjacent neurons. Chemical transmitters do not always have an electrical effect on postsynaptic neurons but may influence metabolism or gene transcription.

2. Organization of the Brain

The anatomy of the brain is quite complex, and only a few terms will be defined here. The *cerebrum,* which is made up of two hemispheres, accounts for the largest part of the brain. The deeply folded outermost layer, the *cerebral cortex,* consists of *gray matter,* a mass of cell bodies, and fine unmyelinated nerve fibers. Beneath this lies a layer of *white matter* made up of myelin-covered axons connecting the cerebral cortex with other parts of the brain. The two

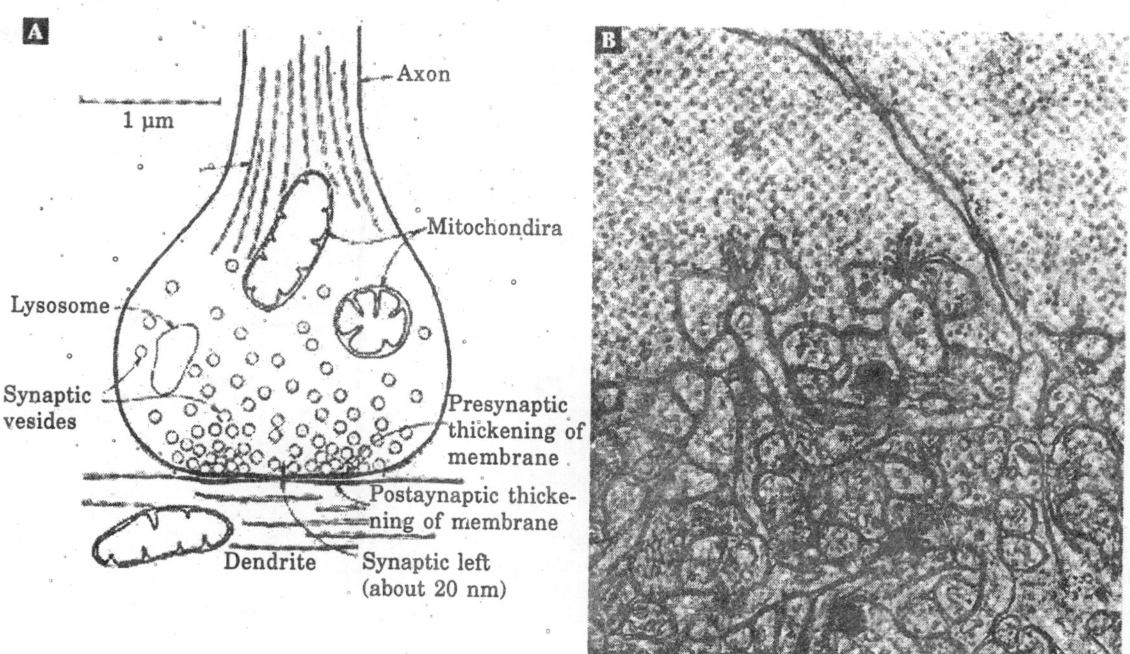

Fig. 7.10. (A) Schematic drawing of a synapse. (B) Electron micrograph showing the synaptic junctions in the basal part (pedicle) of a retinal cone cell of a monkey. Each pedicle contains synaptic contacts with 12 triads, each made up of processes from a bipolar cell center that carries the principal output signal and processes from two horizontal cells that also synapse with other cones. A ribbon structure within the pedicle is characteristic of these synapses. Note the numerous synaptic vesicles in the pedicle, some arranged around the ribbon, the synaptic clefts, and the characteristic thickening of the membranes surrounding the cleft (below the ribbons).

cerebral hemispheres are connected by the *corpus callosum,* a band of ~2×10^8 nerve fibers. Remarkably, these fibers can be completely severed with a relatively minimal disruption of the nervous system. In the past the corpus callosum was sometimes cut to control almost incessant epileptic seizures that could not be prevented by drugs. The "split-brain" patients suffered relatively little disability as long as both eyes functioned normally. Studies of these patients provided some insights into the differing functions of the two hemispheres of the cerebrum.

Deeper in the cerebrum lie the *basal ganglia,* which include the caudate, lenticular, and amygdaloid nuclei. The lenticular nuclei are further divided into putamen (an outer portion) and the globus pallidus. The putamen and caudate nuclei together are known as the *striatum* (Fig. 7.12). The lower lying subthalamic nuclei and substantia nigra are sometimes also included in the basal ganglia.

The outer parts of the cerebrum, including the basal ganglia, make up the telencephalon. Deep in the center of the brain is the diencephalon consisting of the *thalamus* (actually two thalami), *hypothalamus, hypophysis* (Figs. 7.1, 7.13), and other attached regions. A major structure at the back of the brain is the *cerebellum.* Like the cerebrum, its cortex is highly folded. The 30 billion neurons of the cerebellum are organized in a highly regular fashion. The interconnections of the seven types of neurons present in this part of the brain have been worked out in fine detail.

The basal part of the brain or *brain stem* consists of the medulla oblongata and the pons. While the bulk of the tissue consists of myelinated nerve tracts passing into the spinal cord, synaptic regions such as the olivary nucleus are also present.

The brain, which must function in a chemically stable environment, is protected by a tough outer covering, the *arachnoid membrane,* and by the *bloodbrain barrier* and the *blood-cerebrospinal barrier.* Both of these barriers consist of tight junctions similar to those seen in Fig. 7.15 A. They are formed between the endothelial cells of the cerebral capillaries and between the epithelial cells that surround the capillaries of the *choroid plexus.* The choroid plexus consists of capillary beds around portions of the fluid-filled *ventricles* deep in the interior of the brain. They serve as a kind of "kidney" for the brain assisting in bringing nutrients in from the blood and helping to keep dangerous compounds out.

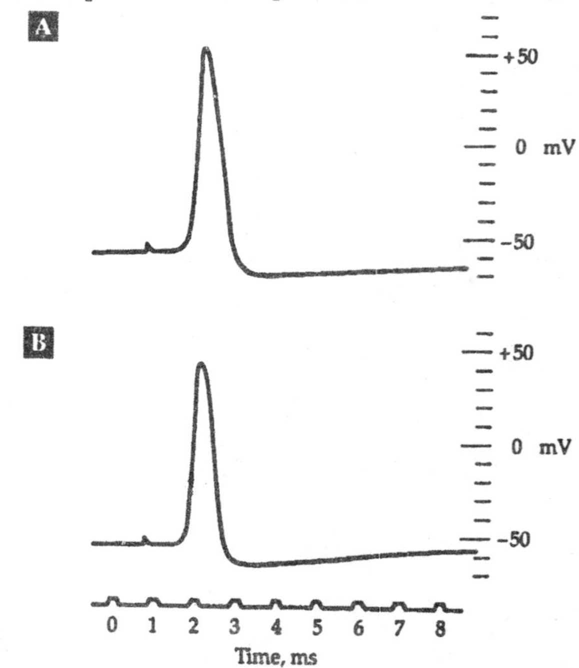

Fig. 7.11. (A) Action potential recorded with internal electrode from extruded axon filled with potassium sulfate (16° C). (B) Action potential of an intact axon, with same amplification and time scale (18° C). The voltage scale gives the potential of the internal electrode relative to its potential in the external solution with no correction for junction potential.

3. Neuronal Pathways and Systems

Consider a message originating with a nerve receptor in the skin or in another sense organ. A nerve signal passes via a *sensory neuron* (*afferent fiber*) upward toward the brain. It may pass through two or more synapses (often through one in the spinal cord and one in the thalamus) finally reaching a spot in the sensory region of the cerebral cortex. From there the signal in modified form spreads through the *inter-neurons* of virtually the entire cortex. In each synapse, as well as in the cortex, the impulse excites inhibitory fibers that dampen impulses flowing through adjacent fibers. Likewise, if a given impulse is not strong enough, it will itself be inhibited before reaching the cortex. Among the important sensory neurons are those from the seven million cone cells and 100 million rod cells of the eye. The nerve signals pass out of the retina by way of a million axons from retinal ganglion cells reaching, among other parts of the brain, the *visual cortex* (Fig. 7.14).

The neuronal events that occur within the cerebral cortex are extraordinarily complex and little understood. In what way the brain is able to initiate voluntary movement of muscles is obscure. However, it is established that the signals that travel out of the brain down the *efferent fibers* to the muscles arise from *large motor neurons* of the *motor cortex,* a region that extends in a band across the brain and adjacent to the sensory cortex (Fig. 7.14). The axons of these cells form the *pyramidal tract* that carries impulses downward to synapses in the spinal cord and from there to the *neuromuscular junctions.* These are specialized synapses at which acetycholine is released, carrying the signal to the muscle fibers themselves. Passing over the cell surface and into the T tubules, a wave of depolarization initiates the release of calcium and muscular contraction.

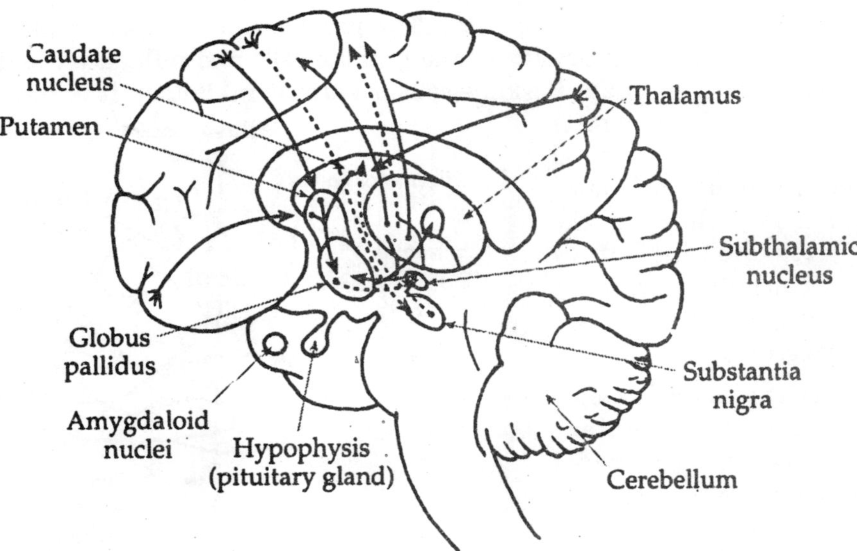

Fig. 7.12. Diagram illustrating some of the major interconnections of the "extrapyramidal system" of the brain. Arrows indicate major direction of projections. The nigrostriatal (substantia nigra to striatum) and related neuronal pathways are indicated with dashed lines.

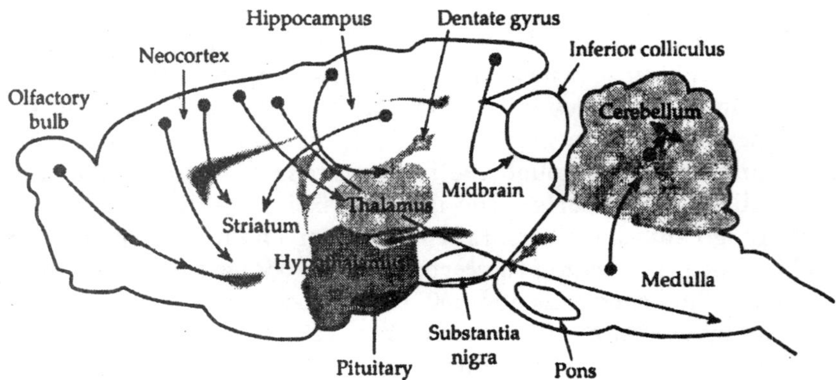

Fig. 7.13. Section through a rat brain. This brain, which has been very widely used in neurochemical studies, appears superficially to be quite different from the human brain, which is characterized by its large cerebral cortex. However, basic pathways are the same. Some major pathways for glutamate-secreting (glutamatergic) neurons are marked by arrows. Most of these originate in the neocortex (outer layers of the cerebral cortex) and the hippocampus.

At the same time that the motor neurons send signals to the muscles, branches travel into other parts of the brain including the olivary nuclei, which send neurons into the cerebellum. The cerebellum acts as a kind of computer needed for fine tuning of the impulses to the muscles. Injury to the cerebellum leads to difficulty in finely coordinated motions. Input to the Purkinje cells arises from the climbing fibers, which originate in the inferior olive of the brain stem. Each climbing fiber activates a single Purkinje cell, but the dendrites of each Purkinje cell also form as many as 200,000 different synapses with parallel fibers that run across the cortex of the cerebellum (Fig. 7.15). The parallel fibers receive input from many sources via a complex series of mossy fibers and granule cells and influence the firing of the Punkinje cells. The output from the Purkinje cells is entirely inhibitory. It is transmitted via synapses in the cerebellar nuclei to neurons that lead back to the cerebral cortex, into the thalamus, and down the spinal cord. The pathway to the cortex completes an inhibitory feedback loop, of which there are many in the nervous system. For details see Llinas and Nicholls.

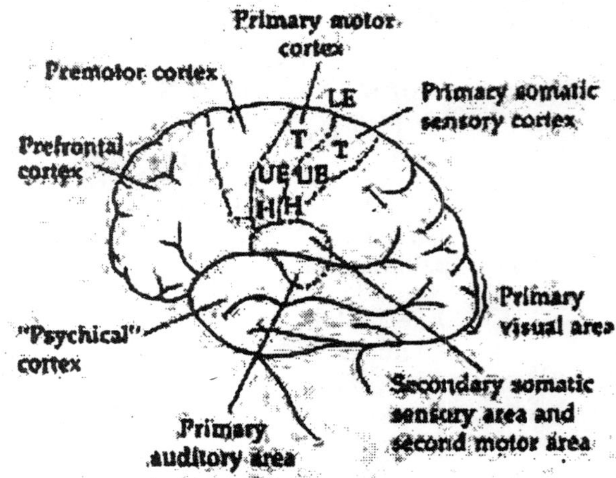

Fig. 7.14. The location of several functional areas of the cerebral cortex. The representation of body parts on the primary motor and somatic sensory cortices include the head (H), upper extremity (UE), trunk (T), and lower extremity (LE)..

In addition to the *somatic motor system* that operates the voluntary (striated) muscles via the pyramidal tract, there is the *autonomic system,* which controls the involuntary (smooth) muscles, glands, heartbeat, blood pressure, and body temperature. This system has its origins in both the cerebral cortex and hypothalamus. It is subdivided into two systems, the *sympathetic* and *parasympathetic* systems, which are anatomically distinct. The sympathetic system is geared to the fight and fright reactions. Its *postganglionic fibers* (those below the ganglia in the spinal cord) liberate norephinephrine (noradrenaline) and include the adrenal medulla, which consists of specialized neurons, the *chromaffin* cells. The parasympathetic system has to do more with homeostasis and maintenance of body systems. Biochemically it is characterized by the release of acetylcholine as a transmitter substance.

The hypothalamus, a four gram portion of tine brain, receives a great deal of biochemical attention because of its function in the autonomic nervous system, in homeostasis, and in endocrine secretion. Its liberation of neurohormonhes that stimulate the hypophysis has already been considered. The hypothalamus is also involved in the regulation of the body temperature, of water balance, and possibly of glucose concentration.

Two other systems of importance in the brain are the *reticular system* and the *limbic system.* The former is the mediator of the sleep-wake cycle and is responsible for characteristic waves in the electroencephalogram. The limbic system is the mediator of *affect* or mood and of *instincts.* It is anatomically complex with centers in the amygdala, other subcortical nuclei, and the limbic lobe of the cortex. The limbic cortex forms a ring lying largely within the longitudinal fissure between the two hemispheres. It includes the olfactory cortex, the *hippocampus,* a region associated with formation of conscious memories, and other evolutionarily older regions of the cerebral cortex. Within the limbic lobe are the *pleasure centers.* When electrodes are implanted in these regions, animals will repeatedly push levers that are designed to electrically stimulate these centers. There are also *punishing centers,* whose stimulation causes animals to avoid further stimulation.

4. The Propagation of Nerve Impulses

Although the chemical basis of the conduction of nerve impulses via an action potential is not entirely clear, the electrical events have been described with precision. If the permeability of a membrane toward sodium ions is increased in a local region, sodium ions flow through the membrane into the cell neutralizing the negative charge inside and depolarizing the membrane. Such depolarization leads to propagation of an electrical signal of diminishing intensity over the surface of the membrane in a manner analogous to the flow of electrical current along a coaxial cable. It is thought that local increases in Na^+ permeability of the plasma membrane often trigger nerve impulses. Other ions such as Ca^{2+} may also play a role. While the kind of passive transmission of electrical signals that results from a local depolarization of the membrane is suitable for very short nerve cells, it cannot be used to send signals for long distances. Most nerve axons employ the more efficient action potential. This is an impulse that passes along the axon and for a short fraction of a second (~0.5 ms in mammalian nerves) changes the membrane potential in the characteristic way shown in Fig. 7.11. Initially, the negative potential of 50 – 70 mV drops rapidly to zero and then becomes positive by as much as 40 – 50 mV, after which it returns to the resting potential. The remarkable thing about the action potential is that it is propagated down the axons at velocities of 1 – 100 m/s without loss of intensity.

To establish the chemical basis of the action potential, A. L. Hodgkin and A. F. Huxley in the 1950s devised the *voltage clamp,* a sophisticated device by which the transmembrane current

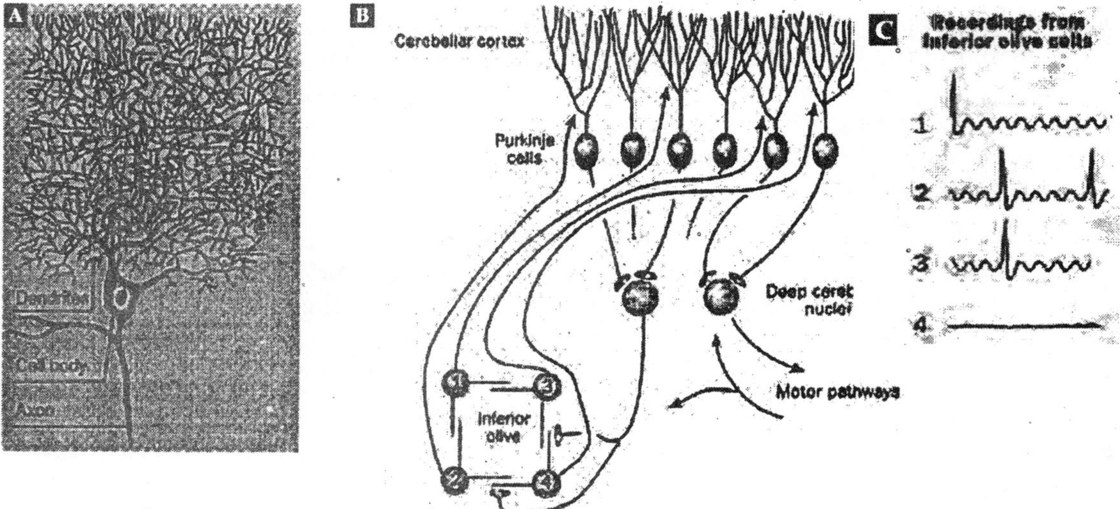

Fig. 7.15. (A) Diagram of the two-dimensional tree formed by dendrites of a single Purkinje cell of the cerebellum. From Llinas. (B) Schematic diagram showing input and output pathways for Purkinje cells. (C) Recordings of output from four different neurons of the inferior olive. These action potentials are thought to arise from oscillations that arise within the neurons or within arrays of adjacent neurons coupled by electrical (gap junction) synapses. These oscillations synchronize the generation of action potentials so that some cells oscillate in synchrony while others (*e.g.*, cell 4 above) do not.

can be measured while using a feedback mechanism to fix the membrane potential at a preselected value. Using the voltage clamp the membrane conductance could be measured as a function of the membrane potential and of time. It was found that immediately after a decrease in membrane potential was imposed with the voltage clamp, the permeability of the membrane toward sodium ions rose rapidly. Since an increased sodium ion permeability automatically leads to depolarization in an adjacent region of the membrane, a self-propagating wave is established and moves down the axon. The voltage clamp studies also revealed that after a fraction of a millisecond the permeability to potassium ions also increases. At the same time the sodium ion permeability decreases again, and the normal membrane potential is soon reestablished. However, during an *absolute refractory period* of ~0.5 ms no other nerve impulse can be passed. The sequence of events during passage of the nerve impulse can be described as the opening of sodium channels followed by the opening of potassium channels, and then by a closing of the channels in the same sequence. The results of these investigations led Hodgkin and Huxley to propose equations that quantitatively describe the action potential and that predict the observed conduction velocities and other features of nerve impulses.

A special feature of nerves that are designed to transmit impulses very rapidly is the presence of the wrapping of *myelin*. As can be seen in this figure, the extracellular surfaces of the consecutive wraps bind tightly together, and the cytoplasm of the cell interior is squeezed out to form the compact myelin sheath. Mutations in the integral membrane proteolipid protein are associated with a variety of defects in myelin formation. Some of these are severe, for example, leading to loosely wrapped myelin. The proteolipid protein is encoded by an X-linked gene. The most abundant protein in peripheral nerve myelin is the integral membrane *peripheral myelin glycoprotein* P_0. It is encoded by an autosomal gene for which 29 known defects account for a variety of human diseases, including an autoimmune inner ear disease. The extracellular domain of P_0, like many other cell adhesion molecules, has a structure related to that of immunoglobulins. Four molecules of P_0, each of which carries a single immunoglobulin domain, associate via these domains in a kind of square donut that protrudes from the outer cell surface. There it can interact with four similar donuts from the apposed cell surface, zipping up the cell-cell interface by a kind of Velcro action. Protein P_0 accounts for 50% of the total protein of peripheral myelin, but the *myelin basic protein,* which constitutes 20% of the total protein, is also essential. This protein exists as a variety of forms that arise from differential splicing of its mRNA and extensive posttranslational modification. Deimination of argine side chains to form citrulline residues has been associated with development of the autoimmune disease *multiple sclerosis. Peripheral myosirt protein* 22 is a 160-residue polypeptide with four membrane-spanning helices. It accounts for 2 – 5% of the myelin protein and is the site of defects that cause the demyelinating *Charcot-Marie-Tooth disease* and other serious human diseases.

The axon is effectively insulated from the surrounding medium by the myelin sheets except for special regions, the *nodes of Ranvier,* which lie at 1-to 2-mm intervals along the nerve. The nerve impulse in effect jumps from one nerve to the next. This *saltatory conduction* occurs much more rapidly (up to 100 m/s) than conduction in unmyelinated axorts. It depends upon Na^+ and K^+ channels that are concentrated in the nodes of Ranvier.

5. Ion Conducting Channels

What is known about the channels through which Na^+ and K^+ flow during nerve excitation? That the channels for the two ions are separate was shown by the fact that *tetrodotoxin* (found in the puffer fish) and *saxitoxin* of dinoflagellates, as well as scorpion toxins, exert their toxic action by blocking the Na^+ channels while having no effect upon conductance for K^+. At the same time the K^+ channels can be blocked by certain quaternary ammonium salts. Since the binding constants for the toxins are high (K_f ~3 × 10^8 M^{-1} for tetrodotoxin), it is possible to titrate the sodium channels. The number is usually quite small, about 10 – 400 Na^+ channels/μm^2 of surface (the same surface area contains 2 × 10^6 phopholipid molecules). However, membranes in the nodes of Ranvier of mammalian nerve fibers contain ~12,000 channels/μm^2. Note that the ion channels described here are not the same as those in the ion pump, *i.e.,* the Na^+, K^+-ATPase (Fig. 7.25). In some neurons the number of conduction channels for Na^+ appears to be ten times less than the number of pumping channels, *i.e.,* of Na^+,K^+-ATPase.

Since the number of ion-conducting channels is small, the rate of sodium passage through the open channels must be extremely rapid and has been estimated as ~10^8 ions/s. This is within an order of magnitude of the diffusion-limited rate. On this basis it is clear that the channels cannot act by means of ionophoric carriers but form pores that can be opened and closed (*gated*) in response to changes in the membrane potential. They are *voltage-sensitive ion channels.* The channels are selective for specific ions and the selectivity parallels that of sites in some cation exchange resins such as those containing carboxylate groups. This suggested that the inside surface of the channel might contain one or more carboxylate groups from protein side chains as well as other polar groups. A Na^+ ion approaching the channel entrance might exchange some of its hydration sphere for ligands from the channel surface. The differing affinity of the "ion exchange" sites for various cations could ensure that it is predominately Na^+ that passes through the channel. Anions could be excluded by electrostatic repulsion. Recent structural studies have allowed these speculations to be replaced with experimental findings as described in the following paragraphs. They have revealed that the selectivity mechanism are similar for Na^+ and Ca^{2+} channels.

The sodium ion channel of the electric eel

Making use of the binding of radioactively labeled specific toxins to identify them, the subunits of the sodium channel proteins were purified from several sources including the electrical tissue of the electric eel *Electrophorus electricus*, heart and skeletal muscle, and brain. In all cases a large ~260-kDa glyco-protein, which may be 30% carbohydrate, is present. The saxitoxin-binding protein from rat brain has two additional 33-36 kDa subunits with a stoichiometry of $\alpha\beta_1\beta_2$. The *Electrophorus* a subunit consists of 1820 residues, while rat brain contains α proteins of 2009 and 2005 residues, respectively, for Na^+ channels designated I and II. In fact, mammals contain ten distinct Na^+ channel genes. In every case the channel proteins contain four consecutive homologous sequences of about 300 residues apiece. Within these the hydropathy plots (see Fig. 7.30) suggest that each homology region forms six membrane-spanning helices as shown in Fig. 7.17A. The four sets may then fold together into a square arrangement that provides a pore somewhat familar to that of the voltage-gated K^+ channel (see Fig. 7.18). The three-dimensional structure of the sodium channel protein, based on cry-electron microscopy, appears to be complex. The central channel may resemble that of Fig. 7.18, but there also seem to be smaller peripheral channels (Fig. 7.17). Bacteria also contain Na^+ channels but they are tetramers of smaller subunits, resembling in this respect bacterial K^+ channels (Figs. 7.18).

How do the "gates" to ion channels open? Presumably some part of the channel protein senses the change in potential and undergoes an appropriate alteration in conformation that opens the gate. The current carried by the ions flowing out through a small number or even a single channel can be measured with tiny *patch electrodes* having openings ~1 μm^2 in area. These are pressed against the nerve membrane, where they form a tight seal. With such a small patch of membrane surface the electrical noise level is low, and it is possible to measure the conductance of the pore. From such measurements it was found that a single pore can allow $>10^8$ ions to pass through in one second. Another thing that is apparently measured with patch electrodes is a small *gating current,* which precedes the opening of the channels by ~0.1 ms. This has been interpreted as a flow of ~6 charges across the membrane or the movement of a larger number of dipoles needed to open the gate. One possibility is that a loss of the electrical field from the surface charges on the bilayer induces a rearrangement of charges on protein side chains within the bilayer or induces changes in interactions between two or more dipoles. Such changes could trigger conformation alterations within the proteins, allowing the channel to switch from open to closed.

Recordings with single channels indicate that after a sodium channel is open for a random length of time it spontaneously closes and passes into a third state, an "inactive" state from which it cannot reopen during the refractory period. After the membrane is repolarized it can function again.

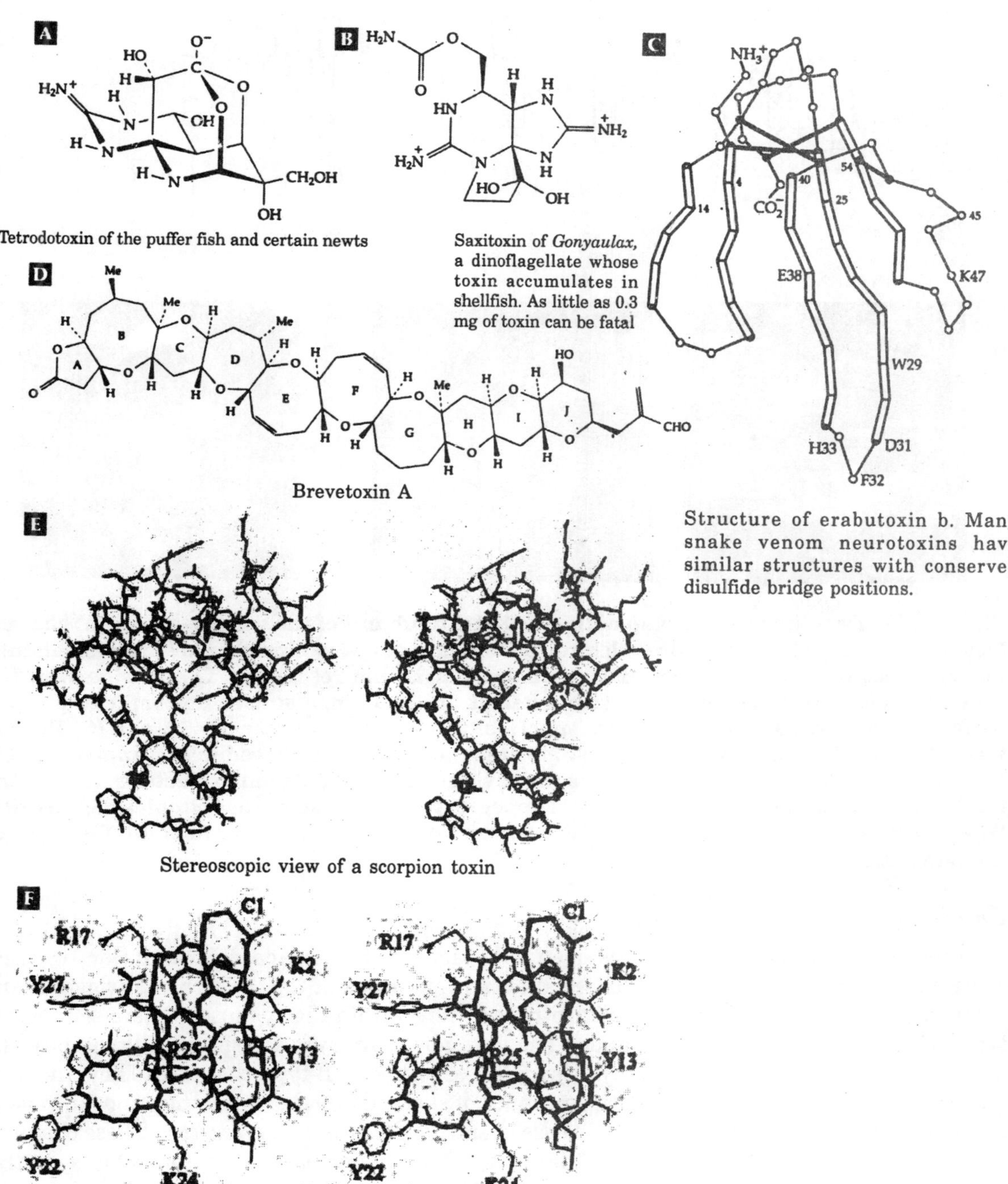

Tetrodotoxin of the puffer fish and certain newts

Saxitoxin of *Gonyaulax*, a dinoflagellate whose toxin accumulates in shellfish. As little as 0.3 mg of toxin can be fatal

Structure of erabutoxin b. Many snake venom neurotoxins have similar structures with conserved disulfide bridge positions.

Brevetoxin A

Stereoscopic view of a scorpion toxin

ω Conotoxin, blocks presynaptic voltage-regulated Ca^{2+} channels

Fig. 7.16. Structures of some neurotoxins that affect ion channels. Other neurotoxins include the Na^+, K^+-ATPase inhibitor ouabain, batrachotoxin, and picrotoxin.

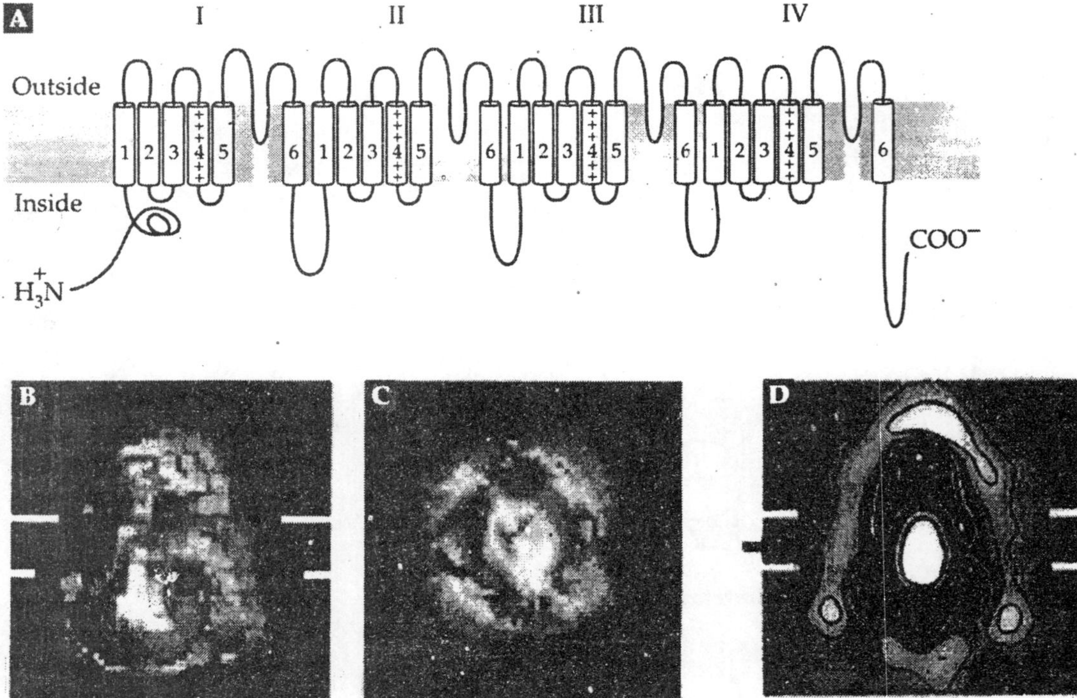

Fig. 7.17. (A) Two-dimensional map of the ~260-kDa α subunit of the voltage-gated Na^+ channel from the electric eel *Electrophorus* electron's. (B) Image of the sodium channel protein obtained by cryo-electron microscopy and image analysis at 1.9 nm resolution. In this side view the protein appears to be bell-shaped with a height of ~13.5 nm, a square bottom (cytoplasmic surface) ~10 nm on a side, and a hemispherical top with a diameter of ~6.5 nm. (C) Bottom view of the protein. (D) Axial section which cuts the bottom, as viewed in (C), approximately along a diagonal. From Sato *et al.* Notice the cavities (dark) and domain structures (light). The black arrow marks a constriction between upper (extracellar) and lower (cytoplasmic) cavities. White lines indicate approximate position of the lipid bilayer. From Sato *et. al.* Courtesy of Chikara Sato.

Calcium ion Channels

Immediately after the Na^+ pores open as a result of membrane depolarization, voltage-sensitive Ca^{2+} channels also open. These allow a rapid influx of Ca^{2+}, which can trigger many processes including the secretion of neurotransmitters within the synapses. There are several types of voltage-sensitive Ca^{2+} channels. The most abundant type are specifically inhibited by dihydropyridines and are called *dihydropyridine-sensitive* or L-type channels. They are most numerous in the transverse tubular membranes of skeletal muscle where they appear to form a complex with the very large calcium release channels, the *ryanodine receptors*. These channels appear to have a structure similar to that of the Na^+ channels. Calcium ions play a central role in cell signaling and there are a large number of different calcium channels in bacteria, plants, and animals. Many of these are coupled to specific receptors. Some are involved in controlling intracellular stores. Some release Ca^{2+} in response to mechanical movement and function in feeling, hearing, maintaining balance, and cardiovascular regulation. Plants sense wind and gravity, and microorganisms sense changes in osmotic pressure with the aid of these channel proteins.

Potassium Ion Channels

Several types of K^+-selective cation channels have been recognized on the basis of electrophysiological and pharmacological studies. More recently, the cloning of channel genes has permitted the study of the proteins by X-ray crystallography. The first structure determined was that of the *Streptococcus lividans* K^+ channel (designated KcsA; Fig. 7.21). There are three large structural families of K^+ channel proteins. One group consists of voltage-regulated (K_v) channels, such as those involved in the action potential of neurons. Like the *S. lividnns* channel, they are tetramers whose predicted structure contains six transmembrane helices per subunit with a pore-forming loop (P region) between helices 5 and 6. This is just what is seen in the *S. lividans* channel and in one-fourth of the much larger Na^+ channel protein (Fig. 7.17A). Furthermore, all known potassium channels, from bacteria to human beings, have the conserved sequence GYGD in the C-terminal half of the P region. A great variety of K_v channels are known. There are ~70 genes for these channels in the *Caenorhabditis elegans* ge-nome. One of the first K_v channel genes to be cloned was from a *Drosophilia* mutant known for its neurological defect as *shaker.* Its structure (Fig. 7.18), which is based in part on modeling from the KcsA channel, has the ion selective filter with the conserved sequence *TVGYG* in the expected location. At the cytoplasmic end of the pore is an additional structure not found in the KcsA channel. This is the *inactivation gate,* so called because it accomplished the rapid self-inactivation of the K^+ channels during passage of the action potential (Fig. 7.18A). This is one of the factors necessary for recovery and repolarization of the axon membrane. The inactivation gate is composed of N-terminal ~130 residue "T1" domains of the α subunits together with parts of the β subunits, which are associated as a tetramer beneath the channel in the cytoplasm (Fig. 7.18B). Various experimental data including mutational analysis suggest that small ball-like domains at the N termini of the β subunits block the channel. Zhou *et. al.*, propose that the N termini unfold into an extended conformation, passing through "windows" between the T1 domains and the channel and allowing the $-NH_3^+$ ends to bind into the central cavity in the channel. The same site can be blocked by well-known quaternary amine inhibitors such as tetraethylammonium, tetrabutylammonium ions, or tetrabutylantimony, an analog used for X-ray crystallography.

The T1 domain of the channel not only participates in control of the ion flux but also stabilizes the pore complex. Among the various K^+, Ca^{2+}, and Na^+ channels the regulatory β subunits are quite variable in their structures and mechanisms of gating. Some β subunits have bound NADH. A speculative possibility is that the rapid interconversion of the positively charged thiazolium ion and negatively charged thiolate ion forms of thiamin plays some role in nerve conduction, *e.g.,* voltage sensing.

Some questions about ion channels have been hard to answer. For example, how are small cations allowed to flow rapidly through a very small opening in a 2 – 3 nm thick nonpolar core of a membrane? From basic electrostatic principles ΔG for transfer of an ion to the center of a membrane has been estimated as –160 kJ/mol, a high thermodynamic barrier to transport. A solution to this problem apparently lies partly in the fact that at the center of the lipid bilayer the ion channel contains a cavity large enough (~0.5 nm diameter) to hold about 50 water molecules. Cations tend to enter this cavity, and X-ray studies have shown that the electrondense Rb^+ does occupy the cavity. A second stabilizing factor is that four helices have their negative (C-terminal) ends pointing toward the cavity. Although the electrostatic effect of these helix dipoles (Fig. 7.20 A) might be regarded as negligible, computations indicate that within the low dielectric bilayer the stabilizing effect of the helices becomes significant.

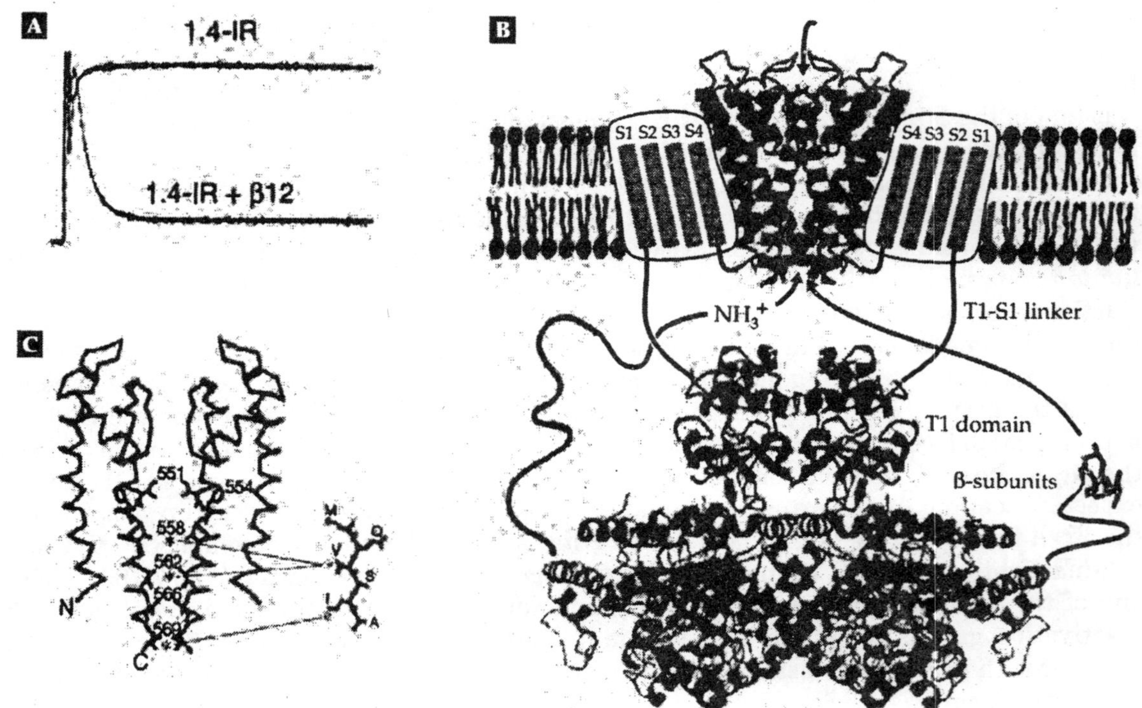

Fig. 7.18. (A) K^+ currents recorded from *Xenopus laevus* oocytes carrying cloned genes of *Drosophila* shaker K^+ channels under two-electrode voltage-clamp conditions. Trace 1.4-IR was obtained from a cell expressing channels that lack the inactivation gate. Trace 1.4-IR + β_{12}, obtained from a cell expressing β subunits as well, shows rapid self-inactivation. (B) Composite model of a voltage-dependent K^+ channel. The pore structure in the α subunit is represented by the KcsA channel. The structure of the T1-β complex is from Gulbis *et. al.* The drawing is modified from that of Zhou. (C) Ball-and-stick view of the selectivity filter showing positions of four bound K^+ ions. Two of the four TVGYG peptide strands of the conduction pore are shown.

How are the pores in these channels opened and closed? Different channels are gated in different ways. The KcsA channel, which is mostly closed at neutral pH, responds by opening at a low external pH. Using methods of spin labeling and EPR spectroscopy, Perozo *et. al.*, found small translational and rotational movements of the helices that form the pore (Fig. 7.18). These may alter the diameter of the pore, opening or closing it. How do the electrostatic sensors control the process? The details are uncertain, but the sensor is thought to lie in a conserved sequence of arginine and lysine residues interspersed with hydrophobic amino acids in transmembrane helix 4 of the channel protein. How do potassium pores select K^+ over Na^+ or Ca^{2+}? One factor is that Na^+ is more heavily hydrated than K^+. This allows K^+ to pass through the channel more readily than Na^+. Potassium ions travel through the 1.2-nm-long selectivity filter at a rate of ~10^8 s^{-1} in consecutive steps of dehydration, movement, and rehydration occurring in ~10 ns. The process is catalyzed by polypeptides and may depend upon competition between a state in which a ring of four hydrogen-bonded peptide groups is formed and a state in which the four carbonyl groups coordinate a K^+ ion.

Belonging to the same structural group as the K_v channels are Ca^{2+}-regulated K^+ channels. Some bacterial channels are controlled by binding of Ca^{2+} ions to a "gating ring" on the intracellular membrane surface. A mammalian channel is controlled by a complex of calmodulin with the intracellular end of the α subunits of the channel and others. A second large group of K^+ channels, containing seven subfamilies, are the *inward rectifying* (Kir) channels. They are tetramers of 360- to 500-residue polypeptide chains, each chain forming two trans-membrane helices with a P region between them. These channels support a large conductance when K^+ ions flow out from a cell but only a small conductance when they flow in. Kir channels are subject to a variety of controls, which include effects of pH.

Some are inhibited by ATP, and others by eicosanoids or inositol hexaphosphate. Some of the ATP-sensitive channels contain an ABC transporter subunit and are binding sites for sulfonylureas and other drugs. A number of human disorders in Kir channels have been identified. The human Kir channels participate in regulation of resting membrane potentials in K^+ homeostasis, control of heart rate, and hormone secretion. A third group of K^+ channels are dimeric, but each subunit contains two tandem P regions and 4-8 transmembrane helices.

Chloride Channels and the Ionic Environment of Neurons

All cells contain voltage-gated chloride channels, which are encoded by the *Clc* genes. Recently crystal structures have revealed chloride channels formed in single polypeptide chains arranged as dimers. The selectivity filter involves stabilization by the positive ends of α-helix dipoles. The importance of the corresponding proteins to the human body is shown by the existence of several specific diseases arising from mutations in their genes. A calcium-regulated Cl^- channel is also present as is the ATP-gated CFTR channel. In addition, other ligand-gated Cl^- channels, such as γ-aminobutyrate receptor channels, are found in the central nervous system. A glutamate-gated chloride channel in invertebrate organisms is the site of action of the antihelminthic and insecticidal compound *ivermectin.*

The significance of ion channels can be better appreciated by considering the ionic environment of nerve axons. Mammalian neurons have roughly the following millimolar concentrations of ions in the cytosol and in the external medium. The concentration gradients for the much-studied squid axon are substantially higher. The membrane potentials that could arise from each one of these concentrations, according to, are also given. In a resting

	Cytosol	*Extracellular*	E_m *(mV)*
K^+	150	5.5	–90
Na^+	15	150	+60
Ca^{2+}	10^{-4}	1.5	+270
Cl^-	9	12.5	-70

neuron the K^+ potential dominates with an observed membrane potential of ~–80 mV. Some K^+ channels are open and the K^+ and $C1^-$ concentrations are nearly in Donnan equilibrium across the membrane. The Na^+ and Ca^{2+} channels are closed, and the sodium and calcium pumps keep the internal concentrations of these ions low.

When an action potential is propagated, a wave of depolarization moves along the axon, changing the membrane potential suddenly to a less negative value. When it reaches ~50 mV the Na^+ channels open, allowing sodium ions to flow into the cell causing further propagation of the wave of depolarization. After ~1 – 2 ms the Na^+ channels begin to deactivate. At the same time the slower K^+ channels open allowing potassium ions to flow out and to repolarize the membrane, the membrane potential sometimes transiently reaching more negative values (hyperpolarization) than the ~80 mV resting potential. Action of the Na^+, K^+-ATPase then restores the original state. The finely tuned properties and sequential opening and closing of the channel proteins are essential to the conduction of nerve impulses.

The existence of voltage-gated ion channels in bilayers are not limited to nerve membranes. They are present to some extent in all cell membranes. Even the paramecium has at least seven kinds of Na^+, K^+, and Ca^{2+} channels. Channels may also be formed by many peptide antibiotics. Among them are the human defensins and the ~20-residue *alamethicin.* Six to eleven of the mostly helical monomers of that antibiotic assemble to form a single voltage-dependent channel. The bacterial toxin colicin E1forms voltage-dependent channels within bacterial membranes.

Receptor-associated Ion Channels

Many neurotransmitters, including acetylcholine and glutamate, act to open ion channels that are part of the receptor protein or of a tight complex of proteins. Such *ionotropic receptors* are

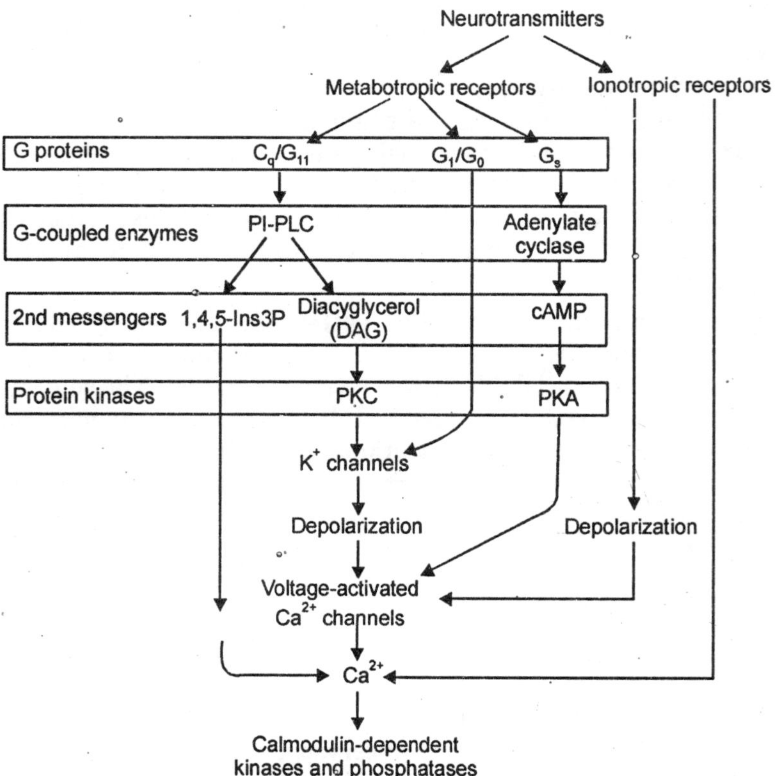

Fig. 7.19. Major signaling pathways from metabotropic and ionotropic receptors in neurons. Various G proteins control the signaling from mutabotropic receptors using phosphatidylinisitol-specific phospholipase C (PI-PLC) and adenylate cyclase or acting directly on K^+ ion channels.

responsible for most rapid neuronal action. For example, binding of acetylcholine to its receptor in the neuromuscular junction causes the release of Ca^{2+} ions from the exterior into the muscle fibers. Binding of glutamate to its ionotropic receptor in a synaptic ending of a dendrite causes an influx of ions into the cytoplasm, initiating an action potential in the dendrite. In most instances the properties of the receptor channel favor the rapid flow of Ca^{2+} ions into the cytoplasm.

Many other receptors are 7-helix transmembrane proteins, which activate guanine nucleotide G proteins. The G proteins couple some receptors directly to Ca^{2+} channels; they couple other receptors to adenylate cyclase and cyclic AMP-activated channels and yet others via phospholipase C to K^+ channels and indirectly to Ca^{2+} channels. All of these G protein coupled receptors are referred to as *metabotropic receptors.* A single synapse often contains both ionotropic receptors and metabotropic receptors. The ionotropic receptors induce a rapid (<1 ms) response, while the metabotropic receptors act more slowly. However, in most cases the final effect is the release of calcium ions into the cytoplasm. The rapid response may be initiation of an action potential, while the slow response may be activation of calmodulin-dependent kinases and phosphatases.

6. A Plethora of Neurotoxins

Bacteria, protozoa, and venomous animals synthesize numerous toxins that are used to kill their prey or to defend themselves. Sea anemones, jellyfish, cone snails, insects, spiders, scorpions, and snakes all make potent and highly specific neurotoxins. Plants form a host of alkaloids and other specialized products, some of which are specifically neurotoxic and able to deter predators. More than 500 species of marine cone snails of the genus *Conus* synthesize a vast array of polypeptide toxins (conotoxins), some with unusual posttranslational modifications. The slow-moving snails are voracious predators that use their toxins, which they inject with a disposible harpoon-like tooth, to paralyze fish, molluscs, or worms.

The targets for natural biological toxins include ion channels and receptors for transmitters. At least four parts of the voltage-gated sodium channels are binding sites for extremely toxic natural products. *Tetrodotoxin* (Fig. 7.16), which is found in the puffer fish, certain newts, and venom of the blue-ringed octopus, and also the shellfish poison *saxitoxin* (Fig. 7.16) block the entry of sodium ions into the channels. *Bactrachotoxin* (Fig. 7.12) and related lipophilic compounds such as *veratridine increase* sodium permeability by blocking the channels permanently open. *Pyrethroid insecticides* prolong the time that the sodium channels stay open after excitation. Some *scorpion toxins* (Fig. 7.16), which all have a hydrophobic core made from a short α helix and a three-strand antiparallel β sheet, and *sea anemone toxins* also the open conformation of the Na^+ channels. Other smaller ~4-kDa scorpion toxins block K^+ or Cl^- channels or other receptors. Some are most toxic to insects and others to mammals. Although their three-dimensional structures resemble those of scorpion toxins, the amino acid sequences of anemone toxins show no homology. The most potent poison produced by the red tide organism, the dinoflagellate *Gymnodinium breve* (Fig. 7.9), *is brevetoxin A* (Fig. 7.16). It selectively opens one class of sodium channels.

Venoms of *cobras, sea snakes,* and pit vipers contain several 6- to 7-kDa proteins that bind to acetylcholine receptors (Fig. 7.23) of the postsynaptic neurons, preventing binding of the neurotransmitter and opening of the ion channels. All of these toxins contain four disulfide bridges and share with certain plant proteins a folding pattern that has been called the *toxin-*

agglutinin fold (Fig. 7.16). These toxins include *erabutoxin a* (Fig. 7.16) from a sea snake as well as the 74-residue toxin *bungaratoxin a* (from the banded Krait). This toxin, which has been used to titrate acetylcholine receptors in neuromuscular junctions, is a member of the *long neurotoxin* group, which contains 71 – 74 residues and five disulfide bonds. Other *short neurotoxins* are 60 – 62 residues in length with four disulfide bridges. Cobra toxins contain both neurotoxins and *cardiotoxins,* which have somewhat similar structures but quite different modes of action. In contrast, *crotoxin* from the venom of a South American rattlesnake and *β-bungarotoxin* consist of 13-kDa phospholipases A_2 complexed with smaller 7.5-kDa proteins. They act at the presynaptic membranes of selected neurons by blocking neurotransmitter release.

The seven types of *botulinum toxin* and the *tetanus toxin* are the most neurotoxic substances known. Only 10^8 molecules are sufficient to kill a mouse. Both toxins are zinc proteases, which block presynaptic transmitter release by cleaving specific synaptic vesicles proteins. They bind initially to ganglioside in the neuromuscular junction, one subunit then being internalized as with the diphtheria toxin. Botulinum toxins specifically enter motor neurons, while tetanus toxin is taken up via synaptic vesicle endocytosis by both peripheral and central neurons. Retrograde axonal transport carries the toxin into the central nervous system and across synaptic clefts into cholinergic interneurons, which are poisoned.

The black widow spider produces the 130-kDa *α-lactrotoxin,* which causes massive release of acetylcholine, norepinephrine, dopamine, and GABA from synaptosomal endings. The small *anatoxin-a* or "very fast death factor" (Fig. 7.22), which is synthesized by various cyanobacteria, antagonizes both muscarinic and nicotinic acetylcholine receptors. Cone snails synthesize mixtures of the 13- to 17-residue conotoxins (Fig. 7.16). They cause rapid paralysis of fish permitting the snails to prey on the much faster fish. They bind to a variety of targets, which include Na^+, K^+, and Ca^{2+} channels, and acetylcholine, and glutamate receptors. One of the toxins is a 17-residue peptide containing five residues of γ–carboxyglutamate and is also notable for the fact that intercerebral injection of less than one microgram of the toxin induces a prolonged sleeplike state in mice. The venom of *Conus geographicus* is so toxic that two-thirds of human stinging cases are fatal.

The most deadly nonproteinaceous toxin known, *palytoxin,* is also the most complex structure ever established without the aid of X-ray crystallography. It is produced by marine zoanthids of the genus *Palythoa* and has the molecular formula $C_{129} H_{223} N_3 O_{54}$.

7. Neuronal Metabolism

The brain has a very high rate of metabolism. Although accounting for only 1/50 of the body mass its utilization of energy amounts to 1/5 of the basal metabolism. This is ~20 watts and is nearly constant day and night. It reflects the unusually active metabolism of neurons, a major part of which can be attributed to the sodium–potassium ion pumps in the membranes and to the maintenance of the excitable state. The source of energy for these processes is the ATP that is utilized to drive the ion pumps and thereby to maintain the membrane potential needed to drive the action potentials. The ATP is formed largely by oxidative metabolism of glucose and, to a lesser extent, of acetoacetate. The large surface area of the axons as well as the frequency with which they transmit nerve impulses accounts for the high rate of metabolism.

Another factor peculiar to neurons doubtless contributes also to their rapid metabolism. The nucleus and most of the ribosomes are found in the cell body. Although few ribosomes are seen in axons and dendrites, many proteins are needed in high concentrations within the axons and synaptic endings. Among these are enzymes catalyzing synthesis and catabolism of neurotransmitters and membrane proteins. If an axon is cut, the separated synaptic endings soon atrophy, an observation that long ago suggested that essential materials, which may include mRNAs, might flow from the cell body. It has now been established experimentally that many materials do move at the rate of 0.3 – 3 mm/day from the cell body down the axon. More remarkable is *fast axonal transport* by which proteins and other materials move at rates of up to 5 μm/s (0.4 m/day). This transport is specifically blocked by vinblastine and batrachotoxin (Fig. 7.12). As has been pointed out in Chapter 20, an ATP-hydrolyzing protein chemically related to the myosin heads functions together with microtubules to provide a kind of miniature railway that moves materials along the microtubules. Transport is sometimes in the opposite direction, *i.e.,* from the synaptic endings to the cell body. This *retrograde axonal transport* may be of importance in altering neuronal properties in response to electrical activity at synaptic endings. It also provides a means of recycling materials originally sent in the other direction.

Brain cells appear to transcribe an unusually large fraction of the genome. About 20% of the DNA of human brain was found to hybridize with mRNA formed by brain cells. In other tissues about half this amount of DNA appears to be transcribed. A related observation that seems surprising is the absence of common electrophoretic variants of enzymes in the brain. However, brain cells synthesize specialized isoforms of many proteins, *e.g.,* of the G proteins, the cytoskeletal protein 4.1 (Fig. 7.14), and transglutaminase. Unusual lipids, such as the cationic acetal of a galactosylcerebroside shown above, are also formed. Adult rat brain contains about 30,000 different kinds of polyadenylated messenger RNA, much of which lacks the poly(A) tail. Many of these mRNAs contain a specific 82-nucleotide sequence within at least one of their introns. Sutcliffe *et. al.,* suggest that this is an *identifier sequence* instructing brain cells to express these genes. However, the sequence is also found in genes transcribed in other tissues, and its significance is not clear.

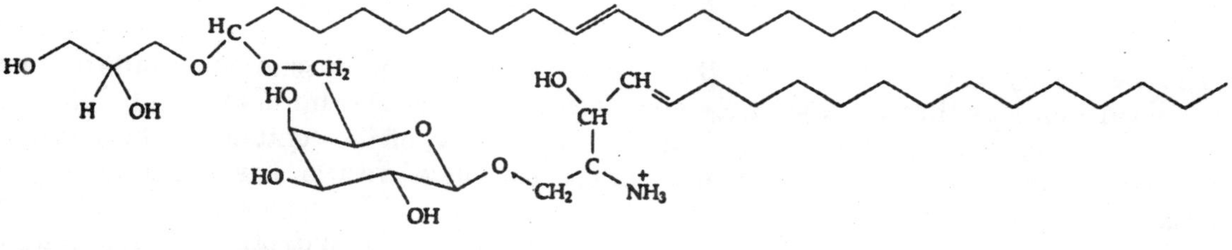

HO H OH O HC O—CH_2 HO HO O OH O CH_2 C $\overset{+}{N}H_3$ HO CH CH

8. Synapses and Gap Junctions

Like the micro-transistors in a computer chip, synapses are the devices by which the brain operates. Synapses process and integrate information from many input channels, send signals on to other neurons, and store information. The information is not stored in digital form, but as chemical alterations in the synapses themselves. Synapses are formed when axons, growing in response to a chemical trail, reach their destinations and send out branches, each with a bulbous terminal knob (*bouton*). When these boutons meet receptive regions on dendrites of another axon, synapses are formed. The synapse is a very firm connection with a thin, tight synaptic cleft through which signaling takes place. It is surrounded in part by astrocytes or other glial cells.

With the advent of electron microscopy, the fine structure of synaptic contacts became evident. The synaptic knobs were often found to contain vesicles of ~30 – 80 nm diameter, which were later shown by chemical analysis and staining procedures to contain the neurotransmitters (Fig. 7.10). In the case of the acetylcholine-releasing synapses *(cholinergic synapses)* each 80-nm vesicle contains ~40,000 molecules of acetylcholine, the concentration in the vesicle being of the order of 0.5 M. To show that the acetylcholine released at a synapse stimulated the postsynaptic membrane to initiate an impulse, the technique of *electrophoretic injection* or *microiontophoresis* was developed. By using ultramicrocapillaries a small pulse of current, *e.g.*, 3×10^8 amp for 1 ms, can be used to inject electrically a compound directly into a synaptic cleft. The results may be observed with separate recording electrodes, one of which is inserted into an axon or a muscle fiber. By this means it was shown that amounts of acetylcholine comparable to those released at the large synapses of the neuromuscular junction do cause muscles to contract.

How does the release of neurotransmitter occur? That the release is "quantal," *i.e.*, involving the entire content of a vesicle, was established from the observation of *miniature end-plate potentials*. These are fluctuations in the postsynaptic potential observed under conditions of weak stimulation of the presynaptic neuron. They reflect the randon release of neuro-transmitter from individual vesicles. Normally, a strong impulse will release on the order of 100 – 200 quanta of transmitter, enough to initiate an action potential in the postsynaptic neuron.

A Synaptic Vesicle Cycle

The number of synaptic vesicles in a single synapse in the brain varies from fewer than 100 to several hundred. In specialized synapses there may be thousands. However, at any moment only a fraction of the total are in the "active zone," often aligned along the presynaptic membrane or in specialized ribbons. The vesicles are normally reused repeatedly, undergoing a cycle of filling with neurotransmitter, translocation to the active zone, ATP-dependent priming, exocytosis with release of the neurotransmitter into the synaptic cleft, coating with clathrin, endocytosis, and acidification. The entire cycle may be completed within 40 – 60 s to avoid depletion of active vesicles. A key event in the cycle is the arrival of an action potential at the presynaptic neuron end.

The accompanying depolarization of the membrane at the synaptic ending permits a rapid inflow of calcium ions through a voltage-gated calcium channel. Within less than 0.1 ms the transient increase in intracellular [Ca^{2+}] triggers the release of the contents of the vesicles. About four calcium ions are needed to release one clathrin-coated vesicle. The membrane fusion required for transmitter release involves cytoskeletal proteins of the synaptic endings as well as specialized proteins that are present in the membranes of the synaptic vesicles. In fact, every step in the cycle depends upon specialized proteins.

Synaptic vesicles can be isolated in large quantities. Their composition is well known, and the proteins have been studied intensively. Indeed, much of what we know about exocytosis and vescular transport has been learned from investigation of synaptic vesicles. A small synaptic vesicle of 35 nm diameter will contain ~10,000 phospholipid molecules in its membrane and only about 200 protein molecules, at least one of which must be a 13-subunit vacuolar type proton pump (Fig. 7.14). This pump acidifies the vacuole, allowing uptake of a neurotransmitter. Although many different proteins may be found in synaptic membranes, only about 15, which are listed in Table 7.6, are found in all synaptic vesicles and appear essential to function.

The synaptic vesicles, which are formed by budding from early endosomes, take up neurotransmitters using one of the transporters. Transmitter uptake is G-protein dependent and is driven by the proton electrochemical gradient generated by a vacuolar type (V-type)

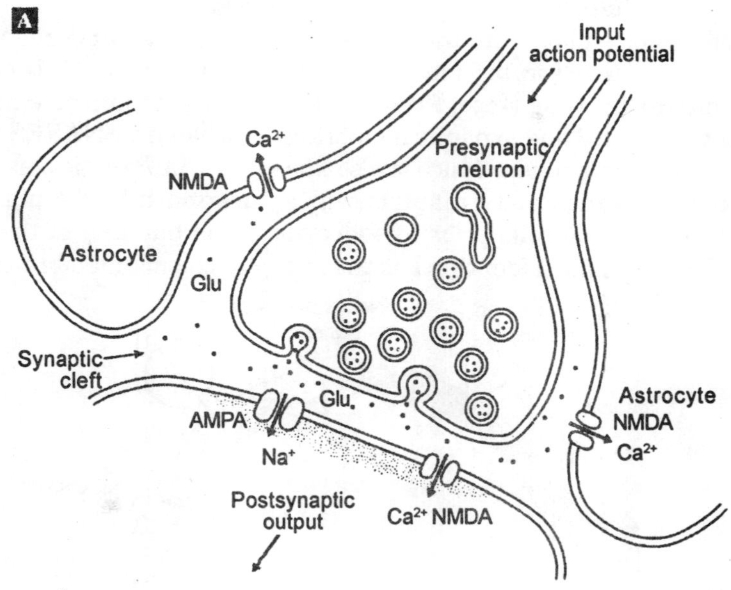
A
Input
action potential
Ca2+
NMDA
Presynaptic
neuron
Astrocyte
Glu
Synaptic
cleft
Glu
AMPA
Na+
Astrocyte
NMDA
Ca2+
Postsynaptic
output
Ca2+ NMDA

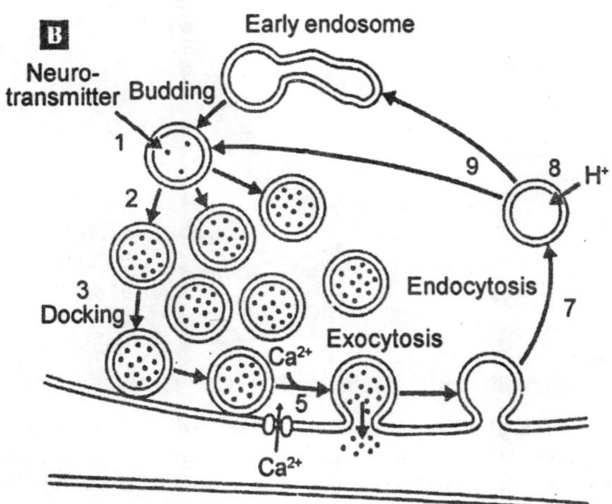
B
Early endosome
Neuro-
transmitter
Budding
1
9
8
H+
2
Endocytosis
3
Docking
7
Exocytosis
Ca2+
5
Ca2+

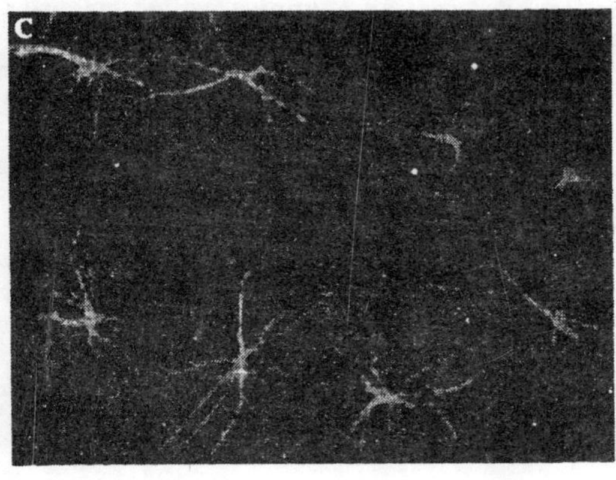
C

ATPase. The filled vesicles move into the active zone where they undergo an ATP-dependent priming of uncertain nature. Exocytosis requires membrane fusion, and it is possible that partial fusion occurs during the priming steps. Priming is also thought to involve interaction between vesicle-associated v-SNARES and synaptic membrane-associated t-SNARES. A major v-SNARE has been identified as *synaptobrevin,* which is also known as *VAMP* (vesicle-associated membrane protein).. The C-terminal-anchored synaptobrevin is inserted into the plasma membrane of neuronal and neuroendocrine cells prior to endocytosis and budding of the synaptic vesicles. The target t-SNARES have been identified as the synaptic plasma membrane proteins *syntaxin*

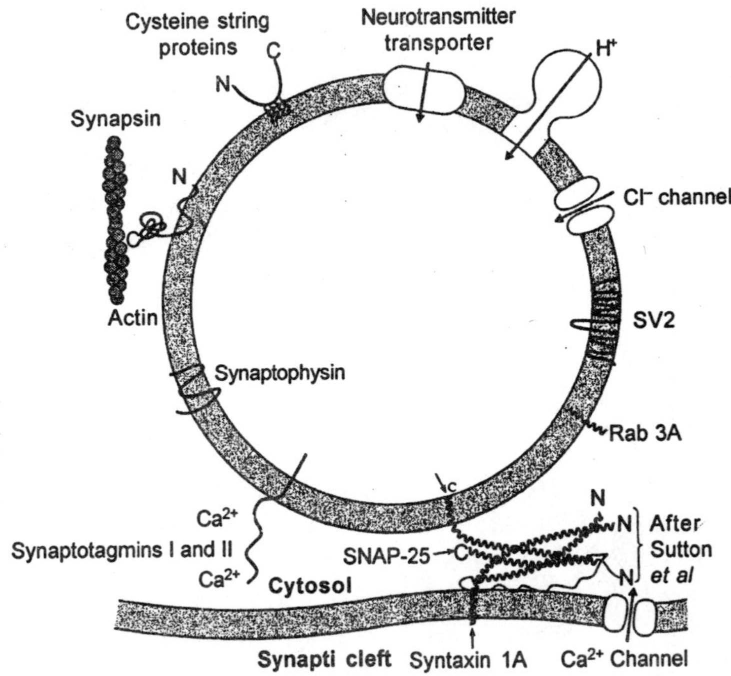

Fig. 7.20. (A) Schematic drawing of a fast glutamatergic synapse. An action potential arrives at the synapse, depolarizing the presynaptic membrane and allowing calcium ions to enter the cytoplasm via voltage-gated Ca^{2+} channels. The Ca^{2+} ions induce exocytosis of small synaptic vesicles from the "active zone" near the membrane, releasing glutamate into the synaptic cleft. After diffusing rapidly across the narrow ~50-nm synaptic gap the glutamate binds to its receptors on the ending on a dendrite from a second (postsynaptic) neuron. Glutamatergic synapses usually have two types of receptor, NMDA and AMPA. Both are ligand-gated ion channels, which release Ca^{2+} and Na^{+} into the cytosol of the postsynaptic ending depolarizing its membrane and possibly initiating an action potential. (B) The synaptic vesicle cycle. The synaptic vesicles, which are formed by budding from an early endosome, are filled with neurotransmitter (1). The filled vesicles are then transported to the active zone near the presynaptic membrane (2), are "docked" on the membrane surface (3), and undergo ATP-dependent priming (4). Binding of four Ca^{2+} ions induces exocytosis and rapid release of the neurotransmitter (5). The empty vesicles receive a clathrin coat (6) and undergo endocytosis (7) and uptake of protons (8) to acidify the content in preparation for a second round of neurotransmitter uptake. Alternatively the vesicle can fuse with an endosome as part of the cycle. After Südhof and Scheller. (C) Small section of brain stained to reveal the astrocytes whose extensions form synapses not only with neurons, as in (A), but also with capillary blood vessels. (D) Illustration of some proteins essential to the synaptic vesicle cycle. Several are integral membrane proteins. Synaptotagmins contain Ca^{2+}-binding domains and may serve as calcium sensors. The vesicle is portrayed as if docked to the presynaptic membrane by interaction of the SNARE proteins synaptobrevin, syntaxin, and synaptotagmin.

and *SNAP*-25. Syntaxin is an integral membrane protein, whereas SNAP-25 is anchored by palmitoylation. These proteins bind together to form a synaptobrevin•syntaxin•taxin-SNAP-25 complex, which forms a four-helix bundle. Synaptobrevin and syntaxin each contribute one helix, while SNAP-25 provides two; all four have a mutually parallel orientation. The helix bundle is so tight that it has a high melting temperature and is resistant to proteolytic cleavage. Nevertheless, the helical domains of both synaptobrevin and syntaxin are sites of very specific cleavage by the zinc proteases of tetanus and botulinin toxins. Cutting of the protein chains by these toxins prevents proper formation of the four-helix bundle and prevents release of neurotransmitter. It is thought that the complex, which probably forms at several points on the periphery of the docked synaptic vesicle, is essential for membrane fusion.

Other proteins are also needed. All cell fusion processes seem to require regulatory proteins that are essential to neurotransmission in the nematode C. *elegans*. Two of these are encoded by the nematode genes *unc*-13 and *unc*-18. The corresponding mammalian proteins *munc*-13 *and munc*-18 interact with syntaxin and are essential for exocytosis of synaptic vesicles. An ATPase is also needed for correct functioning of the SNARE complex as are other additional proteins.

Details of the control of exocytosis are also uncertain. *Synaptotagmin I,* which contains two Ca^{2+}-binding domains, is probably the sensor that detects the rapid influx of Ca^{21+} that initiates exocytosis. It binds several Ca^{2+} ions via a p-sandwich motif that contains five aspartate side chains at its tip. This motif is conserved in a large family of synaptotagmins. A possibility is that Ca^{2+}-Synaptotagmin complexes may self-associate to form a protein ring around the site where the fusion pore forms. Synaptotagmin I also interacts with both syntaxin and with *neurexins,* proteins related to laminin (Fig. 7.33) and present in numerous variant forms in nerve endings. Neurexins are also targets for the α-lathrotoxin of the black widow spider. Other proteins that may participate in membrane fusion include the unique *cysteine string proteins,* which in *Drosophila* contain 13 cysteine residues, 11 of which are palmitoylated. Nitric oxide NO may be involved in a late stage of exocytosis, and phospholipase Dl may also be required.

Presynaptic nerve terminals may contain as few as a hundred vesicles, which must be recycled rapidly after exocytosis in order to allow for repetitive firing. Several proteins are needed for endocytosis. These include *endophilin I,* the vesicle transport ATPase NSF, GTPases, and the soluble NSF attachment protein α-SNAP (which is not related to SNAP-25).

Functions of some other abundant proteins of synaptic vesicles have not yet been accurately defined. The *synapsins* are abundant peripheral membrane ATP-binding proteins with multiple phosphorylation sites and variable C-terminal domains that interact with cytoskeletal proteins such as actin microtubules, microfilaments, and spectrin. Another abundant protein is *synaptophysin,* an integral membrane protein found in all synaptic vesicles. Other proteins are discussed by Südhof and Scheller. The small G protein *rab* 3 together with the Ca^{2+}-binding protein *rabphilin* participate in a G-protein cycle that helps to drive exocytosis. Synaptotagmin, as well as clathrin assembly proteins bind inositol hexaphosphate, which undergoes active turnover in synapses. This suggests a role for $InsP_6$ in the endocytosis steps of the synaptic vesicle cycle. The brain is rich in zinc ions. Much of the Zn^{2+} is bound into zinc finger domains of transcriptional regulators, but much is also present in a relatively free form within synapses of the hippocampus, cerebral cortex, and other regions. Zinc ions may function as a neuromodulator in glutamatergic synapses.

Table 7.6. Some Proteins Important to the Formation and Functioning of Synaptic Vesicles.[a]

1. Synaptic vesicle proteins	
Synapsins Ia, Ib, IIa, IIb	Peripheral, abundant
Rab3, rabphilin	Rab 3 has lipid anchor
Cysteine string proteins (CSP)	Ca^{2+}-binding
Synaptotagmins	Single transmembrane helix; Ca^{2+}receptor N terminus in vesicle
Synaptobrevins (VAMPs)[b]	SNARE proteins, C termini in vesicle
Synaptophysins, synapogyrin	Integral membrane protein
SV2 A, B, C	Integral membrane protein, Cl^- transporter
SCAMPS 1 and 4	Integral membrane protein
SVOP	Integral membrane protein
Vacuolar H^+ pump	13 subunits
Cytochrome 561	H^+ generator
Neurotransmitter transporters	For acetylcholine, glutamate, GABA/glycine, catecholamines, ATP
Ancillary transporters	Zn^{2+}, Cl^-
2. Presynaptic membrane proteins	
Syntaxin[b]	t-SNARE
SNAP-25[b]	t-SNARE
Munc-13	
Ca^{2+} channel	
Agrin	
Neurexin	
Actin and microtubules	In dendrites
3. Postsynaptic specializations	
Receptors	*e.g.*, NMDA, AMPA

What does a neurotransmitter do at the postsynaptic membrane? In the case of acetylcholine in neuromuscular junctions the principal action appears to be one of opening sodium channels and thereby depolarizing the postsynaptic membrane. If enough nerve impulses arrive, an action potential will be initiated in the postsynaptic neuron. In other cases, the first response may be activation of a protein kinase either directly or by opening a channel for Ca^{2+}, which indirectly regulates protein kinases and phosphatases. Thus, a complex cascade may be activated.

The postsynaptic nerve ending, which is usually the tip of an axonal dendrite, has its own set of proteins, which varies to some extent with the nature of the neurotransmitter. In excitatory cells the plasma membrane of the postsynaptic neuron is thickened to ~30 – 40 nm to form the *"postsynaptic density,"* a disc-like structure of clustered receptors of two types, which extends ~30 nm into the cytosol. Only single receptor channels are indicated in Fig.

7.20, but many receptors are present in the clusters as are other specialized proteins. One of these, designated PSD-95, was found to associate with the NMDA receptor using the yeast two-hybrid system. Neuronal nitric oxide synthase may also be present.

The large neuromuscular junctions, which contain clusters of acetylcholine receptors, have wider synaptic clefts (> 40 nm), which contain basal lamina, a dense network of collagen fibrils together with the heparan sulfate proteoglycan *agrin*. Agrin activates a muscle-specific kinase MusK, which phosphorylates the acetylcholine receptors inducing clustering of the receptors together with other proteins embedded in the plasma membrane and binding to the cytosolic protein *rapsyn*. Agrin is also a component of *immunological synapses,* which are important in lymphocyte development. The neuromuscular junction is formed between two cell types, a neuron and a muscle myotube. Both contribute proteins, which include a muscle-specific laminin.

Astrocytes and Other Glia

Although the glial cells greatly outnumber neurons, they were long regarded simply as glue, as implied by the name glia. We know now that the several types of glial cells have functions in many different aspects of brain chemistry. The oligodendrocytes generate myelin sheaths around many brain neurons. Macrophages that invade the brain differentiate into *microglia* that serve as part of the innate immune system (Chapter 31). *Bergmann glia* of the cerebellum help guide axons during brain development. The astrocytes have many processes, which not only contact synapses directly but also form contacts with capillary blood vessels. They often contain receptor ion channels of the same types as are found in post-synaptic membranes and respond to Ca^{2+} influx as do neurons. Glia often take up neurotransmitters and ions from synapses in order to prepare for consecutive nerve impulses. Glia may also control the number of synapses formed, and they may have other roles in brain development. For example, an iodothyronine deiodinase is expressed primarily in neonatal brain, where it supplies thyroid hormone essential to brain development.

Gap Junctions in Synapses

Not all neurons communicate via chemical synapses. Gap junctions, which are found in both neurons, astrocytes, and other cells, serve as *electrical synapses*. Thus, heart cells are all electrically coupled together by gap junctions. Gap junctions are formed with the aid of hexameric *connexons,* which are present in each of the opposed membranes and are aligned one with the other. There may be thousands of connexons in a single gap junction, which resemble ion channels in appearance but contain pores ~1.5 nm in diameter. They are formed from 26- to 43- kDa protein subunits of the multigene family of *connexins*. Each gap junction consists of a pair of hexameric rings of connexins, one ring from each of the two juxtaposed membrane surfaces. Defects in connexins cause inherited deafness, neuropathy, malignancy, and cataract formation. The connexin subunits each contain four transmembrane helices and are related structurally to the peripheral myosin protein 22, the myelin proteolipid, and the protein stargazin, which is involved in synapse formation in the brain.

Another type of channel has been recognized quite recently. An ion channel, which regulates Mg^{2+} ion transport in kidney tubules, forms within the tight junctions that seal the extracellular space between cells. A protein *paracellin* forms channels through the tight junction protein complexes that surround the cells.

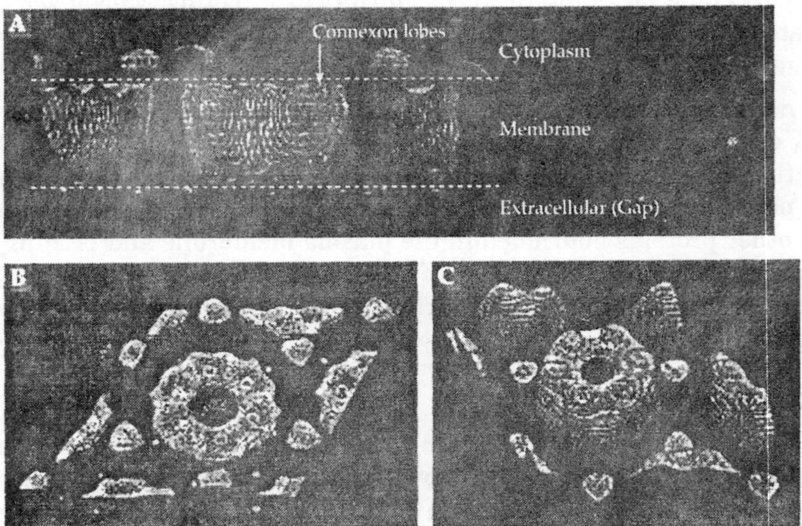

Fig. 7.21. Images of gap junction connexins obtained by electron crystallographic methods at a resolution of 1.6 nm. (A) Cross-section. The thinkness of the (43 × 6) kDa hexameric connexin is 5.0 mm. (B) View of the connexin from the cytoplasmic side. (C) View from the extracellular side.

9. Neurotransmitters

Studies of neuromuscular junctions of the autonomic nervous system as early as 1904 led to the suggestion that adrenaline might be released at the nerve endings. Later it was shown that, while adrenaline does serve as a transmitter at neuromuscular junctions in amphibians, it is primarily a hormone in mammals. Nevertheless, it was through this proposal that the concept of chemical communication in synapses was formulated. By 1921, it was shown that acetylcholine is released at nerve endings of the parasympathetic system, and it later became clear the motor nerve endings of the somatic system also release acetylcholine.

Acetylcholine is an established neurotransmitter because it meets five important criteria: (1) a synthetic mechanism exists within the presynaptic neuron; (2) a mechanism of storage (in vesicles) is evident; (3) the transmitter is released in proportion to the strength of the stimulus (frequency of firing); (4) postsynaptic action of the transmitter has been demonstrated directly by microiontophoresis; and (5) an efficient means for inactivation of the transmitter is present. The same five criteria must be met by other compounds if they are to be considered as transmitters.

At present, in addition to acetylcholine, glutamate, and γ-aminobutyrate (GABA), glycine, noradrenaline (norepinephrine), and dopamine and 5-hydroxytryptamine (serotonin) are regarded as established transmitters. Other probable *(putative)* or possible *candidate transmitters* are also known. Aspartate, taurine, and a large number of peptides are under consideration.

Some transmitters, including noradrenaline, dopamine, serotonin, and various neuropeptides, are sometimes called *neuromodulators* rather than neurotransmitters. These compounds may not initiate a nerve impulse but may act on adenylate cyclase to increase or decrease cAMP levels and protein kinase activity. They may also diffuse through the extracellular space to influence a region of the brain greater than a single synaptic cleft. However, the distinction between transmitters and modulators is not exact.

For many years it was assumed that a single neuron released only a single transmitter. We know now that this is incorrect. For example, enzymes in neuromuscular junctions synthesize not only acetylcholine but also catecholamines, taurine, and GABA. Some synapses in the central nervous system release both glycine and GABA.

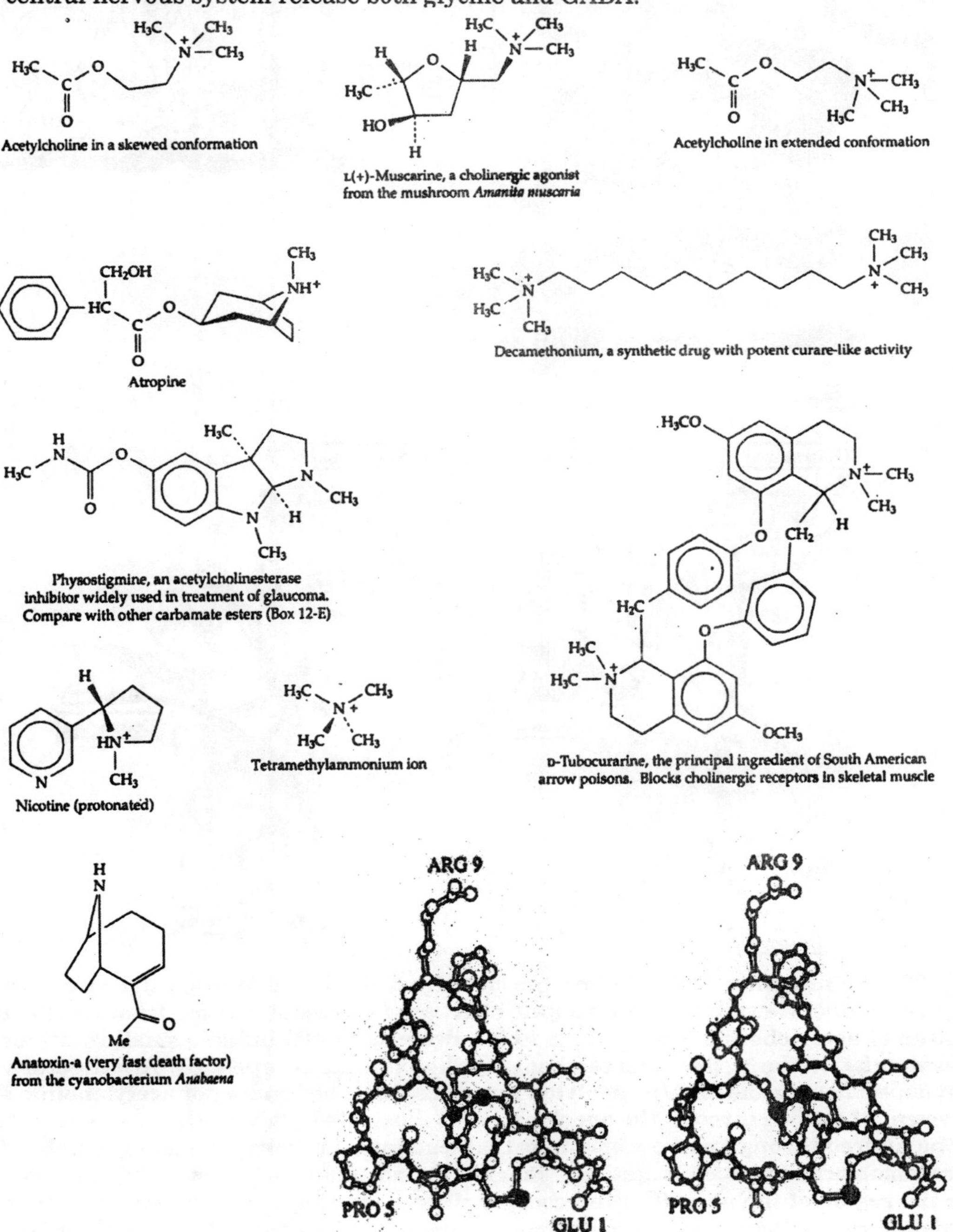

Fig. 7.22. Some inhibitors of cholinergic synapses.

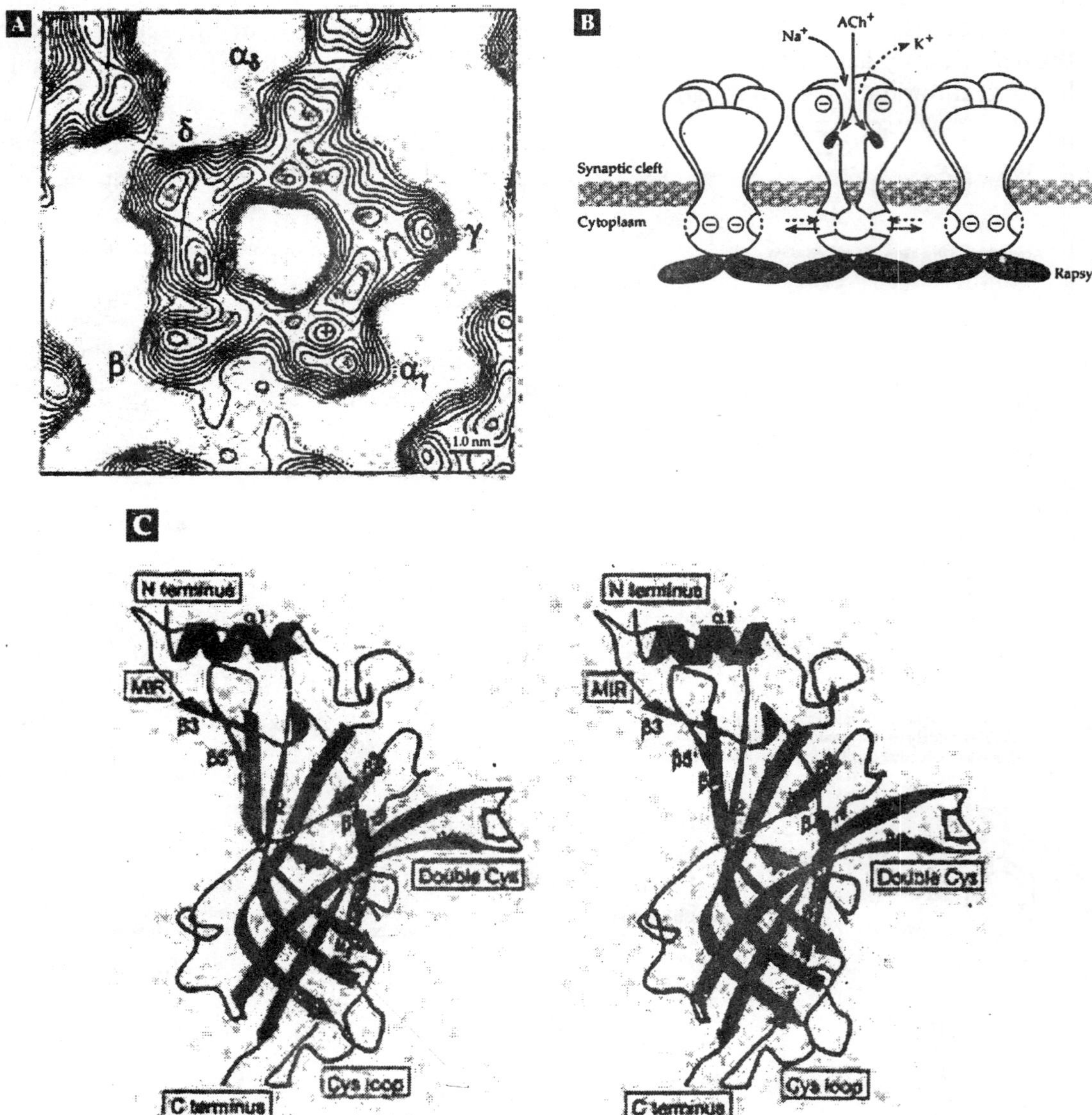

Fig. 7.23. The nicotinic acetylcholine receptor from the *Torpedo ray*. (A) The mouth of the receptor channel viewed from the synaptic cleft based on reconstruction from cryoEm images. Addition of acetylcholine, which binds to the two α subunits, induces small rotations in the five subunits of the $\alpha_2\beta\gamma\delta$ complex causing the channel to open. (B) Architecture of the subsynaptic membrane and the acetylcholine receptor. The binding of acetylcholine and the movement of cations through the open channel is illustrated. Cations that leave the cytoplasm may be filtered through narrow openings that lead into the central channel, which is formed by transmembrane helices. Negatively charged amino acid residues may help exclude anions from the region of the pore. (C) Stereoscopic ribbon drawing of one subunit of a pentameric acetylcholine-binding protein, which mimics the receptor structure. Disulfide bonds are shown in a ball-and-stick form. The N terminus in a receptor would point toward the synaptic cleft and the C terminus would continue at the bottom into the transmembrane helix.

Cholinergic Receptors and their Agonists and Antagonists

Among the acetylcholine-releasing (cholinergic) neurons are the motor neurons that form synapses at neuromuscular junctions, the preganglionic neurons of the entire autonomic system, and the postganglionic neurons of the parasympathetic system. There are also many cholinergic synapses within the brain. In contrast, in insects neuromuscular transmission is mediated by glutamate while acetylcholine is the principal neurotransmitter in the central nervous system.

Important in the study of neurotransmitters is the identification of specific agonists, which mimic the action of a transmitter, and of antagonists, which block the action of the transmitter. Two groups of compounds influence acetylcholine-secreting neurons, leading to the classification of these neurons either as *muscarinic* (activated by muscarine; Fig. 7.22) or *nicotinic* (stimulated by nicotine). The muscarinic receptors, which are found in many autonomic neurons, are specifically inhibited by *atropine* and *decamethonium* (Fig. 7.22). The nicotinic synapses occur in ganglia and skeletal muscle. They are inhibited by curare and its active ingredient *D-tubocurarine* (Fig. 7.22) and by the protein snake venom *α-bungarotoxin*. This toxin has been used to titrate the number of acetylcholine receptors in the motor end plate of the rat diaphragm. About 4×10^7 receptors per end plate (or 13,000/μm^2) were found.

Nicotinic receptors (nAChRs; Fig. 7.23) of the type found in neuromuscular junctions are most frequently isolated from the electric organs of the electric eel *Electrophorus* or from electric fish of the genus *Torpedo*. They have been studied more intensively than any other receptor. They contain four kinds of subunit with a stoichiometry $\alpha_2\beta\gamma\sigma$ and molecular masses of 39, 48, 58, and 64 kDa respectively. The amino acid sequences of the four proteins contain homologous regions, some of which are thought to represent membrane-spanning segments of the peptides. These receptors are ligand-gated ion channels and are closely similar to $GABA_A$ and $GABA_C$ receptors, to glycine receptors, and to 5-hydroxytryptamine (serotonin) receptors of the $5\text{-}HT_3$ type. Parts of their amino acid sequences are also homologous to those of both the voltage-gated Na^+ channels and gap junctions, suggesting that the transmembrane domain may resemble that of Fig. 7.18. However, notice the difference in symmetry. Acetylcholine binds to the two a subunits (Fig. 7.23). Neurotoxins may bind at several sites. Some indication of the possible function of the various subunits comes from studies of the neuromuscular junction in which the different subunits are degraded at different rates with half-lives of from one to ten days. During development fetal γ subunits are replaced by adult γ subunits. Perhaps more rapid changes in receptor composition are sometimes needed.

Similar nAChRs are also found in the brain. However, they are not identical but have at least 17 differing amino acid sequences (α1 – α10, β1 – β4, γ, δ, and ε). The neuromuscular junction receptor (muscle type) from fish is described as $(\alpha 1)_2 \bullet \beta 1 \bullet \gamma/\varepsilon \bullet \delta$. The brain contains homopentamers of subunits α7, α8, and α9 as well as various heteropentamers. The various forms possess different affinities for acetylcholine and for antagonists such as nicotine. In the brain the highest affinity for nicotine is shown by an α4 β2 form, which represents over 80% of the nAChR in mammalian brain. Knockout mice in which the β2 subunit gene has been deleted lose their sensitivity to nicotine.

Conductance measurements showed that the nicotinic receptors contain channels permeable to Na^+ and other cations and that they are acetylcholine-gated ion channels. Construction of a three-dimensional image from electron micrographs at various angles of tilt shows a tube with approximate pentagonal symmetry and a narrow channel through the center (Fig. 7.23). Acetylcholine binds to sites on the two α subunits ~3 nm away from the ion channel. An allosteric change opens the channel, allowing cations (largely Na^+) to flow out, depolarizing the membrane. There are at least four structural states in the channel opening-and-closing cycle. The three-dimensional structure has been modeled using an acetylcholine-binding protein of known structure from a snail as a mimic of the cytoplasmic nicotine-binding domain of the receptor. The structure of one subunit of the binding protein is shown in Fig. 7.23C. This protein which is secreted into synapses by glial cells, may provide a buffering action by binding the acetylcholine. Although the most rapid effect of acetylcholine binding to the nicotinic receptor is depolarization of the postsynaptic membrane, other slower effects follow. Thus, protein kinases are activated and phosphorylate the receptor as well as other proteins.

After a pulse of transmitter is released, it must be removed or inactivated quickly to prepare the synapse for arrival of a new nerve impulse. This is accomplished in two ways in cholinergic synapses. The first is via hydrolytic destruction by acetylcholinesterase. This esterase and the related butyrylcholinesterase are present in the synaptic membrane itself. The second mechanism is energy-dependent transport of acetylcholine into the neuron for reuse. Since much of the transmitter is hydrolyzed, new acetylcholine is synthesized by transfer of an acetyl group of acetyl-CoA to choline.

In the central nervous system muscarinic acetyl-choline receptors are more abundant than nicotinic receptors. They consist of single-chain proteins of mass ~70 kDa. They are not ion channels but are 7-helix receptors homologous in sequence with β-adrenergic receptors (Fig. 7.6) and with rhodopsin. Five different subtypes (M1 – M5) have been characterized. The M1, M3, and M5 receptors are coupled to the G_q/G_{11} family of G proteins, and M2 and M4 are coupled to G_i/G_o proteins. Their effects are slower and of longer duration than those of the nicotinic receptors. It has been difficult to assign functions to the individual types. Most regions of the brain contain more than one type, but they are thought to be involved in locomotion, learning, memory, thermoregulation, and cardiac and pulmonary functions. Many drugs, some of which are used in treatment of Parkinson and Alzheimer diseases, epilepsy, and asthma, affect muscarinic receptors. The M2 receptors predominate in the heart where they help to regulate the beating frequency and atrial contractility. Sudden infant death may sometimes result from a defect in muscarinic receptors. Knockout mice lacking M2 receptors also have problems with movement control, body temperature, and pain responses. Mice lacking M3 receptors are lean with very low levels of serum leptin and insulin. Many of the muscarinic receptors activate adenylate cyclase, while others are coupled to the phosphoinositide cascade. Some indirectly activate K^+ channels. Muscarinic receptors are also studied in insects, but it is difficult to correlate the insect and mammalian receptors.

Amino Acids as Neurotransmitters

The concentrations of *glutamate* and of its decarboxylation product *γ-aminobutyrate* (GABA) are high in all regions of the brain. The two compounds are generated sequentially in the γ-aminobutyrate shunt, a pathway that accounts for a quantitatively significant part of the total metabolism of the brain (Fig. 7.5). Because they are present in all parts of the brain in high concentrations, there was initially reluctance to accept glutamate and GABA as neurotransmitters. However, it is now accepted that L-glutamate is the major excitatory transmitter in the central nervous system. It seems to be responsible for nearly all of the very fast acting nerve impulses in the brain. At the same time GABA is recognized as the most important inhibitory transmitter. The role of glutamate as an excitatory transmitter was first established for the neuromuscular junction of arthropods. Although it is a constituent of all animal tissues, the concentration of glutamate is much higher in brain than in other tissues, and it is higher in neurons than in glia.

Microiontophoretic application of either glutamate or aspartate to the brain cortex leads to very strong excitatory responses.

Three subtypes of ionotropic glutamate receptors (iGluR) are named for the specific agonists *α-amino-3-hydroxy-5-methyl–4-isoxazolepropionic acid* (AMPA), *N-methyl-D-aspartate* (NMDA), and *kain-ate.* The receptors resemble the acetylcholine receptor in containing a cation channel. In addition, there are 7-helix *metabotropic glutamate receptors,* which are coupled to G proteins. The AMPA receptors were in the recent past called *quisqualate* receptors, because they are also activated by the agonist with that name. The toxic domoate (Fig. 7.24) also binds to kainate receptors. Both domoic acid and kainic acid are terrible convulsant toxins. They are formed by two different red algae. Domoic acid accumulates in contaminated mussels and causes shellfish poisoning. The ionotropic glutamate receptors, which may be stimulated by either glutamate or aspartate, are directly linked to the opening of cation channels. Their activation may also induce the inositol phosphate cascade and slower Ca^{2+}-dependent changes. A peculiarity of the high-conductance NMDA channels is that they are blocked by Mg^{2+} in a voltage-dependent manner. They do not open unless the frequency of nerve impulses is high or some other factor causes membrane depolarization.

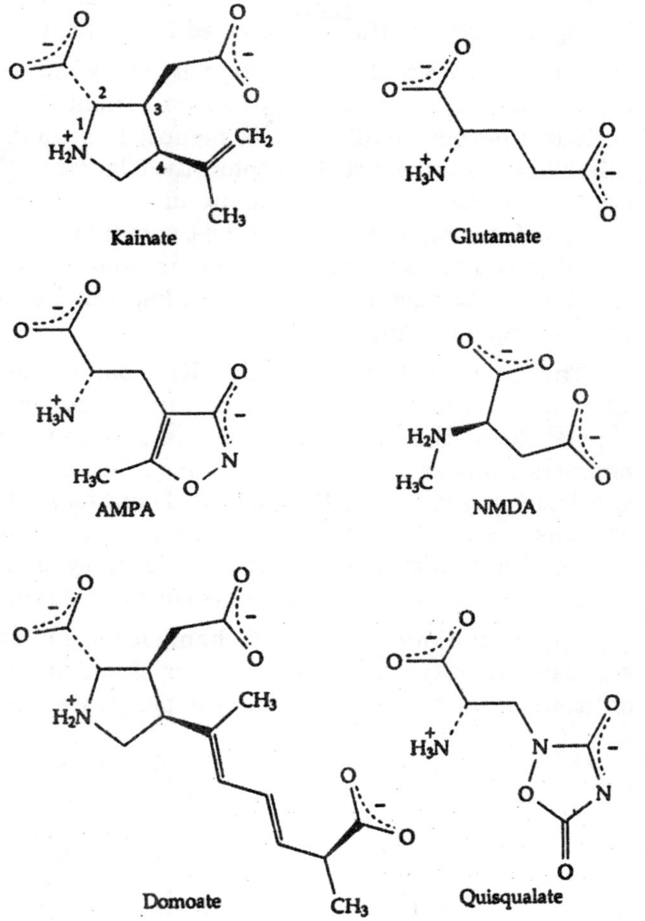

Fig. 7.24. Chemical structures of some agonists of ionotropic glutamate receptors (iGluR).

The AMPA receptors, which are thought to be the predominant mediators of fast excitatory transmission in the brain, are oligomers (probably tetramers) of 950- to 1500-residue protein subunits. These subunits have large N-terminal domains in the synaptic cleft. There are probably three transmembrane helices and a membrane-associated loop similar to those depicted in Fig. 7.17A. A long C-terminal tail protrudes into the cytosol, while a large loop between transmembrane regions extends from the outer membrane surface, joining with the N-terminal domain to form the ligand-binding site, the structure of which resembles those of bacterial periplasmic binding proteins. Four related AMP receptors, designated GluR1, 2, 3, and 4, have been identified. Related kainate receptors, whose properties overlap those of AMPA receptors, are designated GluR5, 6, and 7. Although AMPA receptors are essential for fast signal transmission they lose sensitivity rapidly (on a millisecond time scale) as a result of conformational alterations. Many factors, including inhibition by polyamines, affect these receptors. However, brief high-frequency activation of some AMP receptors leads to a long-lasting increase in efficiency, termed LTP, which is important to learning.

The NMDA receptors are heterooligomers with two type of subunits. The NR1 (or ζ) subunits exist as a series of at least eight splice variants. The NR2A, B, C, and D (ε series) are encoded by four different genes. NR1 is regarded as the principal subunit ard NR2 as a regulatory subunit. As with the AMPA receptors the oligomeric NMDA receptors are anchored at appropriate locations in the postsynaptic membrane by scaffolding proteins containing PDZ domains (Table 7.3). The C-terminal domains of the ε subunits are unusually long and participate in anchoring. NMDA receptors are found not only in neurons but also in astrocytes (Fig. 7.20), where they are thought to have important signaling functions. These include regulation of Ca^{2+} flow, in part via gap junctions.

The N-terminal domain of the NR1 subunit of the NMDA receptor contains a glycine-binding site. Full activity of the receptor requires a *coagonist* bound in this site. Surprisingly, *D-serine* seems to be the normal coagonist, at least in some sites. This newly recognized neurotransmitter is synthesized from L-serine by a pyridoxal phosphate-dependent recemase and is destroyed by the flavoprotein D-amino acid oxidase. Associated with NMDA receptors are clusters of *ephrin receptors,* proteins that bind the glycosylphosphatidylinositol (GPI)-anchored proteins known as ephrins in presynaptic membranes. Binding of ephrins to their postsynaptic receptors activates tyrosine kinases and enhances the influx of Ca^{2+} ions.

Specific inhibitors of NMDA channels include a 27-residue "spasmodic" conotoxin, 2-amino-4-phosphonobutyrate, related longer chain aminophosphonates, and the following potent anticonvulsant drug, which is able to penetrate the blood-brain barrier.

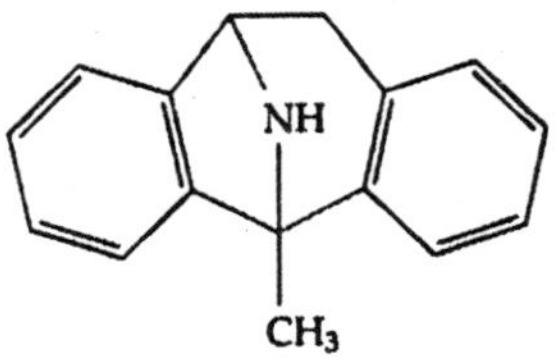

NH

CH_3

(+) 5-Methyl-10, 11-dihydro-5H-dibenzo [a, d] cyclohepten-5, 10 imine.

Metabotropic glutamate receptors have been classified into eight types (mGluRs1–8). Group I (mGluRs1 – 5) are selectively activated by 3, 5-dihydroxyphenylglycine; Group II (mGluR2 and mGluR3) are activated by L-2-(carboxycyclopropyl)glycine; and Group III (mRluR4 and mGluR 6 – 8) are activated by L-2-aminophosphonobutyrate. They are all 7-helix G-protein-coupled receptors with external ligand-binding domains that resemble those of bacterial periplasmic binding proteins. Splice variants for at least mGlnR1 are known. Metabotropic glutamate receptors are neuromodulary but nevertheless play essential roles in the cerebellum and other parts of the brain. For example, mice deficient in the mGluR1 protein have severe problems with motor coordination and learning. Metabotropic glutamate receptors may participate in calcium sensing and signaling.

Synaptosomal particles have a high-affinity proton-dependent uptake system for glutamate. Glutamate and aspartate may also be taken up from the synaptic cleft by neurons or by glial cells, which then transfer the glutamate into neurons for reuse. Five distinct mammalian transporter genes have been cloned. They are driven by concentration gradients of Na^+ and K^+ across the membrane. However, some serve as glutamate-gated chloride ion channels.

Excitotoxicity

As essential as glutamate is for brain function it is toxic in excess. Excessive stimulation of the NMDA receptors, which occurs during convulsions, strokes, or traumatic injury and which can accompany anoxia or hypoglycemia, causes neuronal death. Blocking these receptors with the above-mentioned anticonvulsant drug or aminophosphonates has a remarkable protective effect against the neurotoxicity of the accumulating glutamate. Vitamin E and *tócotrienols* may also be protective.

The Inhibitory Neurotransmitter Gamma-Aminobutyrate (GABA)

Glutamate, aspartate, and cysteic acid are all potent exciters, but their decarboxylation products γ-aminobutyrate (GABA), β-alanine, and taurine are inhibitors as is also glycine. Of these GABA is the most important. Its concentration in the brain is high and varies at least threefold in different parts of the brain. It is hardly present elsewhere in the body. GABA and GABA-binding sites are found in 30 – 50% of the nerve endings. The function as an inhibitory transmitter has also been demonstrated in inhibitory neurons present in the peripheral nervous system of arthropods. Virtually every neuron in the brain is to some extent subject to inhibition by GABA. Glial cells also have GABA receptors.

The receptors for GABA are divided into type A, which are blocked by *bicuculline,* and type B, which are stimulated by *baclofen* (Fig. 7.25). The $GABA_A$ receptors are the major sites of fast synaptic inhibition in the central nervous system. They are structurally related to the riicotinic acetylcholine, glycine, and serotonin type 3 ($5\text{-}HT_3$) receptors. Cloning has revealed 16 different mammalian subunits: $\alpha 1 - \alpha 4$, $\beta 1 - \beta 3$, $\gamma 1 - \gamma 3$, δ, ε, π and Φ. The oligomeric receptors are ligand-gated chloride ion channels as are also glycine receptors. These receptors are clustered in synaptic membranes, apparently anchored in part by their β subunits and scaffold proteins such as the microtubule-binding *gephyrin* (from the Greek word for bridge) and a small ~14-kDa GABA receptor-associated protein. A novel serine protein kinase is also associated with GABA receptors.

Whereas excitatory transmitters lead to depolarization of the postsynaptic membrane, inhibitory transmitters cause *hyperpolarization,* apparently by increasing the conductance of K^+ and Cl^-. The result is that it is more difficult to excite the postsynaptic membrane in the presence of, than in the absence of, these transmitters. GABA-dependent interneurons also contain the calcium-binding *parvalbumin* (Fig. 7.7), which suggests that a Ca^{2+}-dependent process is involved.

The $GABA_B$ receptors resemble metabotropic glutamate receptors. They are 7-helix G-protein coupled proteins, which activate adenylate cyclase.

They tend to dimerize, and maximum activity is observed for heterodimers of $GABA_B$ 1 and $GABA_B$ 2 receptors. They are often coupled to inward rectifying K^+ channels.

The GABA receptors provide binding sites for a great variety of toxins and drugs. These include barbiturates, anesthetics, antianxiety drugs, and the insecticides such as toxaphine, cyclodienes, and pyrethroids. *Diazepam, chlordiazepoxide,* and flurazepam (Fig. 7.25) are antianxiety drugs and muscle relaxants, which, during the 1970s, were the most frequently prescribed drugs in the United States. Binding of benzodiazepines to GABA receptor-chloride channels enhances the effect of GABA. The drugs induce relaxation but can interfere with memory, reduce concentration, and cause physical clumsiness. They may also intensify the effects of alcohol and can be addictive.

Specific antagonists for $GABA_A$ receptors include the alkaloid convulsants bicuculline (Fig. 7.25) and *picrotoxin* (Fig. 7.4) and the convulsant terpenoid compound *thujone* (Fig. 7.3), which is present in the wormwood plant *Artemesia absinthium.* Thujone is present in the liqueur absinthe, which was the national drink of France in the late 19th century but, because of its toxicity, has been illegal in most countries since ~1915.

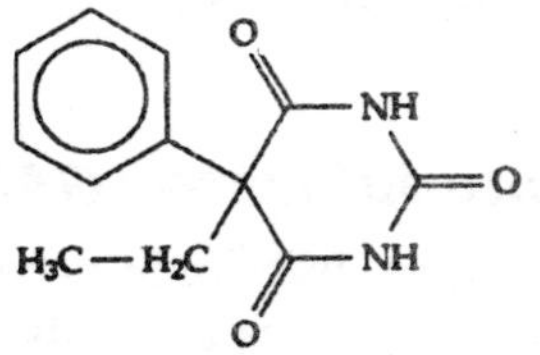

Barbital

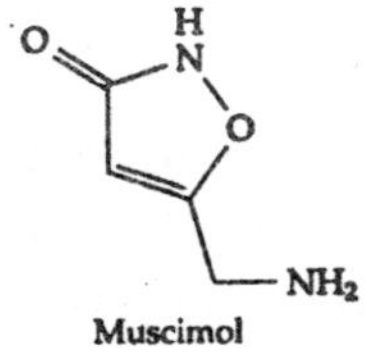

Muscimol

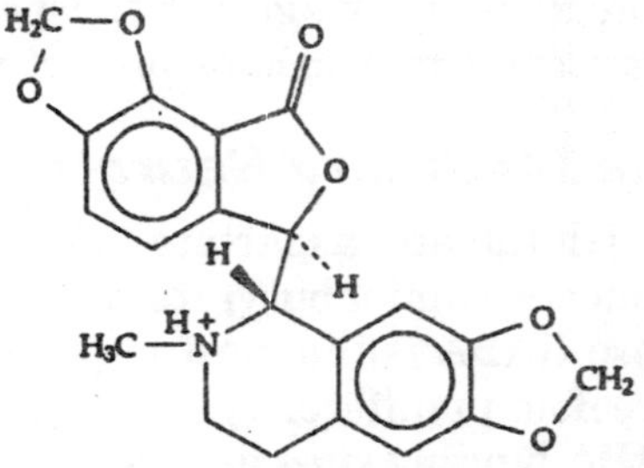

Bicuculline, an antagonist of γ-aminobutyrate (GABA)

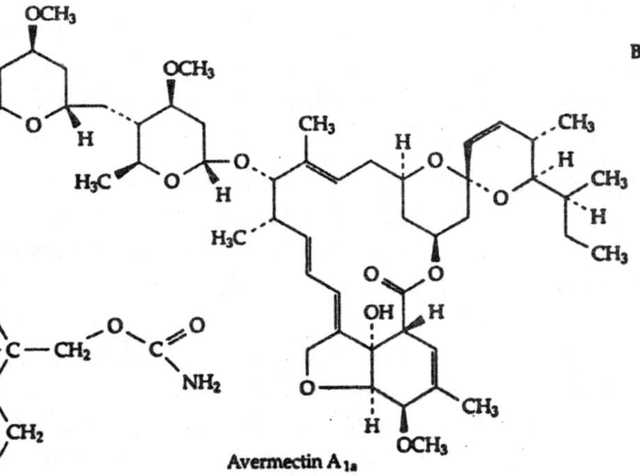

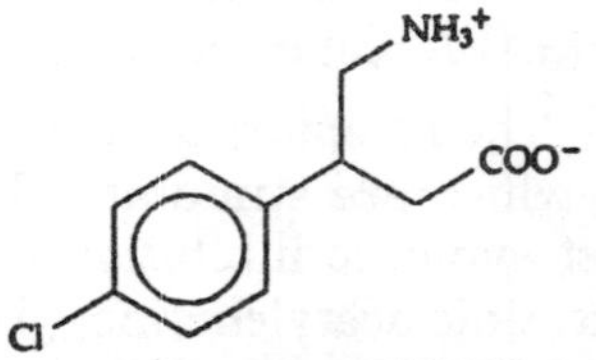

Baclofen, an agonist of $GABA_B$ receptors, muscle relaxant

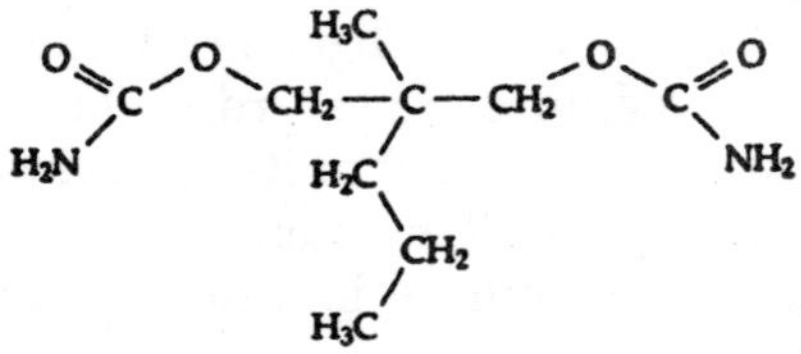

Meprobamate (Miltown)

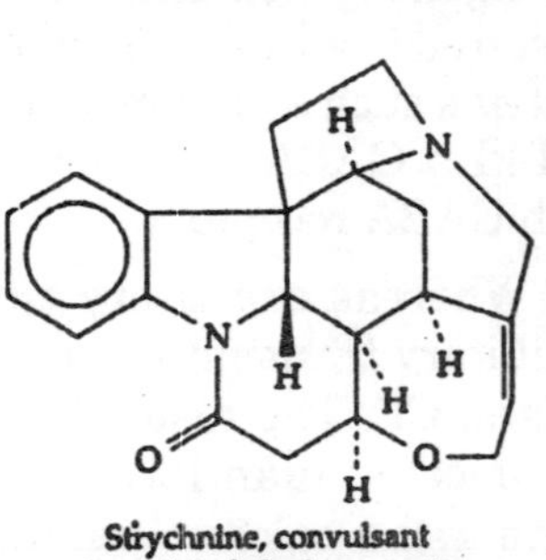

Avermectin A_{1a}
(Ivermectin is a similar semisynthetic compound.)

Strychnine, convulsant antagonist of glycine receptors

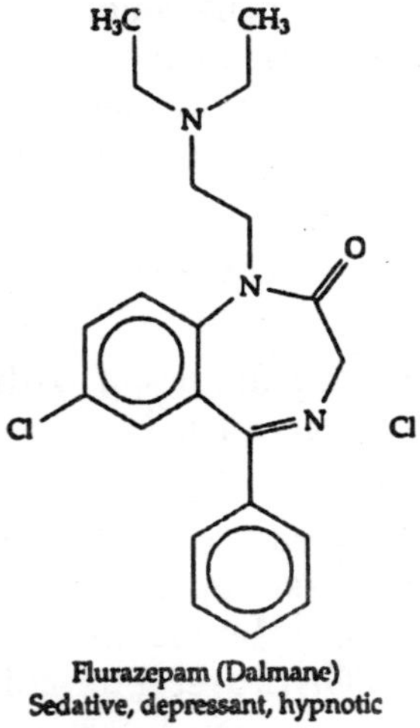

Flurazepam (Dalmane)
Sedative, depressant, hypnotic

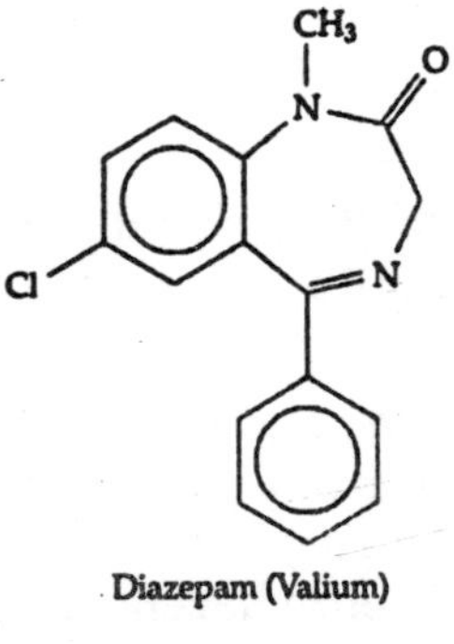

Diazepam (Valium)

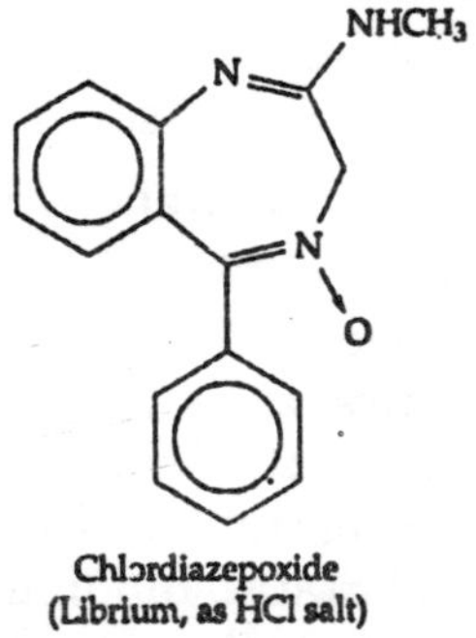

Chlordiazepoxide
(Librium, as HCl salt)

Fig. 7.25. Some toxic antagonists and some useful drugs that bind to receptors of γ-aminobutyrate (GABA) or for glycine.

GABA enters synaptic vesicles via a vesicular GABA transporter, an integral membrane protein whose gene has been found in *Caenorhabditis elegan*. Termination of GABA neuretransmission is accomplished by rapid Na^+-dependent uptake into neurons for reuse and uptake into glial cells. Excess GABA is continuously oxidized to succinic semialdehyde by GABA aminotransferase in the GABA cycle of Fig. 7.4. Notice the manner in which this cycle incorporates synthesis of both of the neurotransmitters glutamate and GABA. Glutamine also functions in neurons, perhaps serving as a buffer for glutamate.

The hereditary triple-repeat disease Huntington's chorea *(Huntingdon disease),* with an incidence of 5 – 10 per 100,000 persons, affects principally persons of age over 40 and is associated with a deficiency of GABA in basal ganglia. The cortex is also affected. Severe neurologic symptoms arise as a result of premature death of neurons in the basal ganglia. Convulsions may also arise because of a deficiency of GABA in the brain.

Glycine

Glycine appears to be the most important neuroinhibitor in the spinal cord and brainstem. It is present at concentration of 3 – 5 mM in the spinal cord and in the medulla but is low in the cerebral cortex. *Strychnine* (Fig. 7.25) is a specific antagonist of glycine receptors in spinal synapses. Ivermectin (Fig. 7.25) also blocks glycine Cl^- channels. A mutant mouse called *spastic* is deficient in glycine receptor function. A small dose of strychnine produces an effect on a normal mouse that resembles the effect of this mutation. A similar disorder affects some Hereford calves. Strychnine-binding studies have suggested a deficit of glycine receptors in human spasticity and in the loss of motor control associated with *Parkinson disease* and *amyotrophic lateral sclerosis.* A human *startle disease,* which causes an exaggerated muscular response to unexpected stimuli, also results from reduced glycinergic neurotransmission.

Most glycine receptors are Cl^- ion channels that open in response to transmitter binding. The strychnine-binding subunit shows significant homology with the nAChR proteins, and the overall structures resemble those of GABA receptors and of nAChRs. Human α1 – α4 and β subunits have been identified. Two integral membrane glycine transporters are known.

Anesthetics

Several types of neurotransmitter receptors provide binding sites for anesthetics. Some anesthetics are molecules of moderate size, *e.g., barbiturate* derivatives, while others, such as *diethyl ether* or *halothane* ($CF_3CHClBr$), are very small. The latter is one of the most widely used inhalation anesthetics. Both Mg^{2+} and Mn^{2+} are also powerful CNS depressants and can cause general anesthesia. It has often been proposed that the effectiveness of anesthetics is related to solubility in lipids, but it has been difficult to pinpoint a site of action. Now it is clear that specific synaptic proteins often provide the binding sites for anesthetics. Important among these are the glycine receptors. GABA receptors and kainate glutamate receptors may also bind anesthetics.

Adrenergic Synapses: The Catecholamines

The three closely related tyrosine metabolites, *dopamine, noradrenaline,* and *adrenaline,* known collectively as catecholamines, are important products of neuronal metabolism. Dopamine and noradrenaline serve as neurotransmitters. Catecholamine-containing neurons are found throughout the brain, including the cortex and cerebellum regions. Very large dopamine-containing neurons are present in the brains of gastropod molluscs. In the human brain a prominent series of dopamine neurons run from the substantia nigra to the caudate nuclei and putamen of the striarum, the *nigrostriatal* pathway (Fig. 7.12). In many invertebrates *octopamine,* which is synthesized via tyramine (Fig. 7.26), apparently functions in place of

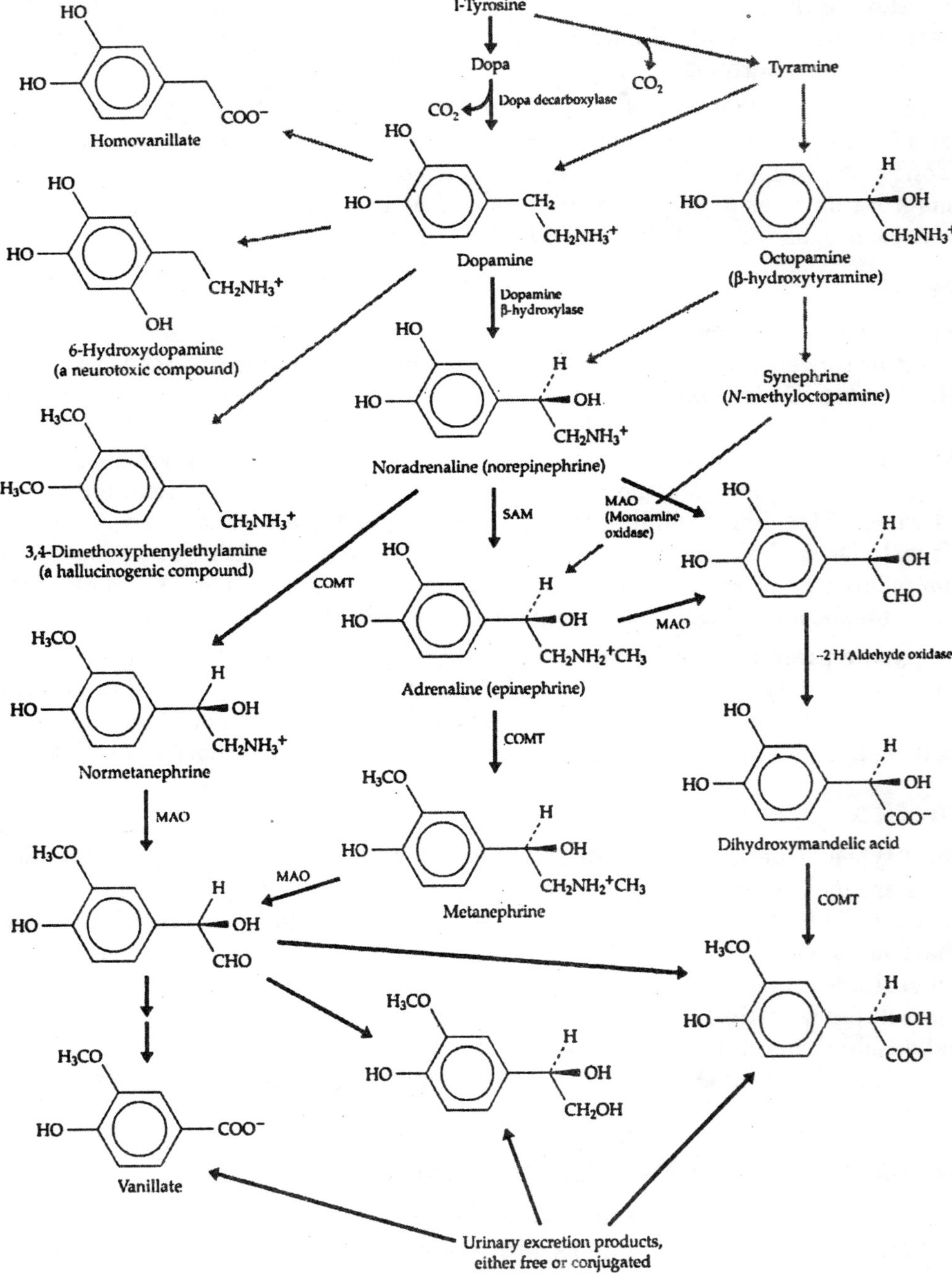

Fig. 7.26. Some pathways of metabolism of the catecholamines.

noradrenaline. Note the precursor-product relationship between dopamine, noradrenaline, and adrenaline. The synthetic pathways to these neurotransmitters involve decarboxylation and hydroxylation, types of reaction important in formation of other transmitters as well. The most important process for terminating the action of released catecholamine transmitters is reuptake by the neurons. High-affinity uptake systems transport the catecholamine molecules back into the neurons and then into the synaptic vesicles. The uptake is specifically blocked by the drug *reserpine*. The dopamine transporter is a major binding site for cocaine. Catecholamine transmitters are catabolized by two enzymes. One is the flavoprotein *monoamine oxidase* (*MAO*), an enzyme present within the mitochondria of neurons membranes as well as in other cells in all parts of the body. The second enzyme is *catechol-O-methyltransferase* (*COMT*; Eq. 7.3), which is found in postsynaptic membranes as well as in liver, kidney, and other tissues. It apparently provides the principal means of inactivating that occurs in all organisms sulfo groups are transferred from PAPS onto hydroxyl groups of catecholamines, steroid compounds, and proteins. Sulfation of catecholamines is relatively specific to humans.

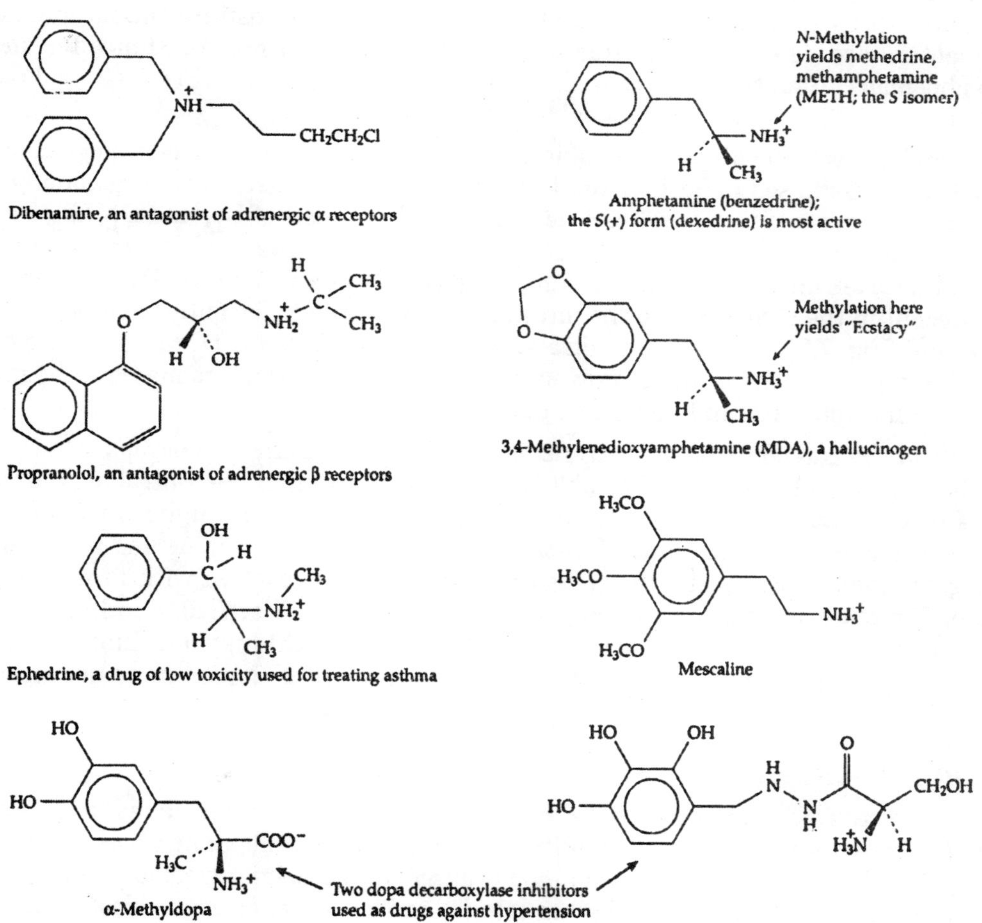

Fig. 7.27. Some agonists and antagonists of adrenergic synapses (shown as cations in most cases).

Both adrenaline and noradrenaline stimulate smooth muscles throughout the body and have a hypertensive effect. Their postsynaptic receptors are 7-helix transmembrane proteins (Fig. 7.6). A comparison of the effects of various analogs led to the classification of these receptors into classes α, α_2, β, and β_2. The α receptors, which are structurally closely related to rhodopsin, are coupled via Gq/11 proteins to a phosphoinositide-activated phospholipase C (Figs. 7.9, 7.19). They usually provoke an excitatory response. However, in intestinal smooth muscles they are inhibitory. Adrenaline is usually more active at a receptors than is noradrenaline. A specific antagonist *is dibenane* (Fig. 7.27). The β receptors usually induce muscular relaxation but cause myocardial stimulation. Noradrenaline is usually more active than adrenaline. In most cases the β receptors of the postsynaptic membrane respond to the neurotransmitter by causing a hyperpolarization of the cell membrane and inhibition of nerve impulses. A specific antagonist is *propranolol* (Fig. 7.27). The β receptors are coupled via proteins of the G_s family. The $\beta 2$ receptors have received special attention because of their importance to heart and pulmonary functions. Both heart failure and asthma are associated with poor $\beta 2$ receptor function. The $\beta 2$ receptors affect many other processes including insulin action. Intense efforts are being made to understand them at the structural level. Of special interest are the mechanisms by which receptors are densensitized after passage of impulses, a process that often involves multiple phosphorylation reactions as well as interaction with *arrestin* (Fig. 7.43) and receptor internalization.

Attention has been focused on dopamine because of its relationship to neurological diseases and to addiction (discussed in Section 10). Dopamine receptors constitute a large family, which are classified into two main subfamilies. The D_1 subfamily consists of D_{1A} and D_{1B} (D_5) receptors and the D_2 subfamily of D_2, D_3, and D_4 receptors. The D_1 receptors, which are prominent in the prefrontal cortex and also in the striatum, are more abundant than the D_2 receptors, which are also present in the striatum and the pituitary and are targets for antipsychotic drugs such as *haloperidol* (Fig. 7.33). The recently discovered and less numerous D_3 receptors are present in only a few regions of the brain. However, a deficiency of D_3 receptors may also be involved in addiction, schizophrenia, and Parkinson disease.

The role of the catecholamines as transmitters in the sympathetic nervous system and in the peripheral ganglia has been well established, but the function in the central nervous system is less clear. Catecholamines are present in varying quantities throughout the brain, and histochemical techniques have made it possible to visualize both dopamine and noradrenaline-containing neurons by the green fluorescence produced from reaction with formaldehyde or glyoxylate. The reactions are presumably analogous to those in Fig. 7.10. Another method for tracing dopamine receptors in the central nervous sytem is through labeling with specific antibodies to dopamine-β-hydroxylase, the enzyme that converts dopamine to noradrenaline, to tyrosine hydroxylase, or to other specific neuronal enzymes.

Parkinson Disease

Neurons of the nigrostriatal pathway degenerate in Parkinson disease, a condition accompanied by severe tremors and rigidity. The significance of dopamine was illustrated by the finding that the precursor amino acid *L-dopa* caused dramatic improvement in many persons with Parkinson disease. Dopamine and other catecholamines do not cross the blood-brain barrier but L-dopa does. This leads to an increase in the dopamine level in the basal ganglia of the brain, which apparently compensates for the deficiency resulting from the neuronal degeneration.

In 1982 a number of young people in California injected themselves with an illegally manufactured opiate drug that was subsequently found contaminated with N-methyl-4-phenyltetrahydropyridine (MPTP). Within a few days they developed irreversible symptoms of Parkinson disease. Subsequent investigation revealed that MPTP itself is not toxic but that it is oxidized by monoamine oxidase B (MAO-B) to the corresponding pyridinium derivative MPP^+ (Eq. 7.4). It is this pyridinium derivative, or perhaps related free radicals, that is toxic. MPP^+ is readily taken up by mitochondria and is apparently concentrated in the mitochondria of the nigrostriatal cells to a toxic level. The MAO inhibitor pargyline (Fig. 7.33) interferes with the

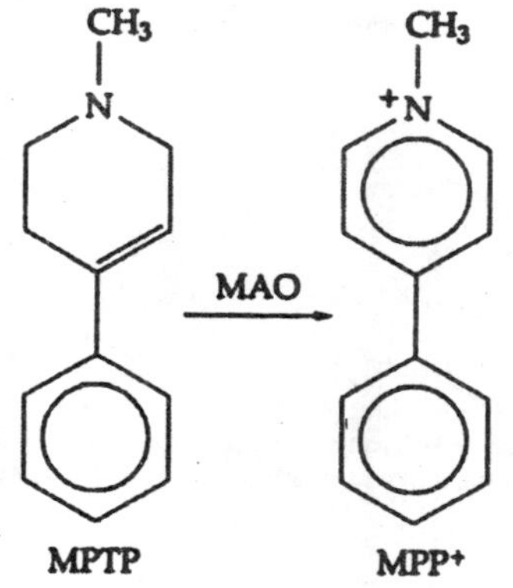

CH_3 CH_3 N $^+$N MAO

MPTP MPP^+

oxidation of Eq. 7.4 and prevents development of Parkinson disease in squirrel monkeys exposed to MPTP. These results suggested possible environmental causes for Parkinson disease and also a new approach to treatment. MPP^+ has been marketed as a herbicide, and it has a close structural relationship to another herbicide, *paraquat.*

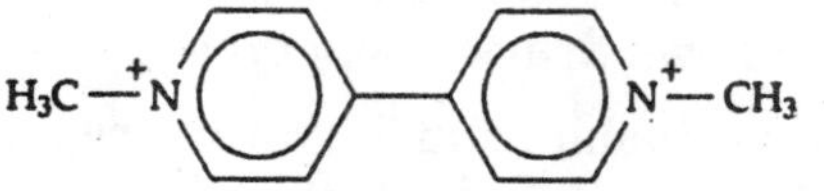

$H_3C—N^+$ $N^+—CH_3$

Paraquat

Many food constituents including peppermint, spearmint, and tea contain 4-phenylpyridine, another close relative. While administration of L-dopa to replace the deficit in the basal ganglia seemed the ideal treatment for Parkinson disease, mental deterioration is not stopped, and for some patients the drug loses its effectiveness in about three years. Based on the new information about MPTP, treatment with extra vitamin E as an antioxidant along with an MAO inhibitor is being tested as a way to prevent further damage from environmental toxins.

Serotonin and Melatonin

The indolealkyl amine serotonin (5-hydroxytryptamine, 5-HT; Fig. 7.28), is found in all mammalian brains and in invertebrates as well. Its distribution in the brain is limited, serotdnin-containing neurons being found in the raphe nuclei of the brainstern from which they ascend into the brain and down the spinal cord. Serotonin-containing neurons have been traced within brains of snails using ^3H-labeled serotonin. Studies with these simpler brains have revealed both inhibitory and excitatory responses to these neurons. Serotonin-accumulating neurons are also found in the retina and are widely distributed in the peripheral nervous system. Serotonin-containing granules are present in blood platelets.

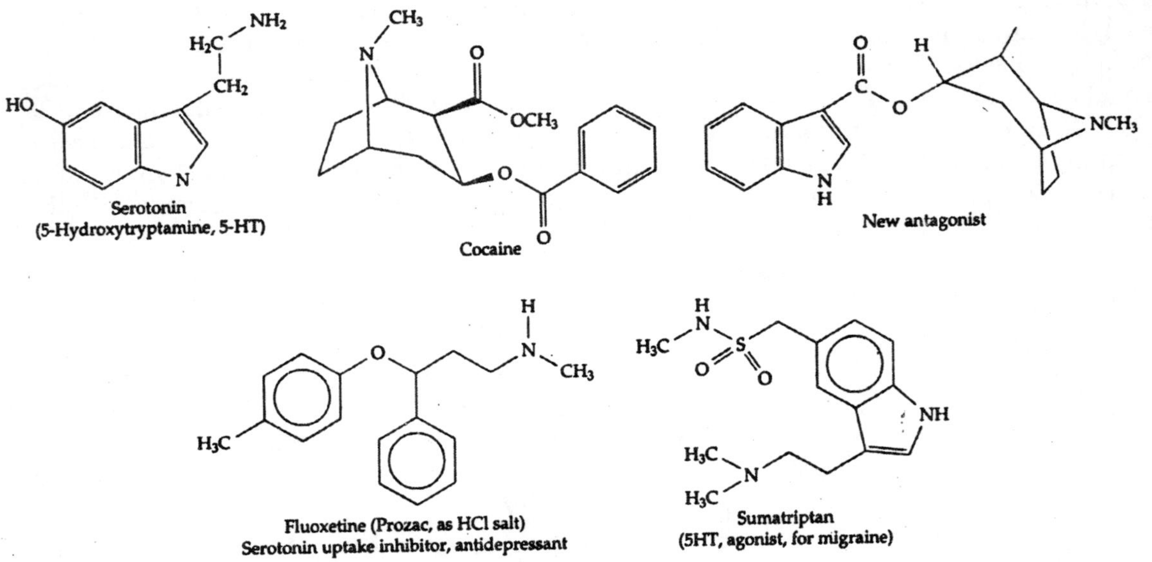

Serotonin (5-Hydroxytryptamine, 5-HT)

Cocaine

New antagonist

Fluoxetine (Prozac, as HCl salt)
Serotonin uptake inhibitor, antidepressant

Sumatriptan
(5HT, agonist, for migraine)

Fig. 7.28. Serotonin (5-hydroxytryptamine) and some drugs that affect receptors and transporters.

Serotonin appears to be involved in activation of pain fibers, when tissues are injured. *Cocaine* (Fig. 7.28) is a powerful pain killer and a weak antagonist of responses to serotonin, a fact that has led to the synthesis of new antagonists, such as the one in Fig. 7.28 whose structure encompasses that of both cocaine and serotonin. It is active at a concentration as low as 10^{-14} M and is among the most potent known drugs of any type.

Serotonin is synthesized via tryptophan and 5-hydroxytryptophan with decarboxylation of the latter (Fig. 7.12). Within the *pineal body* of the brain and in the retina, serotonin is acetylated to *N*-acetylserotonin, which is then *O*-methylated to *melatonin,* the pineal hormone (Fig. 7.12). A specific inhibitor of serotonin synthesis is p-chlorophenylalanine, and studies with this and other inhibitors suggest that serotonin is required for sleep.

At least 14 distinct types of serotonin receptors (5-HT_{1A}, 5-HT_{1B}, 5-HT_{2A}, etc.) have been identified. They are present in the heart, in gastrointestinal tissues, adrenal and other glands, and bone as well as in the brain. Drugs, such as *sumatriptan* (Fig. 7.28), which activate serotonin receptors are important in the treatment of *migraine.* This common disorder of serotonin metabolism is characterized by severe or moderately severe headache and a variety of other symptoms, which are frequently preceded by a visual aura. Serotonin is removed from synapses via a transporter, which also contains the binding site of the widely used antidepressant *Prozac* (Fig. 7.28) and related drugs.

Serotonin and melatonin are evidently involved in maintenance of the 24-h circadian rhythm of the body. Melatonin regulates the sexual cycle in photoperiodic animals and influences the on-set of puberty. The serotonin content of the brain is influenced by the diet, being higher after a meal rich in carbohydrates. Serotonin may serve as a chemical message sent from one set of neurons to the rest of the brain, reporting on the nature of dietary intake. Melatonin, which can readily form free radicals, may function as part of the body's antioxidant system.

Other Neurotransmitters

The abundant glutamate, GABA, and glycine are major neurotransmitters. Do other amino acids also function in the brain? Roles for L-aspartate and D-serine have been identified, but it is very difficult either to discover or to disprove a neurotransmitter function for other amino acids. It is even more difficult for small amounts of various amines and small peptides that are present in the brain. *Taurine* (Fig. 7.25) is one of the most abundant free amino acids in animals and meets several criteria for consideration as both an inhibitory and an excitatory transmitter. However, its function is still uncertain. *Homocysteic acid,* formed by oxidation of homocysteine, is a powerful neuroexcitatory substance, but its concentration in the brain is very low. D-Aspartate is also present at high concentrations in the cerebellum, pituitary, pineal gland, and adrenal chromaffin cells. It appears to be a modulator of melatonin synthesis.

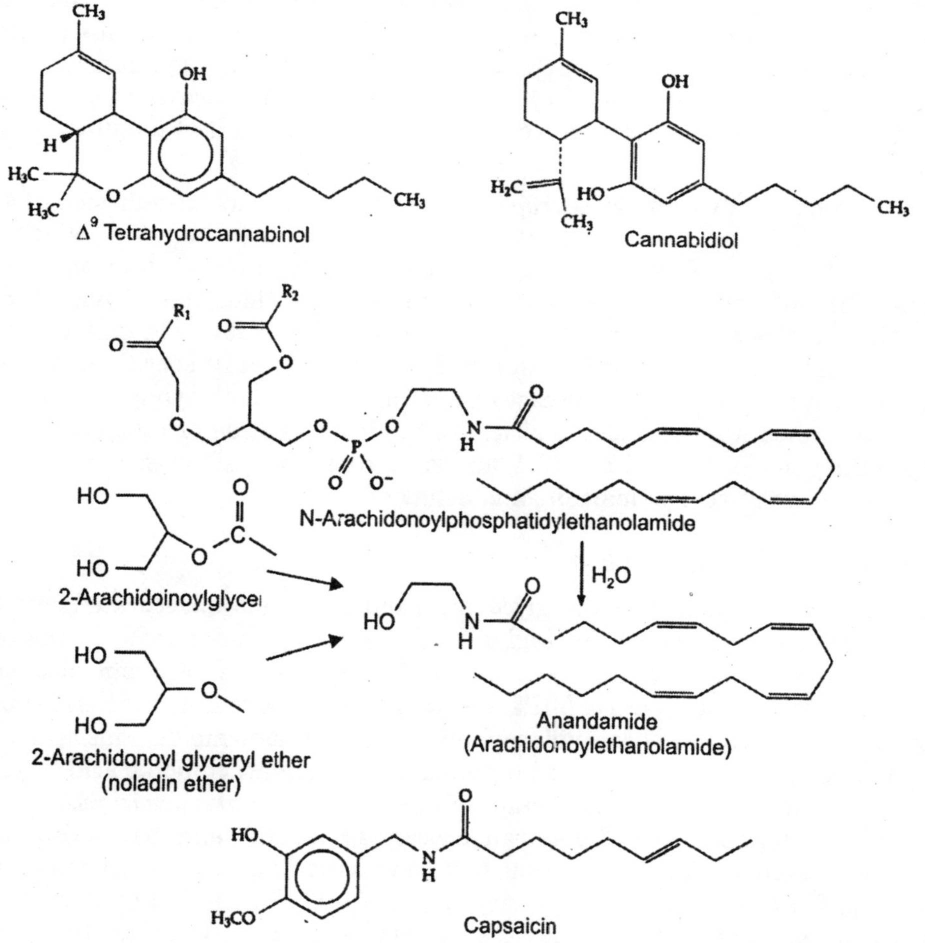

CH_3 OH H H_3C H_3C O CH_3

Δ^9 Tetrahydrocannabinol

CH_3 OH H_2C HO CH_3 CH_3

Cannabidiol

R_1 R_2 O O O O O O P O O^- N H O

N-Arachidonoylphosphatidylethanolamide

HO HO O C O

2-Arachidoinoylglycerol

H_2O

HO N H O

HO HO O

2-Arachidonoyl glyceryl ether (noladin ether)

Anandamide (Arachidonoylethanolamide)

HO H_3CO N H O

Capsaicin

Fig. 7.29. Structures of the active components of cannabis, tetrahydrocannabinol, and cannabidiol, and structures of endogenous cannabinoids and of the vanilloid lipid capsaicin.

Receptors for *histamine,* which probably acts as a neuromodulator, occur in the brain. Histamine is formed by decarboxylation of histidine and is inactivated by histidine

N-methyltransferase. Histamine is best known for its presence in mast cells, components of the immune system that release histamine during inflammatory and allergic reactions. However, histaminergic neurons of the hypothalamus extend throughout the whole forebrain, and specific receptors have been found both in the brain and in peripheral tissues. Several other amines that are formed by decarboxylation of amino acids are present in trace amounts but may have important functions, some of which may be related to psychiatric disorders. These include tyrantine (from tyrosine), β-phenylethylamine (from phenylalanine), and tryptamine (from tryptophan). As previously mentioned, octopamine is also present in trace amounts in. mammalian brains.

ATP, ADP, and *adenosine* are among the purines that are present in some synapses and activate a variety of receptors. Adenosine receptors are blocked specifically by methylated xanthines such as caffeine (Fig. 7.18) and theophylline. A drug almost 10^5 times as potent as theophylline is 1, 3-dipropyl-8-(2-amino-4-chlorophenyl)xanthine. Adenosine receptors, Which are present in large numbers in the hippocampus, form functional complexes with metabotropic glutamate receptors. Adenosine usually has a depressive effect. Craving for chocolate is often attributed to the methylxanthines present, but it may be a result of anandamide and related compounds.

The occurrence of a variety of *neuropeptides* in the brain has been discussed in Section A. The first of these to be discovered was the 11-residue *substance* P (Table 7.4), which was isolated in 1931. Like other neuropeptides it may function either as a transmitter or neuromodulator or perhaps both. Substance P, as well as many other neuropeptides, has been localized to specific neurons. Along with somatostatin, CCK, and enkephalins, it is found in high concentrations in the basal ganglia. Enkephalin and substance P are also found in specific neural elements in the visual system of lobsters. In some cases a neuron contains both synaptic vesicles containing a major neurotransmitter and also vesicles containing a peptide or other cotransmittor. The peptide pituitary hormones ACTH, MSH, and vasopressin as well as the hypothalamic neurohormones may have effects on learning and behavior.

Lipid Mediators in the Brain

The brain is rich in phospholipids, glycolipids, and long-chain unsaturated fatty acids. Many signaling functions seem likely. Prostaglandin D_2 is a major prostanoid in the brain, which induces both hypothermia and sleep. As mentioned in Section A, 7, *oleamide* also induces sleep, perhaps by modulating the effects of 5-HT receptors. *Anandamide* is a lipid derived by hydrolysis of the unusual phospholipids *N*-arachidonoyl-phosphatidylethanolamide. This is one of a recently discovered series of amides, esters, and ethers derived from arachidonic acid (Fig. 7.29). They have been identified as endogenous ligands of the abundant *cannabinoid receptors.* The latter were identified as binding sites of Δ^9-tetrahydrocannabinol and cannabadiol (Fig. 7.29), both of which are constituents of *marijuana.* Anandamide was the first of the endogenous cannabinoids to be isolated. However, the monoglyceride 2-*arachidonoylglycerol* (Fig. 7.29) is much more abundant in brain and also activates cannabinoid receptors. It arises by hydrolysis of a diglyceride. Recently 2-arachidonoyl glyceryl ether *(noladin ether;* Fig. 7.29) has been identified as another endogenous agonist of the CB_1 cannabinoid receptors. A possible alternative pathway for anandamide synthesis is via an energy-dependent coupling of arachidonic acid with ethanolamine. The two known types of cannabinoid receptors are both 7-helix proteins coupled by G_i or G_o proteins to adenylate cyclase and to Ca^{2+} and K^+ channels. The CB_1 receptors are found largely in the brain and are responsible for the psychoactive effects of cannabis, while the CB_2 receptors are more widely distributed. They seem to have a special role in cells of the immune system, *e.g.*, in macrophages and B cells. Palmitoylethanolanide has been proposed as an additional endogenous ligand for CB_2 receptors. Cannabinoid receptors of invertebrate immune system cells and of human monocytes have been found coupled to NO release.

Cannabinoid receptors are present at extremely high levels in the basal ganglia of the brain, but they do not appear to be essential. Knockout mice lacking the CB_1 receptors appear normal in most respects. However, they do not respond to cannabinoid drugs and, curiously, do not become addicted to morphine as normal mice and have less severe withdrawal symptoms than normal after morphine addiction. The CB_1 receptors in the basal ganglia modulate GABA neurons that have outputs to the substantia nigra and the globus pallidus (Fig. 7.30B). The nigrostriatal neurons also secrete substance P and dynorphin, while those extending to the globus pallidus generally contain enkephalin as a cotransmitter. These interconnections affect the dopaminergic neurons. Cannabinoids also have pain supressing and neuro-protective effects. They may have many possible medicinal uses, which are being explored.

The endogenous cannabinoid compounds are lipids and are not stored in synaptic vesicles but are presumably released by enzymatic action following passage of a nerve impulse. Recent evidence suggests that the endocannabinoids are released at a postsynaptic membrane and then diffuse back to a presynaptic surface and outward to other cell surfaces where they affect signaling. This *retrograde signaling* in synapses of the hippocampus is thought to be involved in *long-term potentiation* (LTP), the changes in synaptic properties that occur during learning and in the formation of memories. A monoglyceride lipase participates in inactivation of endocannabinoids. Anandamide is also a substrate for cyclooxygenase-2 (Eq. 7.16), whose action may lead to formation of additional immunomodulatory compounds. Long-chain relatives of arachidonic acid such as docosohexaenoic acid are especially high in brain lipids.

Nitric Oxide and Carbon Monoxide

The gaseous molecules NO and CO have both been found in the brain, and neuronal NO synthase (nNOS or NOS I) has been studied intensively. Complexity in understanding the role of NO in the brain arises from the fact that different isoenzyme forms of NO synthase occur in three different types of cell: nNOS in neurons, iNOS from microglial immune system cells, and eNOS from endothelial cells of capillary blood vessels. All three types of cells are so tightly intermingled in the brain that it is hard to interpret observed experimental effects. Elevated Ca^{2+} concentrations that can arise from stimulation of NMDA receptors in the hippo-campus seem particularly effective in activating the calmodulin-dependet nNOS. This suggests that, like the endogenous cannabinoids, NO may be a retrograde messenger in LTP. The possibility that CO may function in a similar way also remains uncertain, as does any pathway for metabolism of CO. Certainly NO and CO generated in the brain will have some effects that arise from their very tight binding to heme groups. An example is the observed inhibition of dopamine (β-hydroxylase by N_2O_3 with a resulting decrease in noradrenaline synthesis.

10. Some Addictive, Psychotropic, and Toxic Drugs

Humans have a long history of use of stimulant and mind-altering substances. Tea, coffee, alcohol, tobacco, opium, cocaine, marijuana, and a host of modern synthetic compounds have been used as stimulants, as medications, and for pleasurable experiences. Many are also addictive and sometimes lethal. Stimulant drugs such as nicotine, cocaine, methamphetamine (METH), and other amphetamines (Fig. 7.27) can give users feelings of increased energy, well-being, and self-confidence. Nicotine enhances fast excitatory transmission and may sharpen memory. However, all are acutely toxic and are highly addictive. Amphetamines and cocaine act directly to increase the brain dopamine level causing euphoria. However, in response the dopamine receptors rapidly decrease their sensitivity. This leads to mental depression and the desire for more drug. Nicotine appears to indirectly affect the same dopamine neurons. The wisdom and ethics of giving hypoactive children the addictive stimulant *methyl-phenidate* (Ritalin; Fig. 7.33) have been questioned. The depressive drugs, including *morphine* and other narcotics (Fig. 7.30), barbiturates (Fig. 7.22), and ethanol, are all strongly addictive for susceptible individuals. The phenomenon is most striking in the case of the opiates. Addiction leads to physical dependence, a situation in which painful withdrawal symptoms occur in the absence of the drug. At the same time a striking tolerance to the drug is developed. The addicted individual can survive what would otherwise be a fatal dose without ill effect. Aside from the pathological hunger for the drug, an addict can function normally in almost every respect. Dependence develops only from frequent doses of drug over a long period of time and is not observed with cocaine or amphetamines. Marijuana is only mildly addictive, according to some data about the same as caffeine. However, this conclusion is controversial.

Opioid Receptors

Direct binding of highly radioactive opiates has permitted localization of specific opiate receptors of several types. The three major types (μ, δ, κ) are all 7-helix receptors coupled toadenylate cyclase, K^+ and Ca^{2+} channels, and the MAP kinase cascade. The μ receptors bind morphine most tightly. These receptors are found in various cortical and subcortical regions of the brain. Most narcotics are polycyclic in nature and share the grouping indicated in Fig. 7.30. However, the flexible molecule methadone binds to the same receptors. Among antagonists that block the euphoric effects of opiates the most effective is *naloxone* (Fig. 7.30).

What is the natural function of opiate receptors? Opiates are the most powerful *analgesic agents* known. The existence of the *enkephalins, endorphins,* and *endomorphins* (Section A, 5; Table 7.4) in the brain suggests that opiate drugs mimic the normal action of these peptides, which may function in controlling pain. Although opiates are powerful drugs, their efficiency in diminishing pain is directly related to their addiction potential. To date, it has not been possible to design a nonaddictive analgesic drug of the potency of morphine.

Addiction seems to follow compensatory changes in the receptor-agonist system that result from the occupation of the receptor sites by the drug. For example, studies of opiate receptors indicate that morphine acts in an inhibitory fashion, lowering the internal level of cAMP. The neuron then compensates by increasing the number or activity of adenylate cyclase molecules restoring the internal cAMP level. This leads to dependence upon morphine because in its absence the cAMP level rises too high. The increased number of adenylate cyclase molecules and associated receptors also accounts for the observed tolerance. It is now clear that this adaptation is complex. The properties of many synapses in various parts of the brain are altered by phosphorylation or dephosphorylation or other reactions of receptors and other synaptic proteins. Some changes are rapid, but others are slower and involve alterations in transcriptional patterns within neurons. These changes occur in three different neuronal systems: (1) physical control systems, in which changes lead to physical dependence; (2) motivational control systems; and (3) associative memory systems.

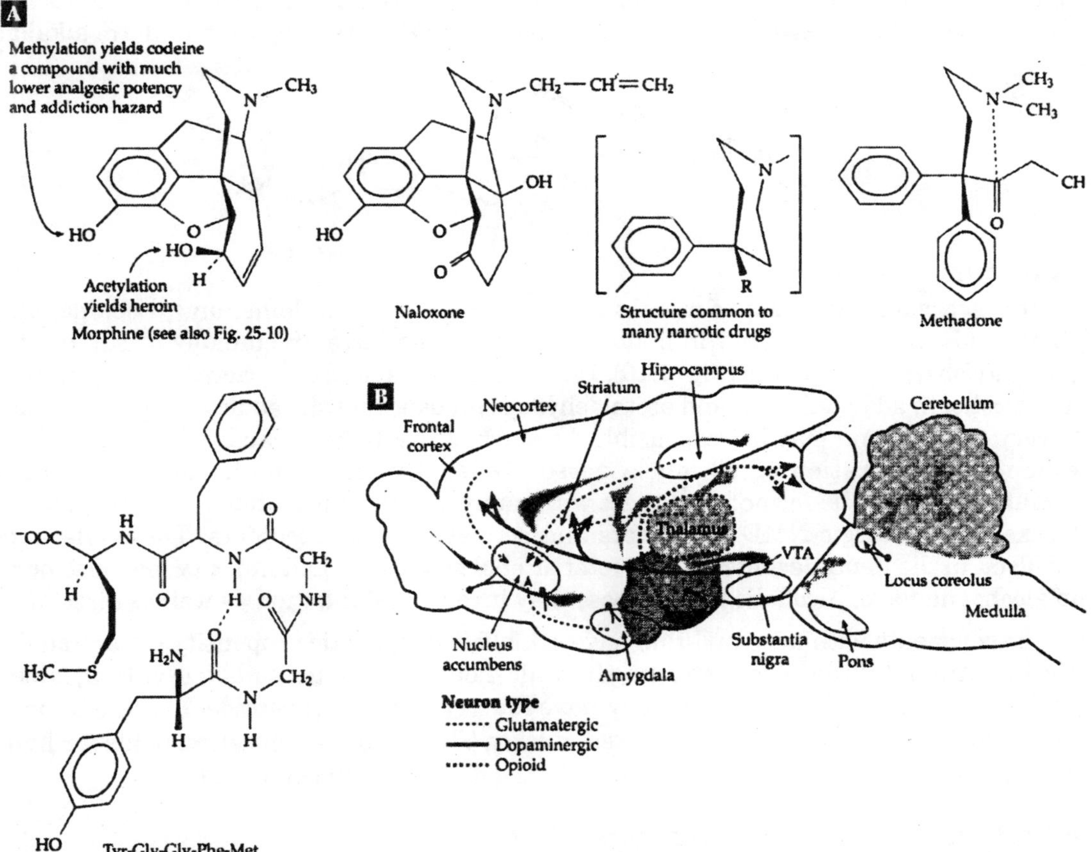

Fig. 7.30. (A) The structures of morphine and of some analogs including the brain peptide Met-enkephalin. Also shown is a structure common to many narcotic drugs. (B) Diagram of a rat brain with some aspects of the mesolimbic dopamine system emphasized.

The *mesolimbic dopamine system* is thought to be involved either directly or indirectly in addiction to many drugs. The dopaminergic neurons of this system have cell bodies in the *ventral tegmental area* (VTA) of the brain (Fig. 7.30B) and extend into the *nucleus accumbens,* a region at the base of the striaturn that is thought to provide the "rewarding effects," *i.e.,* pleasure from drugs such as cocaine or amphetamines. There is direct experimental support for this conclusion. Less certain is the proposal that opiates and other depressive drugs indirectly cause a similar effect in the nucleus accumbens. A more recent view is to regard addiction as an aberrant form of learning. This concept is applicable also to "behavioral addictions.

Ethanol

As with morphine addiction, tolerance to alcohol is developed, and a lack of ethanol produces withdrawal symptoms. The principal route of metabolism of ethanol (both ingested and the small amount of endogenous alcohol) is believed to be oxidation in the liver to the chemically reactive acetaldehyde, which is further oxidized to acetate. Some theories of alcoholism assume that addiction, and possibly also the euphoric feeling experienced by some drinkers, results from a metabolite of ethanol in the brain. For example, acetaldehyde could form alkaloids.

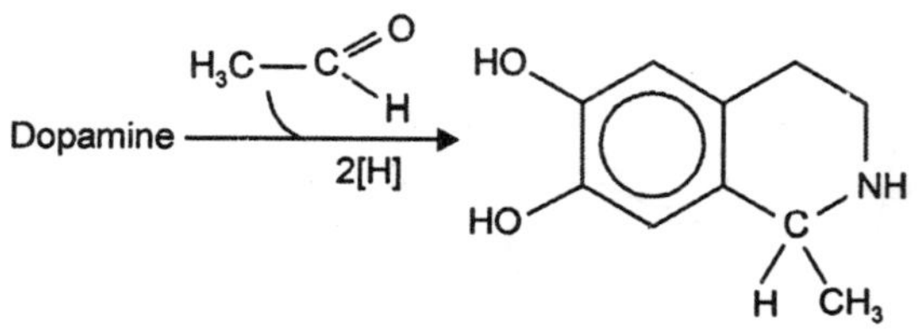

$$\text{Dopamine} \xrightarrow[2[H]]{H_3C-CHO} \text{(6,7-dihydroxy-1-methyl-1,2,3,4-tetrahydroisoquinoline: HO, HO, NH, C, H, } CH_3)$$

In fact, small amounts of morphine, 6-acetylmorphine, codeine, and thebaine, all opiate compounds, have been found in mammalian brain and have presumably arisen by the same pathway observed in plants (Fig. 7.10). However, there is no cross reactivity between morphine and alcohol in addicted mice, and acetaldehyde is probably not the addictive agent. Acetaldehyde is very reactive and may be responsible for much of the damage caused by ethanol. At a blood ethanol concentration of 20 mm a person is legally intoxicated, and large amounts of acetaldehyde may be formed and react with many amines, nucleotides, proteins, etc. Ethanol blocks glutamatergic NMDA receptors and activates GABA receptors. These effects may be involved in the neurodegeneration of fetal alcohol syndrome. Ecitotoxicity may also be a factor in alcohol damage. Alcoholic liver disease may involve malnutrition as well as direct damage.

Experiments with mice and rats have established a genetic propensity toward addiction to alcohol. Animals from some strains shun alcohol and become addicted only if forcefed for prolonged periods. Others, which may have low levels of neuropeptide Y in the brain, accept the alcohol readily and become addicted quickly. That a similar situation holds for humans is quite possible. However, a specific "alcoholism gene" has not been found.

Psychotropic or Mind-changing Drugs

Hallucinogenic compounds have long been a source of special fascination to many people. The presence of the indole ring in the powerful hallucinogen *lysergic acid diethylamide* (LSD; Fig. 7.12) suggests that this compound may mimic the action of serotonin. However, other experiments suggest antagonism of dopamine receptors in the striatum. Other hallucinogens include 3, 4-methylenedioxyamphetamine (MDA; Fig. 7.27), a compound that damages serotonergic neurons, its derivative "ecstasy" (Fig. 7.27), mescaline (Fig. 7.27), and phencyclidine (angel dust), a compound introduced in the late 1950s as a general anesthetic. Unfortunately, it produces a long-lasting condition resembling schizophrenia. A common site of action for a large variety of hallucinogens has been suggested.

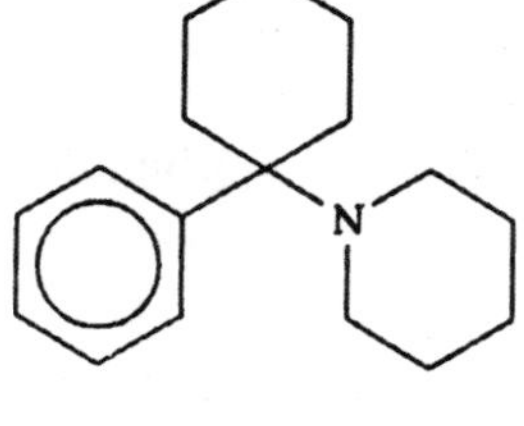

Phencyclidine (Angel dust)
a discredited anesthetic.

Neurotoxins Produced by the Body

Some normal body constituents are neurotoxic in excess. These incluse *quinolinic acid*, 3-*hydroxykynurenine*, and homocysteine. Elevated levels of homocysteine are also associated with vascular disease and stroke. 3-Hydroxykynurenine is a precursor to ommochrome pigments of insects and an intermediate in conversion of tryptophan into the nicotinamide ring of NAD in humans. 6-Hydroxydopamine, which may be formed in the body, is severely toxic to catecholaminergic neurons.

Other neurotoxins can be formed from environmental pollutants. The solvent 1, 4-butanediol is converted to *γ-hydroxybutyrate,* which is also a drug of abuse. Many compounds in commercial use have not been adequately tested as neurotoxins.

11. The Senses: Sight, Smell, Taste, Hearing, Touch, and Others

Our brains receive a continuous stream of impulses from receptors that sense light, taste and odor molecules; sound waves, touch, pain, gravitational pull, etc. Of these receptors those of vision, may be the best known. The photoreceptors consist of rhodopsin and related 7-helix proteins embedded in membranes of the rod and cone cells (Fig. 7.40). A complex series of control mechanisms, some of which are outlined in Fig. 7.43, permit enormous amplification of the initial signal generated by a G protein and a cGMP-gated ion channel. The array of rods and cones in the retina send messages via the optic nerve to the *visual cortex,* an area of ~15 cm^2 on the cerebral cortex surface at the back of the brain. The visual cortex is divided into two halves, but curiously, the right eye sends its signals to the left brain and vice versa. The image viewed by the retina can be mapped to the visual cortex. There it may reside in the form of chemical alterations in the ~40,000 neurons thought to be present in the visual cortex for a short time until it is stored in short-term working memory locations.

Receptors for the other senses, like those for sight, also consist of clusters, often in regular arrays, of 7-helix receptors. Most of these are also G protein-coupled ion channels that are controlled by cAMP or cGMP.

Odor

Even bacteria possess something akin to our ability to taste and smell. As is discussed in Chapter 19, Section A, many bacteria are attracted to L-serine or D-ribose and are repelled by phenol. Receptor proteins in the plasma membrane are involved in sensing these compounds and in allowing bacteria to move toward food and away from danger. Many other examples of chemotaxis are known among the lower invertebrates such as *Euglena.* Chemoreceptors in *Hydra* sense glutathione that flows from the broken tissue of their prey and control the animal's feeding behavior. Related organisms respond to proline. Asparagine induces the bending of the tentacles of the sea anemone *Anthopleura,* while glutathione induces swallowing. Salmon return to their home streams using a memory of specific odors.

Thoughout the animal kingdom the sense of smell is essential for survival. Perhaps it is not surprising that from the nematode *C. elegans* to human beings there is a largely conserved mechanism for sensing odors. A large array of 7-helix G protein-coupled olfactory receptors embedded in an epithelial membrane carry signals directly into the nervous system.

In *C. elegans,* which has only 302 neurons, there are 32 chemosensory neurons and more than 100 genes for 7-helix receptors that are expressed in these neurons. The fruit fly *Drosophila melanogaster* has at least 59 genes for olfactory receptors. Zebrafish and cat-fish have ~100. Mice and rats have ~1000 olfactory receptor genes and human beings at least 500, which account for about 1 – 4 % of the genome. In higher animals most receptors are coupled via G proteins, adenylate cyclase, and cAMP to ion channels in the membrane. Insects utilize both cAMP and Ins 3 – P in their chemosensory receptors. The signaling pathways parallel those of the visual receptors (Figs. 7.40, 7.43), which, however, utilize cGMP. Each gene is thought to give rise to a receptor of a specific *type* able to respond to specific structural features in an odor molecule.

Human olfactory cells are located in the *olfactory epithelium* on the upper surface of the back portion of the nasal cavity. They are neurons with chemosensory cilia similar to the rods and cones of the retina (Fig. 7.40). The cilia, which can be detached and isolated from the olfactory epithelium, contain the odorant-stimulated G-protein-dependent adenylate cyclase. There are ~10 million receptor cells of at least 500 – 1000 different types. The 10 million axons form bundles of ~5000 axons each and pass through small perforations in the skull directly into the *olfactory bulb* (at the front of the brain before the pituitary, Fig. 7.13), a distance of 3 – 4 cm. The cortex of the olfactory bulb is lined with ~1800 *glomeruli.* Each glomerulus is a bundle, ~0.1 – 0.2 mm in diameter, of synaptic endings of the neurosensory nerves coming from the olfactory epithelium with dendrites of neurons that run to the *olfactory cortex* and other regions of the brain. Each sensory receptor sends signals to a single glomerulus, but the glomerulus receives signals from 500 or more sensory neurons, which are not all of the same types. The glomerular cortex of the mouse is divided into four zones, each of which contains only some of the types of receptor. It seems that the cortex contains a crude "map" that relates position to the type of smell. The neural processing involved in the discrimination of odors is not yet clear. Interneurons of the olfactory bulb are unusual, being continuously discarded and replaced by new neurons that arise from neural stem cells. This process seems to be essential for odor discrimination but not for the sensitivity of odor detection.

Most mammals have a second olfactory apparatus, the *vomeronasal organ* (VNO) or "sexual nose," which is located on the lower surface of the nasal cavity. It is a fluid-filled cavity containing chemosensory receptors through which nasal fluid is literally pumped, when the animal seeks to maximize the sensitivity of detection. The VNO is especially important to reproduction, defense, and food-seeking. A specialized set of olfactory sensory neurons that project to. atypical glomeruli in the olfactory bulb utilize cGMP signaling and may also function in reproductive behavior.

The olfactory epithelia are bathed in an aqueous mucus through which odorant molecules must pass. A number of specialized proteins, including *odorant-binding proteins,* are secreted in this fluid. Many odorant-binding proteins are *lipocalins* and presumably assist in transporting lipophilic odorant molecules to the olfactory receptors. They tend to have a low specificity for the odorant and a weak binding affinity, properties that are consistent with this function. Pheromone-binding lipocalins encoded by ~30 genes are also found in rodent urine, where they play a similar role. In contrast, the pheromone-binding proteins of some male moths are largely α helical. Although pheromones are not as important to human physiology, axillary odors from both males and females do apparently carry chemical signals. One well-established effect is the synchronization of menstrual cycles of women living in the same house or dormitory. Alipoprotein D apparently serves as a binding protein that carries odorant precursors that are acted on by bacteria to produce the pheromones.

Virtually all people lack the ability to detect some specific odors. A striking example of such an *anosmia* is the inability to smell the volatile steroid *androsten-one* (5 α-androst-16-en-3-one), a constituent of perspiration, of some pork products, truffles, and celery.

Taste

Less is known about the biochemistry of taste. The taste that we perceive is affected by odor, temperature, and physical contact. However, five primary tastes are recognized.

Salty : apparently perceived by an ion-channel-linked receptor

Sour : apparently linked to an H^+ channel

Bitter : perceived by bitter-sweet G protein-coupled receptors

Sweet : also perceived by bitter-sweet receptors

Umami a recently recognized taste, that of glutamate

An experimental difficulty lies in the fact that there are only a few thousand taste buds in the tongue, with only 50 – 100 cells in a bud. They age rapidly, having a lifespan of only about ten days. There may be only 30,000 – 50,000 hard-to-isolate taste receptor cells on the tongue's surface. However, very recently published reports describe a large family of bitter and sweet receptors in mice and humans and in *Drosophila*. The sweet-sour receptors are thought to activate a G protein called *gustducin*, which plays a role similar to that of transducin in vision and also activates ion channels. Like the odor receptors, taste buds are also bathed in a special fluid. The *von Ebner's glands* in the tongue contain binding proteins, at least some of which are lipocalins.

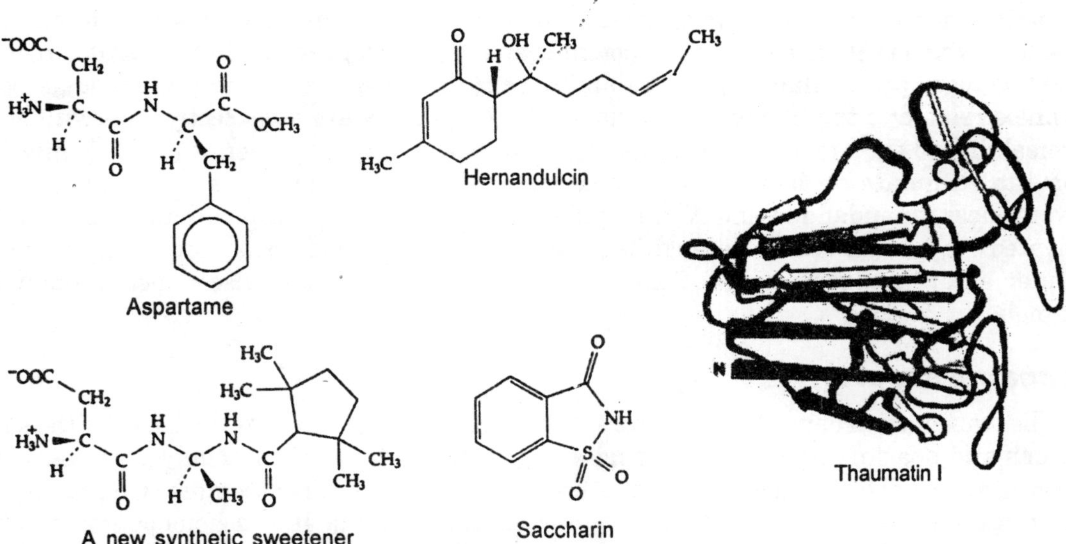

Fig. 7.31. Structures of some very sweet compounds. The backbone structure of the protein thaumatin I is included. The main body of this structure consists of two β sheets forming a flattened β barrel. β Strands in the top sheet are shaded light, and those in the bottom sheet are darker. Open bars represent disulfide bonds, and the regions with sequences homologous to monellin are indicated by the hatched marks.

The relationship between structures and sweet taste in humans has been investigated intensively, but no simple rules have been discovered (see Robyt for a discussion). Sucrose is usually perceived as very sweet. D-Fructose and D-xylose are nearly twice as sweet, but D-glucose is less sweet than sucrose. D-Galactose is usually perceived as not sweet and D-mannose as bitter. Many sucrose derivatives, in which hydroxyl groups have been replaced with Cl or other halogen, are very sweet. One tetrachloro derivative of this type is 7000 times sweeter than sucrose. Some especially sweet materials are depicted in Fig. 7.31. These include peptide derivatives such as Asp-Phe-OCH_3 *(aspartame),* the sesquiterpene *hernandulcin,* and *chemostimulatory proteins.* Among these are some of the sweetest substances known, the 207-residue *thaumatins and monellin* present in certain tropical berries. Thaumatins are ~3000 times sweeter than sucrose on a weight basis and 10^5 times sweeter on a molar basis. Thus, sucrose tastes sweet at a concentration of 10^{-3} M or higher but thaumatin at 3×10^{-8} M.

The proteins *miraculin* and *circulin* from tropical fruits modify taste. Acids taste sweet rather than sour after the tongue has been treated with either protein. Exposure of the tongue to artichokes often makes water taste sweet. Thus, the response of taste receptors can be temporarily altered by binding of other substances, perhaps at adjacent sites on a receptor.

Pain

Receptors for pain *(nociceptors)* are spread over the body in nerve endings found in the skin, muscle, joints, and internal organs. There are several types of receptors, most of which are present in excitatory glutamatergic neurons. Some release substance P. Some activate tyrosine kinases and others ATP-gated ion channels. Some pain receptors are also activated by intense heat or pressure or by irritant compounds. Among the latter is capsaicin (Fig. 7.29), the active ingredient in chili peppers, and an ultrapotent compound, *resiniferatoxin.* Both capsaicin, which is 10,000 times more potent than jalapeno peppers, and resiniferatoxin, which is 20-fold more potent than capsaicin, bind to *vanilloid receptors.* These are ligand-gated ion channels related to the Shaker K^+ channel (Fig. 7.18). They are nonselective but with a high permeability to Ca^{2+} and are members of the *transient receptor potential* (TRP) family. Pain seems to stimulate an increase in anandamide (Fig. 7.29), which has an analgesic effect. Nevertheless, anandamide and *N*-vanillyloleamide activate capsaicin receptors. Because the activated receptors become desensitized rapidly, capsaicin has been used in a paradoxical manner as an analgesic agent. Sensing of temperature changes also depends upon TRP channels.

Mechanoreceptors

The transduction of mechanical force into a chemical signal provides the basis for the senses of touch and hearing. Plants detect wind and gravitational force, and many organisms, even bacteria, respond to changes in osmotic pressure using mechanoreceptors. One of the best known mechanoreceptors is from *Mycobacterium tuberculosis.* It is a homopentamer whose three-dimensional structure resembles that of the nicotinic acetylcholine receptor (Fig. 7.23). A second type of mechanoreceptor is found in the inner membranes of *E. coli* and in plasma membranes of many other bacteria, archaea, and some eukaryotes. These receptors, which are also sensitive to voltage changes, are heptamers of a 282-residue protein that forms a symmetric ion channel in the center. There is also a large cytoplasmic domain consisting largely of β structure. How do such receptors sense mechanical stress? In bacteria they respond to stretch in the membrane induced by an increase in osmotic pressure. One suggestion is that the membrane expansion pulls apart the radially symmetric ion channel in the receptor. In higher organisms transmembrane adhesion receptors and their linkage to the internal cytoskeleton provide a framework for detection of mechanical forces and linkage to mechanoreceptors.

Hearing

Movement of the *stereocilia* of the hair cells of the inner ear activates mechanoreceptors. Each stereocilium contains a core of crosslinked actin filaments, and tens to hundreds of these cilia are connected in hair bundles, which move in response to arrival of sound waves of appropriate frequencies. The movement of the stereocilia induces the opening of receptor ion channels in the hair cell membrane allowing K^+ and other ions to flow inward. The matter is much more complex than this because of the tuning and amplification mechanisms in the cochlea of the inner ear. These mechanisms allow receptors in hair cells to respond to very weak vibrations of specific frequencies. Both mechanical and biochemical mechanisms are involved. A number of specific proteins participate. Among these is a motor protein called *prestin,* which seems to be involved in the rapid changes in length and stiffness of some hair cells in the cochlea.

Other sets of hair cells are formed in specialized parts of the inner ear. The three semicircular canals detect angular acceleration in three directions, while the sac-like utricle and saccule detect linear acceleration including gravitational attraction. These two organs each contain a patch of hair cells whose tips project into a gelatinous layer, which is overlain by a field of small crystals of calcium carbonate. These little stones (*otoliths*) provide an inertial mass, which resists movement causing the hair cell tips to bend and activate mechanoreceptors to send information about balance and orientation to the brain.

While discussing vibrations we may ask whether 60 cycle electromagnetic field fluctuations caused by electrical power transmission can affect the human body? Considerable effort has been expended in addressing this question. The tentative conclusion is that such effects, if they exist, are extremely difficult to detect. However, the possibility has not been disproven.

12. The Chemistry of Learning, Memory, and Thinking

What is known about the chemistry underlying memory, thinking, and the generation of the stream of consciousness within the brain? Nerve impulses originating in sensory receptors are sent to several regions of the brain, among which are the sensory regions of the cerebral cortex (Fig. 7.14). Memory also depends upon other regions of the brain including the hippocampus, amygdala, and cerebellum (Fig. 7.1). Learning, remembering, and thinking all require transfers of information between various neurons and between different parts of the brain. These transfers may perhaps be coordinated via endogenous electrical rhythms (brain waves).

Memory Systems

Memories exist in several forms and are found in various regions that are reached by several pathways. Two major forms of memory are :

Explicit (declarative, episodic) : Conscious recall of facts and events involving people, places, and things

Implicit (associative) : Nonconscious recall of motor skills, conditioned responses, etc.

Explicit memory depends upon the *temporal lobe* of the midbrain, an area that includes the hippo-campus and the nearby subiculum and entorhinal cortex. Implicit associative learning and memory involve the cerebellum, amygdala, and other regions.

Both types of memory possess both *short-term* and *long-term* components. Short-term memory lasts only minutes to hours, but long-term memory lasts days, weeks, and sometimes a lifetime. The difference between the two is clearly seen in individuals who have damage to the hippocampus and impairment of short-term memory. A blow to the head may cause total loss of short-term memory of associations (*amnesia*). Some persons with damage to the hippocampus may never regain their temporary memory, but long-term memories are intact, and new long-term memories may still be formed. An increasingly important tool for study of memory is brain imaging using *PET* or *fMRI*. These tools have become rapid and sensitive with the ability to observe regions of the brain that become activated by visual, auditory, or other stimuli.

The brain often needs to store information for a short period of time. For example, one can recall many details of a visual image after closing one's eyes or shifting one's gaze. The sensory images may be stored in *working memory*. Similarly, if one mentally multiplies two 2-digit numbers the partial product obtained by multiplying the two right most digits is temporarily stored in working memory until the next arithmetic operation is completed, etc. PET and fMRI tomography indicates that regions in the prefrontal cortex may be involved.

Some short-term memory appears to be stored, by neurons that continue to fire after a stimulus has stopped. It has been proposed that such memory consists of reverberations of electrical activity in loops of coupled axons.

Implicit memory can be studied in animals. Much has been learned from the large marine snails *Aplysia* and *Hermissendra* whose simple nervous systems and large neurons have been investigated for over 40 years. The basic chemical mechanism associated with learning in these creatures seems to be similar to those in our own brain. Olfactory memory can be studied in *Drosophila,* even though the organization of the fly's brain differs from ours.

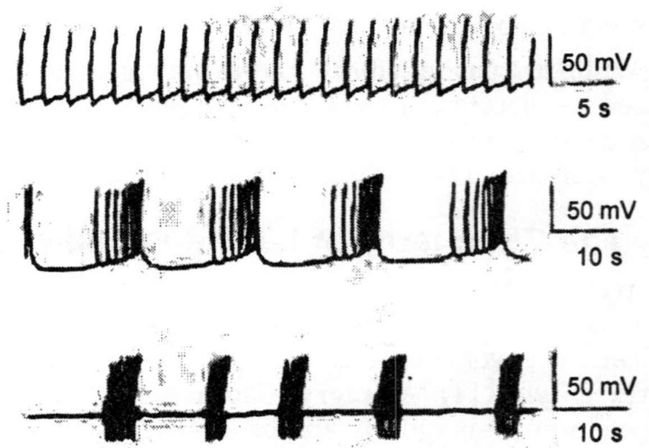

Fig. 7.32. Intracellular recordings from isolated neurons of the mollusc *Aplysia:* (A) beating pacemaker,(B) bursting pacemaker, and (C) oscillating pacemaker.

To be useful for more than a few minutes stored information must be transferred from the temporary to more permanent forms. We know that even temporary memory depends upon chemical changes in synapses. Long-term memory involves both stable chemical changes and also changes in the physical connections between neurons. Before discussing these changes let us consider briefly the waves of nerve impulses that drive the necessary alterations.

Brain Rhythms

A live brain displays characteristic oscillatory activity. Using electrodes placed on the scalp of and awake but relaxed individual, a rhythmic change in the recorded voltage with a frequency of ~10 Hz (*alpha waves*) can be detected. Such *electro-encephalograms* (EEGs) contain other rhythms at ~5 – 6 (theta), ~40 (gamma), and ~200 (high frequency) Hz. More recent studies employ microelec-trodes placed on individual neurons. Some cells generate spikes at frequencies as high as 800 Hz. However, the significance is uncertain. The 40-Hz frequency, which is prominent in the hippocampus, has aroused the most interest because of its probable relationship to learning and memory. Psychophysical experiments have suggested that humans can store only 7 ± 2 items, such as digits in a telephone number. To remember more digits usually requires a conscious effort to place them in longer-term memory. One proposal is that the seven items in temporary memory are stored as 40-Hz oscillations and that ~7 such items can be stored within a single 5-Hz theta oscillation. Thinking rates have also been estimated as 7 ± 2 thoughts per second. This is also the same as the syllable rate in speech. This allows us to speak at the same rate that we think. Shattering may be a result of lack of synchronization of thinking and speaking.

Individual cells or groups of cells are able to initiate rhythms. Examples are provided by the slower Ca^{2+} oscillations shown in Box 6-D, by the periodic release (at –155 intervals) of cAMP by cells of *Dictyostelium*, and by the 24-h circadian cycle observed for virtually all living cells. In simple invertebrates the source of neural rhythms appears to reside in *pacemaker neurons* that fire spontaneously at regular intervals. Their cell membranes apparently undergo a cyclic series of changes in ionic permeabilities sufficient to initiate action potentials. Three types of pacemaker output from molluscan neurons are illustrated in Fig. 7.32. In lobsters three-neuron *pacemaker groups* provide a pyloric rhythm. In these groups the oscillation period of a pacemaker neuron is adjusted from its intrinsic value by feedback through inhibitory and electrical connections to the other cells. Electrical coupling seems to be basic to oscillatory cell networks. Individual neurons or small groups of neurons in our own bodies act as pacemakers for the heartbeat rhythm. The slow 3- to 8-Hz rhythm observed in the EEG apparently originates in pacemaker bursts from the basal ganglia. Rhythms from these endogenous pacemakers may combine with pulses from sensory neurons to evoke *conscious thought.* However, the basis of consciousness is still poorly understood.

As can be seen from Fig. 7.32, neurons send "trains" of spikes down their axons. These form synapses with dendrites, usually on dendritic spikes, of a postsynaptic cell. However, each such cell typically receives input from thousands of other neurons. At any moment most of these are probably "silent," but others are sending trains of impulses. Among the important questions are "How does the postsynaptic neuron know whether to fire or not?" and "What kinds of information, if any, are encoded in the trains of impulses both in the presynaptic inputs and in the output of the postsynaptic neuron?" Part of the answer to the first question is probably that firing occurs if two or more input impulses arrive synchronously, and if there are not too many inhibitory impulses that damp the response. In the hippocampus a network of neurons electrically coupled via gap junctions may be synchronized to the theta and gamma brain rhythms by high-frequency (150 – 200 Hz) oscillations.

Chemical Changes in Synapses

It has long been recognized that the synapses are the probable sites of alterations that lead to memory, whether long-term or short-term. Study of individual synapses has demonstrated the phenomena of *potentiation* (facilitation) and *depression* (habituation). Potentiation refers to the fact that a second impulse will often be transmitted through a synapse more effectively than the first, while depression refers to a decreased response to repeated stimuli. Memory may consist of potentiation and depression at specific synapses. The underlying chemical changes in the synapses are referred to collectively as *synaptic plasticity.* Chemical changes associated with short-term memory are often transient. Those associated with long-term memory are described as *long-term potentiation* (LTP) and *long-term depression* (*LTD*).

Many experimental results have confirmed the chemical basis of memory. For example, learning is facilitated by administration to animals of small doses of strychnine. Puromycin and other inhibitors of protein synthesis disrupt the transfer of information into long-term memory. They are especially effective during the first hour after the initial learning event. Increased synthesis both of mRNA and of proteins within the cell bodies of neurons is observed.

Short-term memory is not affected by inhibitors of protein synthesis, but alteration of synaptic proteins and membranes may be induced by covalent modification of existing macromolecules. One way in which this happens has been described for *Aplysia*. As the snail learns a simple gill-withdrawal reflex, the duration of the action potentials in sensory neurons is increased, and there is a greater release of transmitters. This change comes about because stimulation of the sensory neuron causes simultaneous activation of interneurons that synapse with the sensory neurons. The interneurons release the neuromodulator serotonin, which binds to receptors in the membrane of the sensory neurons. This activates adenylate cyclase, which in turn activates a protein kinase that phosphorylates a class of open K^+ channels. Phosphorylation causes the channels to close with a consequent strengthening of the action-potential. Thus serotonin brings about presynaptic facilitation. The peptide FMRF amide has the opposite effect. It causes hyperpolarization and a decrease in the duration of the action potential. It also binds to a receptor on the neuronal membrane, and presumably via a different second messenger than cAMP causes the K^+ channels to stay open a longer fraction of the time.

Evidence that LTP is essential to learning in rats was provided by the observation that the antagonist 2-aminophosphonovalerate, which blocks the NMDA class of glutamate receptors (Fig. 7.20A), impairs both LTP and learning. The potentiation is thought to result, in part, from Ca^{2+} influx through the nonselective NMDA cation channels. The increased intracellular calcium may then induce phosphorylation of various proteins with associated long-lasting changes in the postsynaptic endings. A large amount of evidence favors this interpretation of LTP. However, it is a great oversimplification. Most studies of LTP in mammals (usually rodents) have focused on the CA1 region of the hippocampus and nearby brain regions. The excitatory axons in this organ are largely glutamatergic, and, as shown in Fig. 7.20A, the postsynaptic (dendritic) membranes contain both fast AMPA receptors and the slower NMDA receptors. Both are ionotropic. The AMPA receptor channels allow mainly K^+ and Na^+ to pass and are responsible for most of the nerve transmission. However, the NMDA receptors have an important controlling influence.

A generally accepted theory is that no LTP arises unless both the presynaptic and postsynaptic neurons are activated. This can happen if a presynaptic action potential activates many AMPA receptors in a synapse allowing enough flow of $Na^+ + K^+$ to depolarize the postsynaptic membrane and possibly to initiate an action potential in the postsynaptic neuron. (However, many factors, probably including influences from neighboring neurons, will affect this outcome.) The NMDA receptors are usually blocked by extracellular Mg^{2+} ions, and their ion channels remain closed. However, when the postsynaptic membrane becomes depolarized, the Mg^{2+} dissociates, and if the NMDA receptors are also occupied by glutamate, their channels will open, permitting Ca^{2+} to enter the neuron (Fig. 7.20A). This not only enhances the probability of developing a postsynaptic action potential but is also the trigger for LTP. Ca^{2+}

ions have a variety of effects; one of which is to bind to calmodulin. This activates a calcium-calmodulin-dependent protein kinase, which phosphorylates postsynaptic structural and signaling proteins to increase the synapse strength.

Modifications in existing proteins, such as are induced by Ca^{2+} and calmodulin, can provide LTP for a few hours, but other mechanisms must provide for longer-term effects. These require transcription of genes and protein synthesis, processes that occur in the cell bodies of neurons and may depend upon axonal transport mechanisms. Among the experimentally observed results of LTP are ultrastructural changes in synapses and in dendrites. Long-term memory is also thought to involve changes in the neocortex. Again, NMDA receptor activation seems to be involved.

LTP has been demonstrated experimentally, but does it really influence memory? Evidence that it does has been provided by clever experiments with transgenic mice. Using the Cre recombinase the NR2B subunit of the NMDA receptor was overexpressed in the hippocampus and in the forebrain of mice. This was expected to provide better synaptic strengthening than for receptors with the similar NR2A subunit. It was found experimentally that these transgenic mice were more intelligent than normal mice.

LTP is also thought to affect the presynaptic as well as the postsynaptic neuron. One way in which this may happen is for a *retrograde messenger* to pass across the synapse and induce alterations in the presynaptic cell. One proposed retrograde messenger is nitric oxide, NO. Neuronal NO synthase (nNOS; NOS1) contains a calmodulin binding site and is activated by Ca^{2+}. However, other substances as simple as K^+ might also be the messenger.

Long-term depression (LTD) is the *loss* of synaptic strength after passage of an impulse. There is evidence that during brain activity, including that in the hippocampus, both LTP and LTD are essential. LTD may depend upon cyclic ADP-ribose. LTD, like LTP, may also spread via retrograde signaling.

Does learning affect a few specific neurons or a large number of neurons? The rate of glucose utilization in different parts of the brain can be estimated from the rate at which labeled 2-deoxyglucose is taken up. From changes in this rate (obtained using ^{14}C-labeling) in brains of splitbrain cats performing visual tasks it was estimated that from 10^{10} to 10^{11} neurons are activated. This supports the idea that memory is distributed over a large area of the brain, just as information about an image is stored in all parts of a hologram.

An alternative to the idea that synaptic potentiation and depression provide the chemical basis for learning is *molecular coding.* Thus, it was reported that a 15-amino acid peptide isolated from rats trained to avoid the dark carries behavioral information. When this peptide was injected into brains of untrained rats, they also avoided the dark.This was one of several reports of transfer of learned behavior through chemical substances extracted from the brain. These ideas are hard to accept in the light of our present knowledge of the brain. However, in view of the large number of different neuropeptides known the possibility that some aspects of long-term memory may be associated with transcription of specific amino acid sequences within specific neurons should perhaps still be considered.

The Complexity of the Brain

A major obstacle to our understanding of the human brain is its enormous complexity. This problem can be appreciated if we consider the small nematode *Caenorhabditis elegans.*

All of the synaptic connections among its 302 neurons had been mapped by 1986. There are 5000 chemical synapses and 600 electrical (gap junction) connections. There are 80 different types of K^+-selec-tive channels, 90 types of ligand-gated receptors, and ~1000 G-protein-linked receptors. Twenty-six of the neurons are GABAergic and are involved in three distinct behavioral motions that involve muscular contractions. Despite intensive efforts the system has been hard to understand. The brain of the macaque monkey has been described in great detail. Over half of its cerebral cortex is devoted to vision, and this can be subdivided into 20 functional areas. However, the human brain, with its extremely large cerebral cortex, cannot be compared accurately with the monkey brain. Whereas anatomical studies are done on postmortem human tissues, *in vivo* studies rely largely on fMRI and PET imaging. The resolution of these images is now less than 1 mm, but 1 mm^3 of human visual cortex contains more than 40,000 neurons! The microcircuits in the neocortex are still largely unknown and the tissue is of "apparently impenetrable complexity." There may be several hundred different classes of neocortical neurons. The tissue is rich in GABAergic interneurons. Some fast-spiking GABAergic neurons are also connected by electrical synapses and may be involved in detecting and promoting synchronous activity.

Intelligence

We must all agree that there is such a thing as intelligence; but can it be measured? In 1904, Spearman proposed the existence of a general intelligence factor g that could be measured as the IQ (intelligence quotient). Since then various tests have been devised that attempt to measure IQ. Most recently use of PET scan data has indicated that various types of analytical analysis lead to brain activity in the lateral frontal cortex in one or both cerebral hemispheres suggesting that this is a region important to IQ. A question that has been raised is whether analytical intelligence, creative intelligence, and practical intelligence are correlated?

Is intelligence hereditary? Both logic and observation say that heredity must be a major factor. However, it is hard to know how to measure the hereditary component. Also hard to understand is why IQ scores have been increasing about one standard deviation unit per generation. Is this really true? Does environment also influence IQ? The fact that new hippocampal nerve cells are formed continuously provides one mechanism by which learning, nutrition, and other influences may alter intelligence.

A difficult-to-explain aspect of the brain is the existence of rare *savants,* persons with amazing mental abilites in music, art, or computation but who are unable to communicate (autistic) and mentally retarded. One boy at age four could play Mozart piano sonatas flawlessly after a single hearing. A three-year old girl drew horses with lifelike perspective from memory but was unable to communicate. Some mathematical savants can instantly state the day of the week for any arbitrary date such as June 12, 1929; others rapidly identify prime numbers. They evidently use the same strategies as mathematically trained persons. Do we all have these abilities but can't have access to them? How can we explain the fact that rarely a blow to the head will convert a person into a savant?

Behavior

It may seem impossible to interpret complex behavioral patterns at the molecular level. However, the genetics of behavior is a well recognized field of investigation, and some behavioral traits have been linked to single genes. If a gene can be located cloning, sequencing, and biochemical studies may follow quickly. The behavioral genetics of lower organisms, *e.g.*, of *Drosophila,* have provided many insights. Recently, however, the mouse has become a major object of behavioral studies. Its genome is well known, and a very large number of mutations have been mapped. The ability to prepare *knockout mice* and to carry out gene transfer experiments on such animals makes them very attractive for study.

Some behavioral traits are based on simple alterations, often defects, in motor skills. For example, the following traits in mutant mice have been traced to specific brain structures and often to specific biochemical alterations.

Staggerer Purkinje cell defect

Vibrator Phosphatidylinositol transfer protein gene

Tottering Mutation in voltage-gated Ca^{2+} channel

Lurcher Abnormality in cerebellum

Weaver Gly → Ser mutation in K^+ channel

Knockout mice lacking oxytocin or vasopressin have altered social behavior toward other mice. Those lacking galanin seem less intelligent than normal mice, as if they had Alzheimer disease. Mice lacking neuronal NO synthase became aggressive. Human personality, language abilities, and sexual behavior all have a genetic component. However, claims that a "gay gene" has been found are not generally accepted.

13. Circadian Cycles and sleep

In mammals an approximately 24-h *(circadian)* rhythm controls behavior and affects many physiological functions. As previously mentioned, the brain has its own rhythms, which originate with pace-maker neurons. The heart beats with another neurally established rhythm. The circadian rhythm has a much longer period and, therefore, seems more mysterious. It is observable, even with single cells and for virtually all organisms. In most instances the cycle becomes synchronized with the daily light-dark cycle with the aid of suitable light-absorbing pigments often crytochromes. However, the cycle can be observed in various ways under conditions of constant light intensity and temperature. For example, the unicellular marine alga *Gonyaulax* undergoes dramatic circadian changes in the intensity of its bioluminescence. Over one 10-day period the luminescence peaked every 22.99 ± 0.01 hours. It is more difficult to measure the period for human beings, but under suitable conditions during which time cues were missing a precise period of 24.18 hours was observed for the level of melatonin in the blood, the body temperature, and other quantities. From cyanobacteria, fungi *(Neurospora)*, insects (*Drosophia*), and frogs to mice and people, the circadian cycle affects the organism's chemistry and behavior. Green plants likewise observe a circadian cycle.

The cycle is thought to originate in feedback loops that control transcription of a small set of genes. In *Drosophila* the set includes seven genes: *period (per), timeless (tint), dock (elk), cycle (eye), double-time (dbt), vrille,* and *cryptochrome(cry).* Many corresponding genes have been found in mammals. For example, the mouse NPAS2 is a close relative of the *Drosophila* CLOCK protein, and the period proteins PERI and PER2 and the cryptochromes CRY1 and CRY2 are also related to the *Drosophila* proteins. In *Drosophila* the heterodimers PER-TIM and CYC-CLK are thought to serve as DNA-binding transcription factors that repress transcription of their own genes when they reach a high enough concentration in the nucleus.

Because some time is required for transcription and protein synthesis, this inhibitory feedback can lead to oscillations in the concentrations of the circadian clock proteins. Proteosomal degradation of the TIM protein may also be a factor. The need for proteins encoded by other genes indicates that the matter is more complex. Individual cells or individual tissues, *e.g.*, mammalian retinas, may independently set up circadian cycles. However, these normally become *entrained* by the daylight cycle and are reset daily. Other factors such as temperature, activity, and food may also affect the resetting. One factor, which may be influenced by food, is the NAD^+/VADH and $NADP^+$/NADPH ratios within cells. The circadian cycles for mammalian tissues are synchronized by a *master clock* that originates in neural tissues and specifically in a region of the hypothalamus containing the *supra-chiasmatk* nuclei.

Positron Emission Tomography (PET), Functional Magnetic Resonance (FMRI), and other Imaging Techniques

In the widely used technique of transmission computerized tomography (CT) an image of a slice through the body of a patient is obtained using X-rays. An X-ray source moves in a ring around the patient while detectors measure the intensity of the transmitted radiation and send it in digital form to a computer, which generates the desired image. A chemically more sophisticated view of the body can be obtained by positron emission tomography (PET). This technique makes use of a metabolite or drug labeled with a short-lived radioisotope that decays by emission of positrons (antielectrons). Among these are ^{11}C, ^{13}N, ^{15}O, and ^{18}F with half lives of 20 min, 10 min, 2 min, and 110 min, respectively. The isotopes are produced in a cyclotron, and are rapidly introduced into suitable compounds, which can be injected into a bloodstream. An emitted positron travels only a few millimeters before undergoing annihilation with an electron to produce two high-energy (50 keV) photons (γ-rays) that travel in opposite directions and are detected by an array of scintillation detectors.

$$\beta + e^- \rightarrow 2\,h\nu$$

Present-day PET technology allows images to be formed in a few seconds, and in some cases in a fraction of a second. Among the useful compounds for PET imaging is [^{18}F]2-fluoro-2-deoxy-D-glucose. This compound, which contains the longer-lived ^{18}F, is phosphorylated by hexokinase, and the resulting phosphate ester is effectively trapped in the brain. 3-Deoxy-3-fluoro-D-glucose is another useful tracer. One of the most useful PET measurements has been blood-flow monitored by ^{15}O-containing H_2O, which is administered into a vein in the arm. The ^{15}O has a half-life of only two minutes and is almost completely gone in ten minutes. However, very low doses of radioactivity are used, and several images can be obtained before the radioactivity has decayed. A common practice is to subtract images obtained after the isotope has decayed from those obtained at various times while it was still present. The technique is also useful for study of the binding and transport of hormones, other metabolites, drugs, and other inhibitors.

The NMR technique *magnetic resonance imaging (MRI)*, so called to avoid the word nuclear, is rapidly displacing many applications of PET scanning. MRI uses proton NMR spectroscopy to generate very sharp images based largely on the water present in tissues. These images can be made to depend upon variations in T_1 and T_2 (Chapter 3) as well as upon differences in the water content. The first MRI scans required 20 minutes, but the use of more powerful magnets and more sensitive instruments has reduced the acquisition times in ultrafast MRI to ~0.1 s. The decay of the NMR signal from a single RF pulse is observed at several different times, The dynamics of blood flow and neural activity can be followed. Every technique has disadvantages as well as advantages. MRI does not use radioisotopes, but overheating of the brain must be carefully avoided. In addition, patients may suffer from uncomfortably loud noises generated by rapidly changing magnetic gradients. As with PET scans isotopic tracers may be used. However, most MRI scanning is done with ^1H from the solvent water. As with PET, MRI is often used to measure blood flow but with an indirect method. The Fe of deoxyhemoglobin (Hb) is paramagnetic, but upon oxygenation to HbO_2 it becomes diamanetic, and the ^1H signal of the solvent H_2O becomes sharper. In metabolically active regions of the brain the demand for oxygenated blood is greatly increased. Perhaps surprisingly, the ratio [$FIbO_2$]/[Hb] is greater in these areas than in less active areas where a greater fraction of the hemoglobin remains unoxygenated. However, the exact interpretation of the ultrafast MRI images is uncertain. In *functional MRI* (flVfRI), differences in images acquired after some physiological change are recorded. For example, after a visual or other sensory stimulus a change in the MRI image of some region of the cortex will be observed (see figure). The technique is allowing many deductions about learning, memory, and communication pathways in the brain and is being used to investigate many aspects of brain disease.

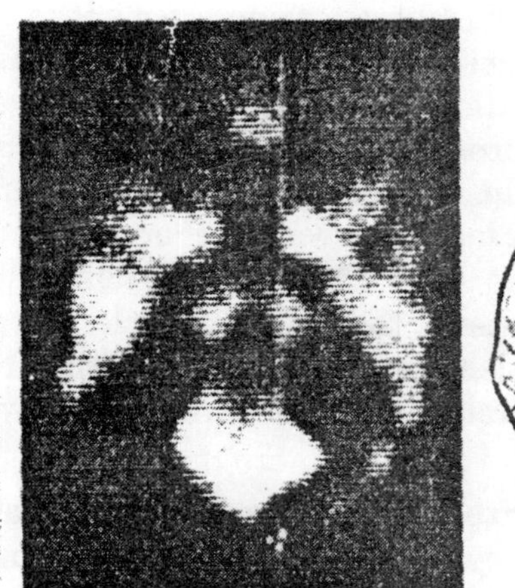

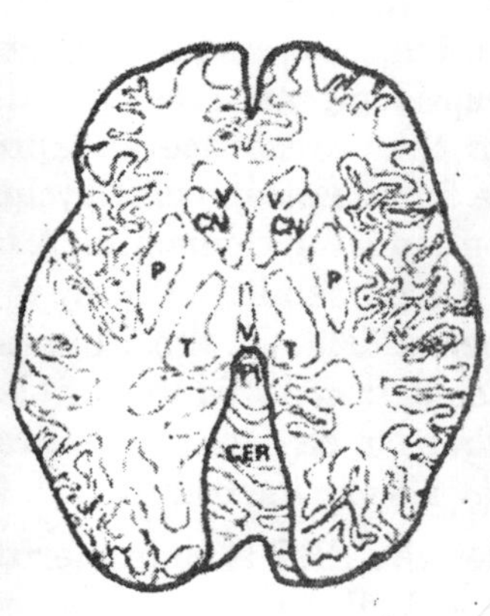

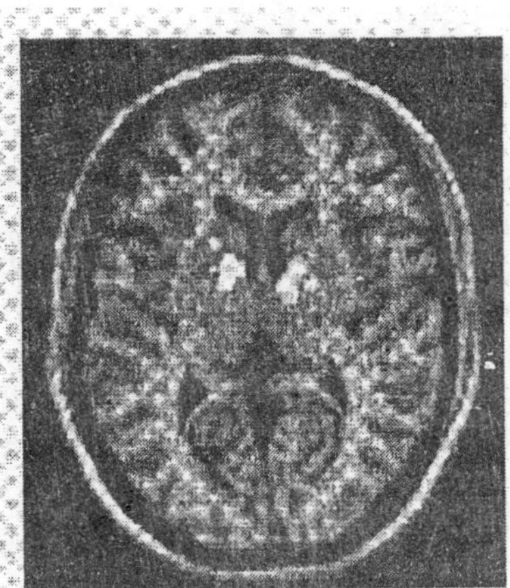

Left: PET image of a human brain obtained using 2-[^{18}F]fluoro-2-deoxyglucose. This tomographic brain slice at the level of the basal ganglia shows the cortical gray matter and subcortical white matter. As marked on the drawing on the right: V, ventricles; CN, caudate nucleus; P, putamen; T, thalamus; PI, pineal gland; CER, cerebellum. From Rottenberg and Cooper. P Right: fMRI image illustrating modulation of neural activity in the ventral stria turn, an area of the brain associated with reward, when eye contact was made with an attractive face. The activation map shown is derived in a complex manner and is based on recorded brain activity of persons viewing images of a series of faces. It portrays the differences in neural activity when viewing images of attractive faces of either sex with the eye gaze directed at the subject and with the eyes averted. See the report of Kampe *et. al.* for details.[q]

Another brain imaging technique is *magne-toencephalography* (MEG).[c, n] It has been uniquely valuable in mapping the sensory regions of the human cerebral cortex. Looking ahead, optical methods, which include use of infrared radiation, are also under development.[o] They may not be adequate for study of the human brain but can be used for smaller animals, for studies of embryonic development, etc.

The pineal gland appears also to play a role in maintaining the mammalian circadian cycle. The concentration of the pineal hormone melatonin (Fig. 7.11) as well as.its precursor N-acetylserotonin and the enzyme serotonin *N*-acetyltransferase (Eq. 7.4) all fluctuate far more than do the concentrations of other metabolites during the 24-h cycle. These metabolites increase over 10-fold concentration at night and decrease by day. During the daytime the serotonin N-acyltransferase, which forms the precursor, is rapidly and apparently irreversibly inactivated, perhaps through a disulfide exchange reaction. Bright light will reset the circadian cycle keeping it approximately (circa) daily. The effect of light is apparently a result of signals sent to the hypothalamus from the optic nerves. In chickens and in lower vertebrates the pineal gland may directly sense light passing through the skull.

The circadian cycle is not the only timing device used by animals. A short-term interval timer helps male doves to know how long to sit on a nest and helps all of us in timing everyday tasks.

We spend a third of our lives asleep, but our understanding of sleep from a molecular viewpoint is minimal. Sleep is essential for the life of mammals, which die if completely deprived of sleep. It has been shown that during prolonged sleep deprivation sleep-inducing material accumulates in the brain. One such substance, isolated from human urine, appears to be a peptide containing glutamate, alanine, diaminopimedic acid, and muramic acid. Thus, it resembles a fragment of bacterial peptidoglycan. Prostaglandin D_2 also induces sleep. Hayashi proposed that a balance between this substance and prostaglandin E_2, which induces wakefulness, is in part responsible for the sleep-wake cycle. More recently oleic acid amide (*oleamide*) was identified as a sleep-inducing compound. A fact observed by everyone is that the longer one is awake the higher the probability of going to sleep. The accumulation of sleep inducers is part of a homeostatic mechanism. On the other hand, the circadian cycle probably provides the signal to awake and tends to consolidate our sleep into the characteristic 8-hour period. The melatonin level, which drops in daylight, plays a role. Release of adrenocorticotropin (ACTH) one hour before waking may also be important.

During much of the night's sleep the EEC is characterized by the slow 5- to 6-Hz waves. However, after ~90 min there is an ~10-min period of *rapid eye movement (REM) sleep* during which the EEG resembles that of an awake person and dreaming occurs. The closed eyes move rapidly in unison, breathing is irregular, and the heart rate increases. Motor neurons are inhibited allowing only minimal body movement. Three more periods of slow-wave sleep, each shorter than the preceding one, are followed by REM sleep. The REM sleep periods become successively longer. The fourth period lasts 20 – 30 min and is followed by awakening. All placental and marsupial mammals follow a similar sleep pattern and all dream. The importance of dreaming is not obvious but is often thought of as a reprocessing of memory, a means of ridding the mind of unneeded memories, a process of *unlearning*. However, this is uncertain as is the relationship of sleep to learning and memory.

A number of disorders of sleep are known. Among these is *narcolepsy,* uncontrollable, sudden daytime sleepiness. It affects 1 in 2000 individuals. The same occurs in dogs. After a 10-year effort at great expense the narcolepsy gene of dogs (*canarc*-1) was located by positional cloning. The corresponding human (and rat) gene was independently discovered by other investigators. It encodes a receptor for neuropeptides produced by the hypothalamus and named *hypocretins* or *orexins* for their stimulation of appetite. It seems probable that the hypocretin/orexin neuropeptides are involved in promoting wakefulness. Another sleep disorder is *familial advanced sleep phase syndrome.* Persons with this trait are "morning larks" who tend to fall asleep at ~7:30 p.m. and awake suddenly at ~4:30 a.m., about four hours in advance of a typical sleep period. A missence Ser → Gly mutation in the human period gene (hper2) has been found.

Some mammals hibernate. Special blood proteins that induce hibernation apparently control the process.

14. Mental Illness

Whereas many metabolic defects affect only a small number of individuals, emotional illnesses including depression, *schizophrenia,* and other *affective disorders* at one time or another afflict a large fraction of the population. Autism affects thousands of children. Parkinson disease and *Alzheimer disease* are just two of a number of degenerative neural diseases attacking older people. Less commonly, young persons contact *multiple sclerosis* and *muscular dystrophy,* which is often a disease of neuromuscular junctions.

Depression

Depression is our most common mental problem. One in four women and one in ten men will have a major depression during their life-time. More than 15 million people in the United States are affected by severe depression in any given year and more than 30,000 may commit suicide. Worldwide psychiatric problems, mostly depression, account for 28% of all disabilities. The *biogenic amine hypothesis* states that depression results from the depletion of neurotransmitters in the areas of the brain involved in sleep, arousal, appetite, sex drive, and psychomotor activity. An excess of transmitters is proposed to give rise to the manic phase of the bipolar (manic-depressive) cycle that is sometimes observed. In support of this hypothesis is the observation that administration of reserpine precipitates depression, which may be serious in 15-20% of hypertensive patients receiving the drug. Similar effects are observed with the dopa decarboxylase inhibitor *α-methyldopa* (Fig. 7.27). The fact that L-tryptophan has some antidepressant activity, but L-dopa does not, was one clue that a low concentration of serotonin (5-hydroxy-tryptamine) might be responsible for depression. Excessive formation of histamine and decreased formation of tyramine and octopamine have also been suggested as causes of depression.

Strong support for the biogenic amine theory of depression is provided by the powerful antidepressant effect of inhibitors of monoamine oxidase. An example is *pargyline* (Fig. 7.33), which forms a covalent adduct with the flavin of MAO. Although effective, this drug is somewhat dangerous. Because their monoamine oxidase activity is so low, patients taking pargyline have been killed by ingesting compounds such as tyramine, which occurs in cheese. Less easy to understand but clinically more important are tricyclic antidepressants such as *imipramine* (Fig. 7.33), whose antidepressant action was discovered accidentally. Notice the close similarity to chlorpromazine but the greater flexibility of the central ring. Imipramine was found to block transporters of both noradrenaline and serotonin. In 1986, the less toxic serotonin reuptake inhibitor fluoxetine was introduced and is now used by many millions of people. Nevertheless, its mode of action is not entirely clear. For example, it blocks nicotinic acetylcholine receptors and may have many other effects. Interestingly, depression sometimes responds to a placebo just as well as to an antidepressant drug. In addition to newer drugs related to Prozac, antagonists of substance P are also effective antidepressants. MRI images of brains of depressed patients show that hippocampal volume has decreased and suggest that formation of new neurons is inhibited. Antidepressants seem to stimulate growth of new cells as does exercise, which also has an antidepressant effect. Dietary treatment can also help. Among older people depression may be caused by deficiency of vitamin B_{12} and can be treated by injection of the vitamin. An *anxiety peptide* that may be the natural ligand for benzodiazepine receptors has been reported.

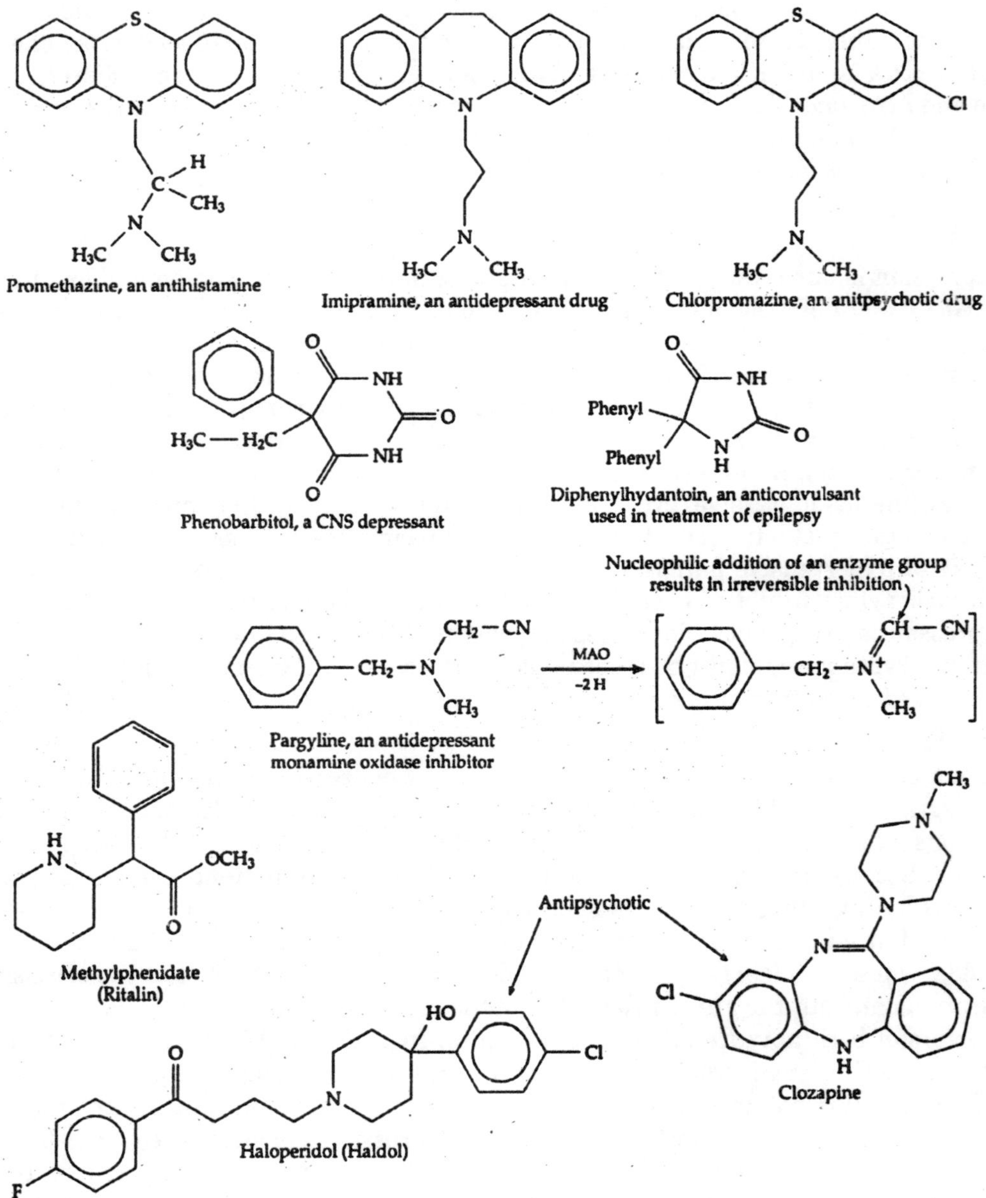

Fig. 7.33. Some drugs used to treat psychiatric disorders. See also Figs. 7.25 and 7.28.

Another recognized type of depression is *seasonal affective disorder* (SAD). People in far northern or southern latitudes develop this condition in the winter, apparently from lack of sunshine needed to lower the melatonin level in the morning. Light therapy is beneficial. Persons with the SAD syndrome also tend to crave carbohydrates and to stay in bed for 9-10 hours.

An effective treatment for *bipolar disorder* (manic–depressive illness) is the administration of lithium salts. Inhibition of the hydrolysis of inositol phosphate by Li^+ (Fig. 7.9) may be related to its therapeutic effect. Reduced phosphatidylinositol turnover may dampen responses to neurotransmitters. Li^+ may affect gene expression in neuropeptide-secreting neurons. Bipolar disorder apparently has more than one cause. There are strong indications of genetic susceptibility, and genes that increase susceptibility have been located on chromosomes 4, 12, 13, 18, 21, and X.

Schizophrenia

Among the most baffling of mental illnesses are the group of diseases known as schizophrenia. They involve thought disorder, disturbance of the affect, and withdrawal from interactions with other people. Hallucinations and paranoid feelings are common. In some cases a striking loss of gray matter in some areas of the brain is revealed by MRI scans. The schizophrenias are of varying degrees of severity and shade continuously into the affective or mood disorders, which include manic-depression and depression. As many as one person in a hundred is affected by schizophrenia. There is a complex genetic susceptibility. One theory about the persistence of the genes favoring schizophrenia is that they are also associated with creativity.

A revolution in the treatment of the schizophrenias, as well as in thinking about mental illnesses, took place following the synthesis, in 1950, of the antipsychotic drug *chlorpromazine* (Fig. 7.33). At about the same time the effect of the *Ranwolfia* alkaloid reserpine (Fig. 7.12) in calming mentally disturbed persons was rediscovered. The Indian plant *Rauwolfia* had been used for centuries in Hindu medicine for the same purpose. The tricyclic phenothiazines such as promethazine (Fig. 7.33) earlier had been found to have powerful *antihistamine* activity. It was the search for better antihistamine drugs that led to the synthesis of chlorpromazine. As many as 250 million people throughout the world were treated with chlorpromazine and related drugs in the 20 years following its discovery before newer and safer drugs (*e.g., clozapine;* Fig. 7.33) were developed. What does chlorpromazine do? A possible clue comes from the fact that it sometimes induces serious "extrapyramidal" side effects including tremors and other symptoms of Parkinson disease. This suggested that chlorpromazine may block dopamine receptors in the corpus striatum, thereby precipitating a functional deficiency of dopamine. If so, it is possible that schizophrenia may result from an overactivity of dopamine neurons, perhaps including some of the same neurons that are hypoactive in Parkinson disease. Supporting this view is the observation that amphetamines (Fig. 7.27), which may substitute for dopamine, worsen the symptoms.of schizophrenia and in very high doses induce striking schizophrenialike symptoms in normal individuals.

A stereotyped compulsive behavior is induced both in humans and in laboratory animals by amphetamines. This provided the basis for a method that has been used to measure the action of drugs on amphetamine-sensitive centers of the brain. A lesion in the nigrostriatal bundle on one side of a rat brain was made by injection of a neurotoxic compound such as 6-

hydroxy- dopamine. This caused degeneration of dopamine-containing neurons on one side of the brain. When rats that had been injured in this way were given amphetamines, they developed a compulsive rotational behavior. Administration of chlorpromazine and several other antipsychotic drugs neutralized this behavior and in direct proportion to the efficacy in clinical use, an observation that also supports the theory that schizophrenia involves over-activity of dopamine neurons.

If schizophrenia results from an elevated dopamine content of the brain, the fault may lie with either an oversupply or a reduced rate of metabolism of dopamine. The possibilities of reduced activity of monoamine oxidase or of dopamine β-hydroxylase have both been suggested. The plasma level of the dopamine metabolite homovanillic acid (Fig. 7.26) is elevated in schizophrenia and is correlated with the severity of the illness, suggesting the hypothesis of a decreased rate of metabolism. Possible defects in dopamine receptors may be at fault.

Chlorpromazine may also act on brain cholinergic neurons. Blockage of muscarinic acetykholine receptors in the brain by belladonna alkaloids such as atropine (Fig. 7.22) has often been used in treatment of Parkinson disease. Apparently antagonizing acetylcholine action is to some extent functionally equivalent to increasing dopamine concentrations. There is evidence that suggests a role for cholecystokinin (CCK) in development of schizophrenia. CCK-containing neurons interact with dopaminergic neurons in the midbrain. GABA neurons in the prefrontal cortex may be faulty. Excessive glutamate may also induce schizophrenia. The schizophrenia-like symptoms induced by phenylcyclidine are eased by antagonists of metabotropic glutamate receptors. This suggests another possible therapy. Among other suggested causes of schizophrenia are dysregulation by *retinoids* and action of retroviruses. Demyelination in portions of the prefrontal cortex may disrupt neural connectivity. Recent genetic evidence points to a possible defect in proline dehydrogenase which reduces Δ^1-pyrroline-5-carboxylate to L-proline (Fig. 7.9).

Numerous theories of mental illness have embodied proposals that a toxic metabolite is produced in abnormal quantities. An example is *6-hydroxydopamine* (Fig. 7.26), which is known to damage dopamine-containing neurons. Overactive methylation of catecholamines has also been suggested as a cause of mental disorders. The hallucinogen 3, *4-dimethoxyphenylethylamine* (Fig. 7.26) has been identified in urine during acute schizophrenic attacks, but the variability is so high that no definite conclusion has been reached. N-Methylation of serotonin yields *bufotenin* (N-methylserotonin) and N-*dimethylserotonin,* known hallucinogenic agents. Enzymatic synthesis of the latter by human brain and other tissues has been demonstrated, and administration of tryptophan and methionine to schizophrenic patients exacerbates their illness.

Another theory of mental illness postulates endogenous alkaloid formation. Aldehydes formed by oxidation of catecholamines as well as formaldehyde and acetaldehyde are present in tissues in small amounts. Condensation with amines could generate Schiff bases and alkaloids as in Fig. 7.10. This "plant chemistry" is spontaneous and can apparently take place in the brain, where it may have a potent effect. Incubation of tryptamine derivatives with 5-methyl-tetrahydrofolic acid and an enzyme preparation from brain gives *tryptolines.* Dopamine and its derivatives form related tetrahydroisoquinolines such as the product that arises from reaction with acetaldehyde. This product has been found in elevated amounts in alcoholics (who synthesize excess acetaldehyde), in phenylketonurics, and in L-dopatreated patients with Parkinson disease.

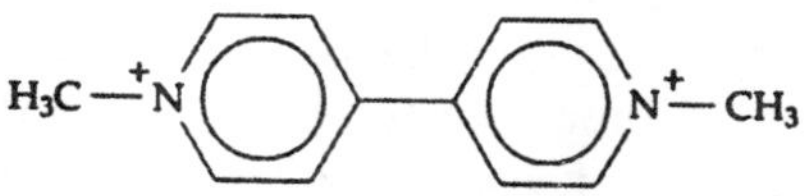

$H_3C-\overset{+}{N}$ ... $\overset{+}{N}-CH_3$

Paraquat

Epilepsy

The brain disorders known as epilepsies affect 1 – 2% of the population worldwide. Characteristic of epilepsies are recurrent *seizures,* sudden brief changes in behavior caused by the simultaneous, disordered firing of large numbers of neurons in the brain. Many seizures are thought to be initiated in specific areas of the cerebral cortex. For example, seizure-induced firing of neurons in the thumb area of the right motor cortex will be accompanied by rhythmic jerking in the left thumb. More than 40 different types of epilepsy are known.

GABA is the principal inhibitor neurotransmitter, and one cause of epilepsy may be a deficiency in GABA formation from glutamate. The brain contains two isoforms of glutamate decarboxylase, designated GAD65 and GAD67, in accordance with their molecular masses in kDa. They are encoded by separate genes. GAD67 is formed mainly in cell bodies of neurons, binds its cofactor PLP tightly, and is essential to survival of young mice. GAD65 is associated mainly with nerve termini, where it is anchored, apparently by association with other proteins to the membranes of synaptic vesicles. It binds PLP weakly. Some convulsive agents such as 1, 1-dimethyl-hydrazine are thought to act by interfering with PLP-dependent enzymes among which is GAD. Convulsions are one of the most striking symptoms of a severe vitamin B_6 deficiency. A zinc deficiency can also cause convulsions, apparently because pyridoxine kinase is a Zn•ATP-requiring protein and the rate of synthesis of PLP is too slow to supply apo-GAD with the PLP needed for GABA synthesis. The PLP in GAD65 undergoes rapid substrate-dependent transamination to pyridoxamine phosphate, which must be replaced by new PLP.

Epilepsy may arise also from defects in a GABA transporter or receptor. One form of epilepsy is a triplerepeat disease of cystatin B. Mutation in potassium channels, glutamate receptors, absence of neuropeptide Y, and absence of L-isoaspartyl/D-aspartyl O-methyltransferase have all been associated with epilepsy.

Neurodegenerative Diseases

As many as 5% of persons of age 65 and 20% of those of age 80 are afflicted with the progressive senile dementia known as *Alzheimer disease.* The condition is characterized by a gradual loss of memory and of the abilities to speak, think, or take care of one's self. Histologically Alzheimer disease is marked by the accumulation within neurons of *paired helical filaments.* These filaments, of 10 nm diameter, twist about each other to form a helix with an 80-nm pitch. The helices aggregate to create *neurofibrillary tangles.* The tangles are composed largely of a highly phosphorylated form of the microtubule-associated protein *tau* together with phosphorylated neurofilaments, apolipoprotein E, and other materials. The tangles are found in the cell bodies, axons, and dendrites of neurons in the hippocampus, amygdala, cerebral cortex, and other areas of the brain. Tangles may also be present in Parkinson disease, in the nearly extinct *Guam* disease, and in some types of prion disease. Outside of the diseased neurons are numerous, spherical *amyloid plaques.* Their principal component is a 40- to 43-residue fragment called *amyloid β-protein* (AB) which appears to be toxic to neurons. Aβ is cut from a larger *amyloid precursor protein* (APP). The APP gene is a member of a family of 16 related genes found in many organisms including nematodes, flies, and mammals. In humans the APP gene is found on chromosome 21, the chromosome that is present in three copies in *Down syndrome.* People with Down syndrome who live into their late thirties or beyond develop Alzheimer disease, presumably from excessive synthesis of APP. Both APP and its cleavage product Ab are formed by nonneuronal cells throughout the body. However, the AP plaques form only in the brain, and the APP gene is essential for life. The rare *familial British dementia* resembles Alzheimer disease in producing amyloid plaques and neurofibrillary tangles. They appear principally in the cerebellum and arise from a different precursor protein.

There are many other neurodegenerative diseases, some with a high incidence, and others rare. They include *Parkinson disease, Huntington disease, spinal muscular atrophy* (SMA; a leading hereditary cause of infant mortality), amyotrophic lateral sclerosis *(ALS)*, prion diseases, *ataxias,* and other diseases caused by triple-repeat DNA sequences and X-linked adrenoleukodystrophy (ALD). In the last, membrane function is disrupted. Although these diseases arise from a variety of causes many of them have in common amyloidosis, the deposition of insoluble proteins in or around neurons.

Parkinson disease, some cases of Alzheimer disease, and some types of prion disease are accompanied by the presence of *Lewy bodies* within the cytoplasm of neurons and also in nearby glia. These deposits consist largely of a dense core of fibrils of *α-synuclein,* a small 140-residue protein abundant in various parts of the brain. Mutations in the α-synuclein gene are associated with autosomal-dominant inheritance of early-onset Parkinson disease. Just as tau tends to be associated with microtubules, α-synuclein may function in cooperation with microfilaments. Studies of an autosomal-recessive form of inherited juvenile Parkinson disease led to mutations in a large (>1 Mbp) gene on chromosome 6. It encodes the 465-residue *Parkin.* Parkin is an E3 ubiquitin ligase, which ubiquitinates a-synuclein. This finding suggests that abnormally slow degradation of synuclein may be an important cause of Parkinson disease.

One of the triple-repeat polyglutamine diseases is Huntington disease. The defective *Huntingtin* is a cytosolic protein that normally protects neurons but fails when the polygultamine sequence becomes too long. Neurons of the cerebral cortex and striatum die, apparently by apoptpsis. Huntingtin interacts with p53, with a CREB-binding protein, and with an EGF receptor suggesting that it functions in regulation of transcription. One of the genes whose transcription is regulated is that of the neurotrophin known as *brain-derived neurotrophic factor (BDNF).*

Many approaches have been taken in therapy of Parkinson disease. As mentioned already that enhancing dopamine production by administration of L-dopa or by use of MAO inhibitors is a standard treatment. Experimental gene therapy with a glial cell line-derived neurotrophic factor also appears promising.

Another aspect of neurodegeneration involves oxidative damage. A clue comes from amyotrophic lateral sclerosis (*ALS*), which struck down the New York Yankees baseball player Lou Gehrig, after he had started 2130 consecutive games over a 15-year period. ALS (Lou Gehrig disease) is the most prevalent of more than 70 diseases, that cause loss of motor neurons. The cause of a rare hereditary form of ALS is a defect in superoxide dismutase, which appears to promote excessive formation of free radicals. However, this interpretation is uncertain. Parkinson disease induced by the compound MPTP may also arise as a result of free radical damage. Among possible effects, MPTP may induce apoptosis. Both oxidative damage and apoptosis may be factors in Alzheimer and other neurodegenerative diseases as well.

In every disease in which an abnormal protein is found there must be pathways of processing the protein to generate its functional form and pathways for degradation. These pathways are being investigated for all of the neurodegenerative diseases and none more intensively than for Alzheimer disease. The amyloid precursor protein APP is an integral membrane glycoprotein with a large ~687-residue extracellular N-terminal portion, which resembles a cell surface receptor. It contains both a protease inhibitor-like domain and a zinc-binding region, which can be phosphorylated. It binds heparin and collagen as well as other proteins. Rare mutations that cause early-onset familial Alzheimer disease are found in the APP gene. Some of these mutations alter the regulation of pre-mRNA splicing. Splicing generates eight different APP isoforms vaying in length from 677 to 770 residues. The properties of the isoforms vary. For example, if the 18 residues of exon 15 are spliced out a new motif for posttranslational modification is created by fusion of exons 14 and 16. The newly created sequence ENEGSG is recognized by a xylosyltrans-ferase, which initiates formation of the terminal unit for glycosaminoglycan formation. The resulting proteoglycan is known as *appican*. The precursor protein APP is transported down axons to the nerve endings and is proteolytically cleaved to form the insoluble amyloid deposits. Alzheimer disease may occur when there is excessively rapid proteolysis of the precursor or if there is a failure to metabolize the amyloid protein. The folding and glycosylation reactions of APP occur in the ER and the Golgi, but the major problem in Alzheimer disease appears to be in subsequent proteolytic processing, some of which may occur as the APP is being transported through the Golgi to the cytoplasmic membrane. The protein may be cleaved at three sites by enzymes known as α-, β-, and *γ-secretases* as is indicated in Fig. 7.34, in which the protein is represented as an unfolded "stick." Most of the mutations in APP that cause Alzheimer disease are near these three cleavage sites. As indicated in Fig. 7.34, cutting at the α site liberates into the extracellular space the large N-terminal portion as a soluble protein called APPsα. It is thought to have a protective effect on neurons. If this cleavage occurs the protein is not cut at the β site and fragment Aβ is not formed. However, if cleavage by β-secretase (beta-site APP-cleaving enzyme or BACE) occurs first and is accompanied by or followed by cleavage at the γ-site, AB is liberated (Fig. 7.34). The β-secretase is an integral membrane protein, which carries a pepsinlike domain in its lumenal (or extracellular) part. Since the AP peptide in an aggregated form appears to be toxic to neurons, a logical therapy for Alzheimer disease may be to block either the β-or γ-secretase.

The γ-secretase has been difficult to locate but has been identified as a result of other rare familial forms of Alzheimer disease. These are caused by mutations in genes for proteins known as *presenilin*-1 (on chromosome 14) and *presenilin*-2 (on chromosome 1). The presenilins are integral membrane proteins with multiple transmembrane helices.

They have been regarded as regulators of γ-secretase, but there is much evidence that the presenilin molecules may be cleaved proteolytically and that the C- and N-terminal domains formed in this way may associate to form an unusual aspartyl protease. It, too, is a target for inhibitors. The picture is made more complex by the fact that presenilins form complexes with other proteins. These include a newly discovered protein *nicastrin,* a large 709-residue transmembrane glycoprotein. Nicastrin not only seems to modulate presenilin action but also participates in an important developmental process via the highly conserved *Notch pathway*. Many other proteins are found in the amyloid plaques of Alzheimer disease. Among them are acetylcholinesterase, proteoglycans, hydroxyacyl-CoA dehydrogenase, GM1 ganglioside, apolipoprotein A-1, and lithostatine.

What are the possible adverse consequences of accumulation of the A β protein? It may cause inflammation by activation of *microglia,* which may cause damage by release of NO. Aβ may induce death of neurons by apoptosis. A defect in protesomal degradation may be a factor. Both Aβ and the prion protein may promote oxidative damage. The brain derives most of its energy from oxidative metabolism, a major source of damaging radicals. Mitochondria are found in dendrites as well as cell bodies. Methionine residues in glycinerich parts of the Aβ and prion proteins are suspected as centers of free radical formation.

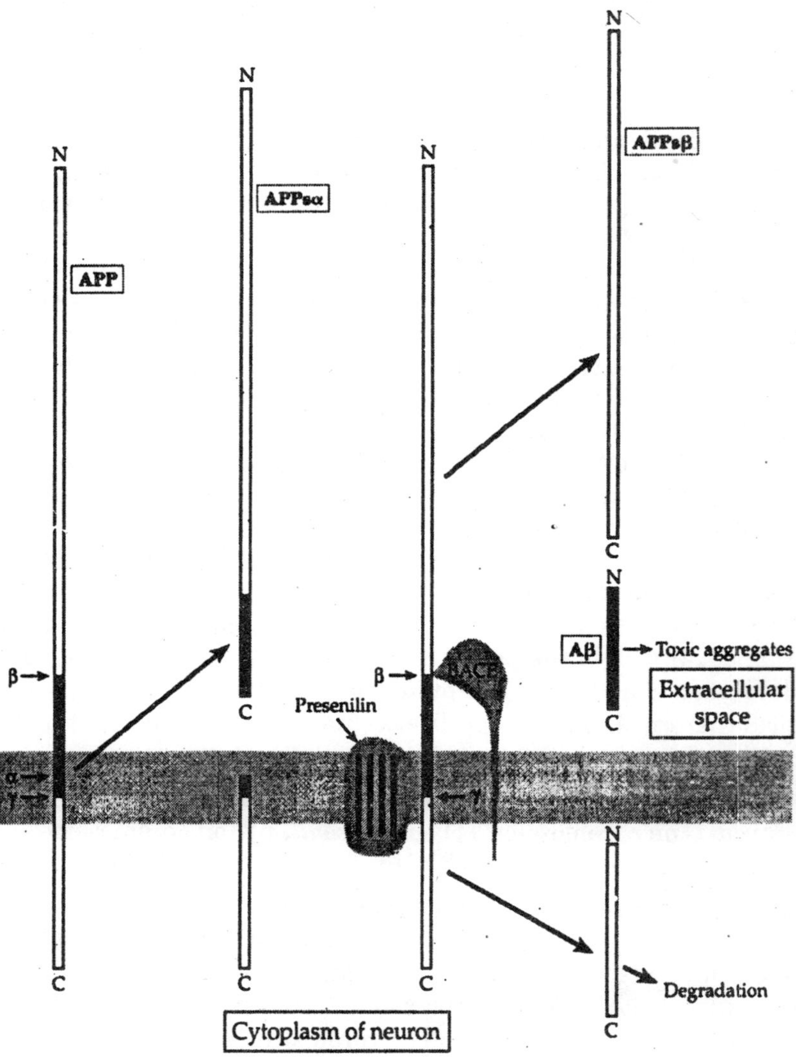

Fig. 7.34. Cleavage of the amyloid precursor protein APP with liberation of amyloid Aβ protein. The proteins are represented as sticks (not to scale) but in reality contain both intracellular and extracellular globular domains.

Both amyloid plaques and the tangles of protein taucontaining paired helical filaments are typically present in Alzheimer disease. Which comes first? Some hereditary neurodegenerative diseases are known in which tau filaments are present in neurons and sometimes also in glia. Since mutations in tau don't lead to Alzheimer disease whereas mutations affecting APP do, it is often assumed that the primary defect in the disease is with APP and that accumulating AP induces the observed changes in tau. However, this is by no means certain. Six different isoforms of tau (the longest with 441 residues) are created by alternative splicing of the mRNA. During its normal functioning tau is phosphorylated and carries an average of 2 – 3 phospho groups. In Alzheimer disease the level of tau is greatly increased (4- to 8-fold) and the molecules carry 5 – 9 phospho groups. It is this hyperphosphorylated tau that forms the paired helical filaments and tau tangles, which appear to clog the slender neurons.

What does tau do normally? Although it has been studied for many years, its exact functions are elusive. However, the role of the microtubules in axonal transport is well established. The tau isoforms may play a functional role in this process. The hyperphosphorylated tau of Alzheimer disease doesn't promote proper assembly of microtubules and may interfere with axonal transport of materials along the microtubules. Alzheimer disease may reflect an imbalance between the phosphorylation and dephosphorylation processes. Another possible problem with tau may be slow isomerization of prolyl linkages because of a deficiency of a prolyl *cis-trans* isomerase.

Until 1993 *apolipoprotein E* was best known for its central role in plasma lipoproteins and cholesterol transport (Fig. 7.1). However, one of the three common alleles of the apoE gene confers a significant risk of development of Alzheimer disease. A high blood cholesterol level is also correlated with increased risk. Membrane abnormalities in mitochondria have been associated with Alzheimer disease. Also related to membranes and lipid metabolism, *vitamin E* appears to combat Alzheimer disease.

Environmental and nutritional factors may also affect the development of Alzheimer disease and other mental illness. Aluminum frequently accumulates in the neurons containing neurofibrillary tangles. Copper and zinc ions can cause the amyloid AP to aggregate. However, Zn^{2+} may actually protect against neurotoxicity. The amino acid β-N-methylamino-L-alanine, a constituent of the toxic seeds of a type of palm *(Cycas circinalis* L.), may have induced both ALS and Guam disease, a condition resembling Parkinson disease, in a population in Guam that traditionally used these seeds as food.

Can neurodegenerative diseases be prevented or delayed? Much evidence suggests that the answer is yes. Rare early onset forms pose a special problem, but for most of us maintaining an active life style, using our minds, and choosing a good diet with adequate amounts of vitamins and essential ω3 fatty acids may be very helpful. New methods of treatment are being tested. Antiinflammatory drugs are helpful, and even vaccination against Aβ and other amyloid proteins appears possible. Is it possible that antibodies and phagocytic cells can clear the cobwebs from our brains?

8 The Life History of Hormones

The use of hormones for the purpose of coordination involves a complex series of physiological events. Such a life history begins with the formation of the excitant by the endocrine glands and concludes with the response of a target, or effector tissue, and the hormone's ultimate destruction or its excretion from the body. The events that determine the action of a hormone are shown in Fig. 8.1. This basic pattern persists throughout the vertebrates, though, as will be described, certain differences exist.

The Formation of Hormones

Although the formation of all hormones is determined at the genetic level, it can be either a relatively direct translational procedure or, alternatively, occur as a result of the prior formation of enzymes that mediate synthesis. Enzymes are also involved in the former process, however.

TRANSLATIONAL FORMATION OF HORMONES

It seems likely that the sequences of amino acids in the polypeptide and protein hormones directly reflect genetic translation via messenger RNA. Even when a direct translation of genetic material into the amino acid sequence of a hormone is made, the active hormone product does not result. Many peptide hormones, including glucagon, corticotropin, MSH, the neurohypophysial peptides, and calcitonin, contain relatively few amino acids and are formed as a result of a process of disassembly from much larger protein molecules. In addition, some polypeptide hormones are not simple strings of amino acids but consist of subunits as seen in insulin, the gonadotropins, and TSH. The assemblage of these subunits into a hormone must occur subsequent to the formation of the individual parts. The formation of polypeptide hormones thus usually involves the initial formation of a parent molecule called a*preprohormone.* As a result of post-translational changes, involving cleavage by enzymes, the pre-prohormone is broken up to form the hormone itself. The pre-prohormone has only a transient existence (usually less than 2 minutes) and is rapidly converted to the*prohormone.* There is good evidence that most peptide hormones are formed in this way.

Such a process of hormone formation is well illustrated in the production of insulin (Steiner et al., 1974; Chan, Keim, and Steiner, 1976; Lomedico et al., 1977). The genes that are responsible for the formation of insulin have been isolated and allowed to translate their coded amino acid sequence *in vitro.* The initial product is a protein with a

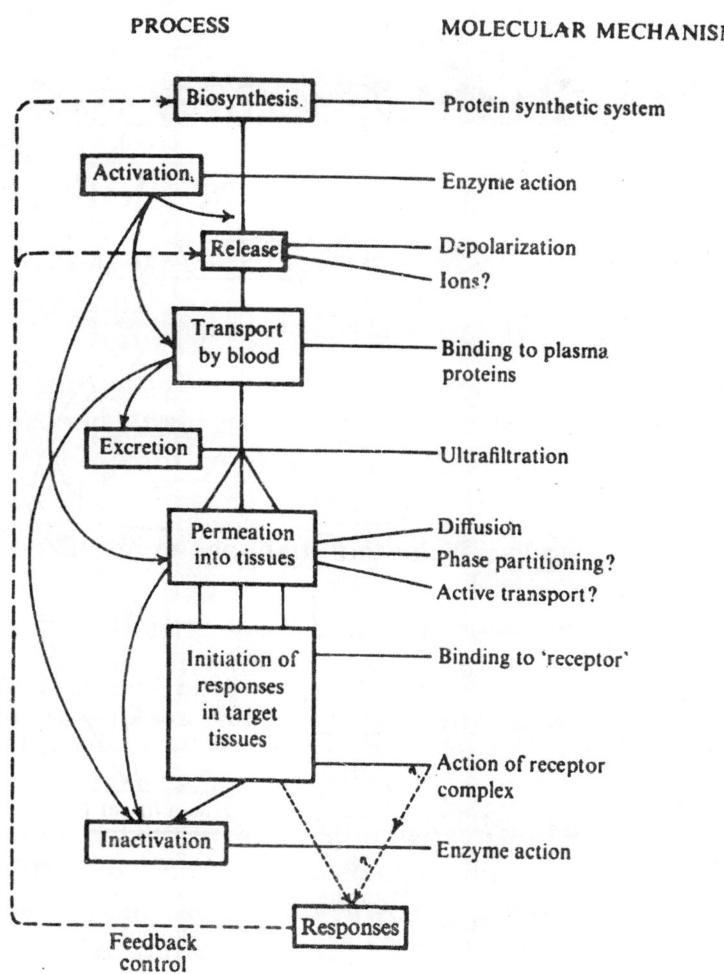

Fig. 8.1. Diagrammatic summary of the life history of a hormone commencing with its biosynthesis and concluding with the response and its inactivation.

molecular weight of about 11,500 (compared with about 6,000 for insulin). It is a linear peptide with a special 23 amino acid segment at its N-terminus. This "signal peptide" attaches the molecule, and the ribosome on which it is being formed, to the endoplasmic reticulum. The molecule is called pre-proinsulin and it only has a brief existence as it passes into the Golgi apparatus where its 86 amino acid C-terminal segment is incorporated into storage granules. This proinsulin is then broken down under the influence of a carboxypeptidase-like and tryptic enzymes that fragment the molecule in the region of certain arginine and lysine residues. The A-chain and B-chain are released but as a result of the folding of the prohormone they remain aligned and joined to each other by two disulfide bridges. The "C"-peptide fragment is released with the hormone. Its amino acid sequence varies quite considerably in different species. The structures of several other prohormones, and these are also converted to hormones in a similar manner to that of proinsulin.

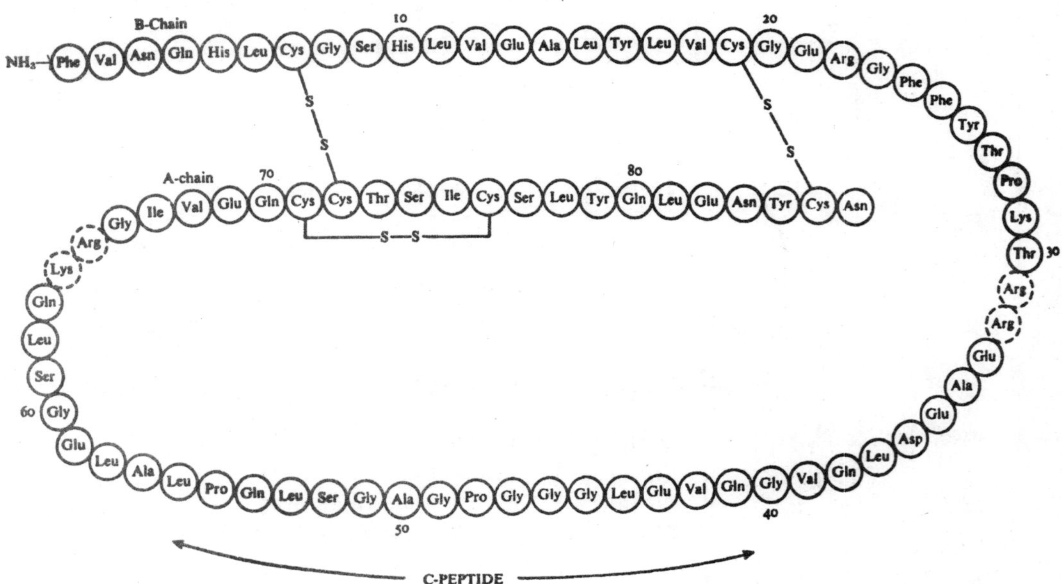

Fig. 8.2 The proposed amino acid sequence in proinsulin from man. The basic residues indicated by the broken circles have been assigned as they are known to occur in bovine and porcine proinsulin.

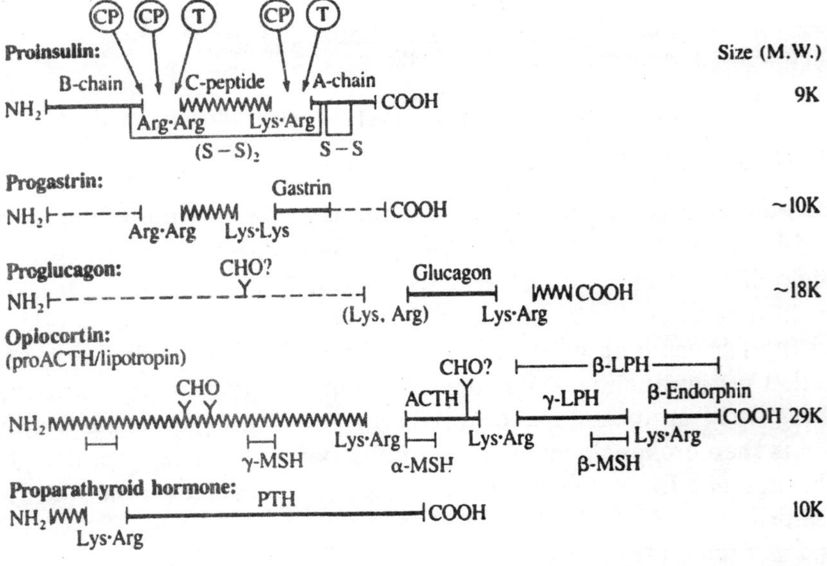

Fig. 8.3. Structures of some prohormones. The heavy lines indicate the biologically active fragments, or hormones. The dashed lines are the regions in which the amino acid sequences have not yet been determined. The parent molecules are broken down as a result of the activities of carboxypeptidase B-like (CP) and trypsin enzymes (T). These reactions occur in the regions of the arginine and lysine residues, as indicated for the pro insulin.

Many polypeptide hormones are stored in granules present in the endocrine cells. These structures are bounded by membranes and are 0.1-0.4 μm in diameter. They appear to originate in the Golgi apparatus of the cell. The precursor, or prohormone, becomes associated with the granules and it seems likely that conversion to the hormone takes place in these. The granules can travel to the peripheral regions of the cell, and, in response to releasing stimuli, combine with the plasma membrane and discharge their contents into the region of blood vessels. A summary of this process as it is thought to occur for insulin.

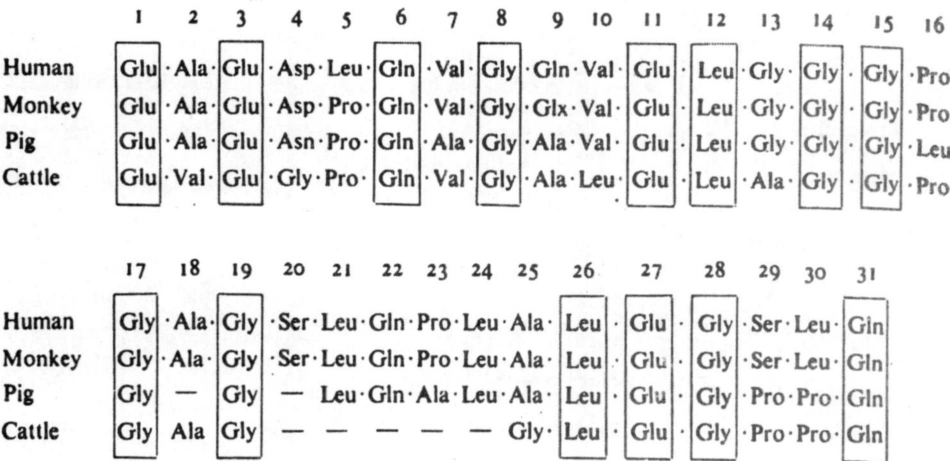

	1	2	3	4	5	6	7	8	9	10	11	12	13	14	15	16
Human	Glu	Ala	Glu	Asp	Leu	Gln	Val	Gly	Gln	Val	Glu	Leu	Gly	Gly	Gly	Pro
Monkey	Glu	Ala	Glu	Asp	Pro	Gln	Val	Gly	Glx	Val	Glu	Leu	Gly	Gly	Gly	Pro
Pig	Glu	Ala	Glu	Asn	Pro	Gln	Ala	Gly	Ala	Val	Glu	Leu	Gly	Gly	Gly	Leu
Cattle	Glu	Val	Glu	Gly	Pro	Gln	Val	Gly	Ala	Leu	Glu	Leu	Ala	Gly	Gly	Pro

	17	18	19	20	21	22	23	24	25	26	27	28	29	30	31
Human	Gly	Ala	Gly	Ser	Leu	Gln	Pro	Leu	Ala	Leu	Glu	Gly	Ser	Leu	Gln
Monkey	Gly	Ala	Gly	Ser	Leu	Gln	Pro	Leu	Ala	Leu	Glu	Gly	Ser	Leu	Gln
Pig	Gly	—	Gly	—	Leu	Gln	Ala	Leu	Ala	Leu	Glu	Gly	Pro	Pro	Gln
Cattle	Gly	Ala	Gly	—	—	—	—	—	Gly	Leu	Glu	Gly	Pro	Pro	Gln

Fig. 8.4. A comparison of the amino acid sequences of the human, monkey, porcine, and bovine C-peptides. The solid bars indicate residues that are identical in all species.

Such granules apparently furnish sites for the formation and storage of many hormones. If released into the cytoplasm of the cell the hormones may be destroyed, as has been observed for the catecholamines when they are exposed to the mitochondrial enzyme monoamine oxidase (MAO). In addition, storage granules may afford convenient vehicles in which hormones can be transported for considerable distances along nerve cells.

Some neurons form hormones by a process called *neurosecretion*. These are like ordinary nerve cells and consist of a cell body with an extended axon and they can also be depolarized and so convey electrical information. The axon, instead of terminating at another neuron or an effector tissue, like a gland or muscle, lies near a capillary into which it can discharge certain of its products. These products may be hormones, the formation of which is initiated some distance away in the cell body. The hormones, parceled up in their granules, travel along the nerves to the peripheral sites in the axon where they can be released into the blood.

Hormones that are formed as a result of neurosecretion are those of the neurohypophysis and hypothalamus including vasopressin, oxytocin, and the various releasing hormones that control the adenohypophysis. In mammals, vasopressin and oxytocin are formed in the supraoptic and paraventricular nuclei that are situated at the base of the brain. These hormonal products pass down the axons in granules to the neural lobe. Inside the granules they are attached to protein molecules of neurophysin, the synthesis of which seems to be closely associated to that of the hormone. The neurophysin and each hormone (vasopressin and oxytocin) are synthesized from two prohormones, each with a molecular weight of about 20,000. They have been called propressophysin and prooxyphysin. Amphibians and fishes have a single preoptic nucleus where the neurohypophysial hormones originate. The putative hormones of the urophysis in fishes apparently formed by a process of neurosecretion.

Beta Granule Formation

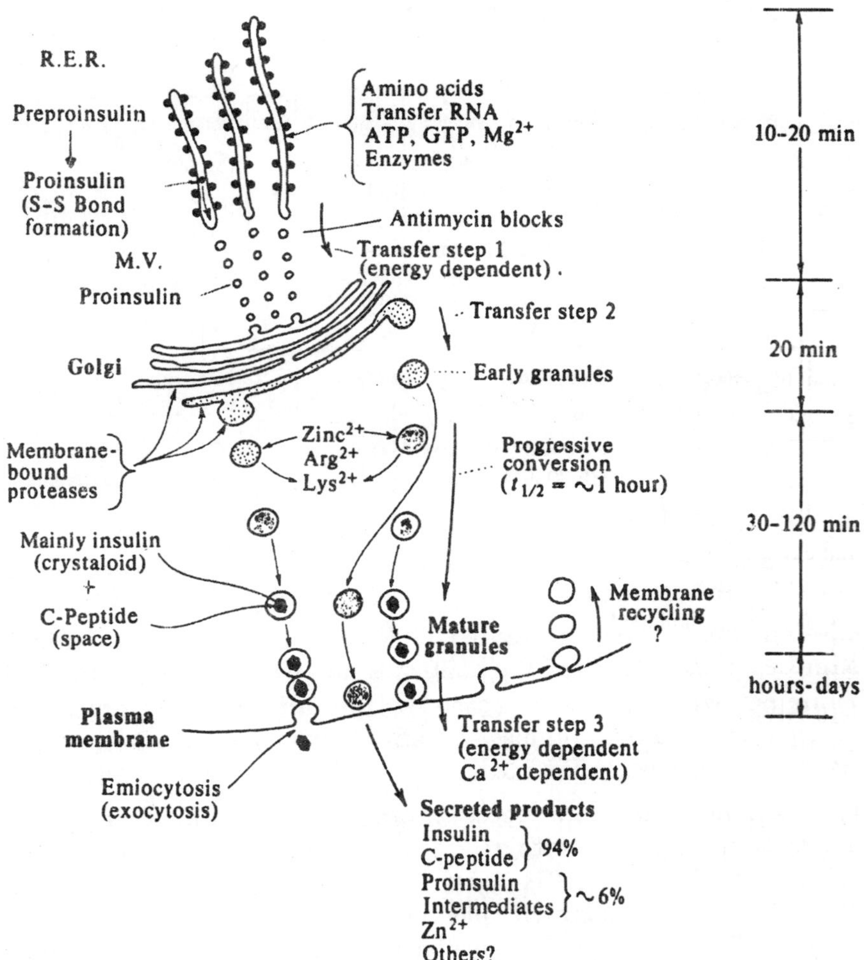

Fig. 8.5. Schematic summary of the insulin biosynthetic mechanism of the pancreatic B-cells. The time scale on the right side of the figure indicates the time required for each of the major stages in the biosynthetic process. R.E.R. = rough endoplasmic reticulum; M.V. = microvesicles.

THE FORMATION OF HORMONES BY ENZYMICALLY CONTROLLED SYNTHESIS

Thyroid Hormones

The endocrine secretions of the thyroid, the adrenal medulla, the gonads, and the adrenal cortex are the result of biosynthetic processes controlled by enzymes. Although the enzymes themselves are the result of genetic translational processes, the hormones are synthesized in chemical reactions controlled by the enzymes.

Thyroid hormones contain iodine and the thyroid gland has a special ability to concentrate inorganic iodide from the blood. This ability to transport iodide actively against an electrochemical gradient is shared by some other tissues, including the intestine and salivary glands. This ability may be controlled by a single gene: In man a congenital inability to accumulate iodide in the thyroid is accompanied by a parallel deficiency at the other iodide transport sites. The accumulated iodide is oxidized to iodine, which combines with tyrosine to form the precursor of thyroxine and triiodothyronine. The latter reaction occurs with the tyrosine residue that is part of a large glycoprotein, thyroglobulin (molecular weight 660,000) which is stored extracellularly in the thyroid follicles. It is synthesized as two subunits, each with a molecular weight of about 300,000 (Van Herle, Vassart, and Dumont, 1979). Each molecule of thyroglobulin binds two of thyroxine and on the average, less than one of truodothyronine. Thyroglobulin itself does not appear to be a particularly remarkable protein; it contains about 30 molecules of tyrosine (or about 2% by weight), and about 0.5% iodine. It nevertheless provides a site for the synthesis and storage of the thyroid hormones. The biosynthetic process for the thyroid hormones appears to be common to all vertebrates and was apparently attained early in their evolution. Nevertheless, most of our information has been derived from studies of mammals.

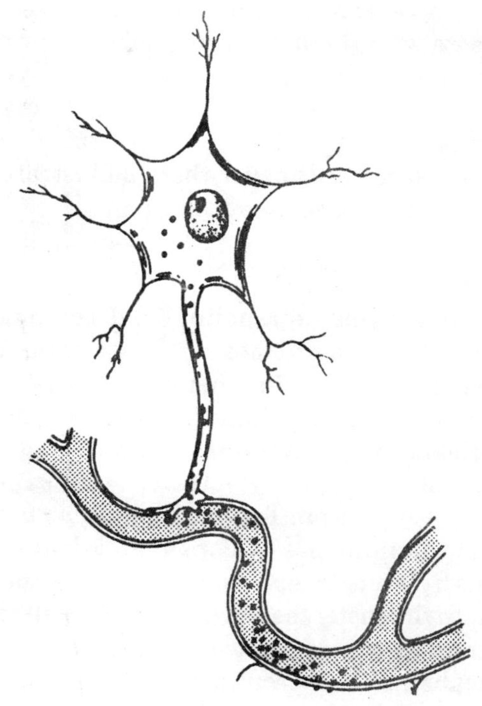

Fig. 8.6. A neurosecretory cell. The hormonal products are transported from the cell body down to the axon from which they can be released into capillaries. In contrast, ordinary nerve cells have axons that about onto other neurons (instead of capillaries).

Thyroglobulins have been identified in thyroid tissues of species from most groups of vertebrates, even including larval cyclostomes (lampreys) where it is present in the subpharyngeal gland or endostyle. These proteins exhibit many similarities with respect to their molecular size (though a few differences have been observed), as determined by centrifugation in sucrose gradients, but their amino acid constitutions may differ. In hagfish, however, a smaller protein may have assumed the role of thyroglobulin. Thyroglobulins also exhibit different immunological behaviour. Antibodies to specific thyroglobulins have been prepared and these react, *in vitro,* with the homologous protein, which can be radioactively labelled. Thyroglobulins from different species may compete with this labelled protein for binding to its antibodies. The relative ability to do this suggests the degree of immunological similarity to the homologous thyroglobulin. Considerable interspecific differences have been observed in such radioimmunoassays. Sheep thyroglobulin readily displaces its labelled form from anti-sheep thyroglobulin antibodies, but thyroglobulins from other mammals, such as pigs and rabbits, are much less effective. Thyroglobulin from a python and a crocodile also compete with the homologous labelled protein for such binding, but this is also much less so than that for the sheep protein. Bird thyroglobulin, from ducks, has no ability to bind with the sheep antibodies. Although it is tempting to construct phylogenetic trees with such information the paucity of species examined makes such predictions of doubtful significance. The measurements never-theless illustrate the diversity that can occur among thyroglobulins from different species.

Catecholamines

Epinephrine (adrenaline) and norepinephrine (noradrenaline) are formed in chromaffin tissues. These hormones are present not only in the adrenal gland but also are associated with nervous tissue in other parts of the body. Norepinephrine is also formed in certain nerve endings in the sympathetic nervous tissue and the brain. The original precursor of these catecholamines is tyrosine, which by a series of enzymically controlled reactions is converted to 3,4-dihydroxyphenylalanine, or dopa, and thence to dopamine. These reactions occur in the cell's cytoplasm. The dopamine is accumulated by storage granules in which it is converted, under the influence of dopamine β-hydroxylase, to norepinephrine. Norepinephrine can be *N*-methylated to epinephrine under the influence of the enzyme phenylethanolamine-*N*-methyltransferase (PNMT), which, in mammals, can be induced in the presence of high concentrations of corticosteroids. There is evidence to suggest that norepinephrine and epinephrine are stored in different granules and even different cells in the adrenal medulla. This could be determined by regional differences in the access of corticosteroids to the medullary tissue, thus influencing the local levels of PNMT.

Steroid Hormones

The formation of the steroid hormones also appears to be basically the same in all vertebrates. All of these are formed from cholesterol, which is present in high concentrations in the steroidogenic endocrine glands. This parent molecule may be formed *in situ* from acetate or be accumulated from the plasma. The enzymic conversion of cholesterol to pregnenolone and progesterone is common to all of the steroidogenic endocrine glands. Sandor has suggested that the use of steroids as hormones may have been determined by a primeval mutation that invented the enzyme systems that determine this transformation of cholesterol. The conversion of progesterone to the androgen, estrogen, and adrenocorticosteroid hormones involves the successive actions of diverse enzymes that hydroxylate, oxidize, aromatize, and reduce the steroid at some of the 21 carbon positions present. The ability of a species to synthesize steroid hormones with differing structures depends on the presence, absence, or activities of the enzymes that mediate these changes. The chondrichthyean fishes that secrete 1α-hydroxycorticosterone possess an enzyme, 1α-hydroxylase (which converts corticosterone to the hormone), that has not been found in other vertebrates. The formation of aldosterone in tetrapods is determined by the presence of 18-hydroxylase (it converts corticosterone to aldosterone). Rats and mice cannot form cortisol and lack 17α-hydroxylase (that con verts progesterone to 17α-hydroxyprogesterone) in their adrenal cortex. Mutations may also arise that influence the ability of a species to form certain hormones. In man, a congenital condition known as the adrenogenital syndrome is due to the complete or partial block of the 21-hydroxylating system (the conversion of progesterone to deoxycorticosterone and 17α-hydroxyprogesterone to 17α-hydroxy-11-deoxycorticosterone). Such enzymic differences determine the presence or absence of the various steroid hormones and furnish the raw materials of evolutionary change.

The steroidogenic enzymes are associated in the cytoplasm of the cell and the mitochondria. The pathways and sites of the enzymes determining the formation of adrenocorticosteroids in the frog's interregnal.

THE RELEASE OF HORMONES FROM THE ENDOCRINE GLANDS

Nature of the Stimuli

The role of the endocrine glands in the regulation of bodily functions is dependent on the release of their secretions on appropriate occasions. Secretion is initiated upon the receipt, by the gland, of a suitable stimulus, which may increase or decrease the discharge of its hormone. The message may arrive either by way of a nerve or be carried in the blood that perfuses the tissue. The primary event that initiates this stimulus may arise either from the external environment (exteroceptive stimulus) or inside the body (interoceptive stimulus).

Exteroceptive stimuli that may affect the endocrine glands include the receipt of light, a change in temperature or of the osmotic concentration (of an aqueous environment), and the acquisition of food, water, and salts. Social situations such as the proximity of prey, a predator, a mate, or the young may evoke psychogenically mediated responses in the endocrine glands. Climatic events such as rain, temperature and even, possibly, humidity and atmospheric pressure can also influence a hormone's release. The receipt of and endocrine response to such external stimuli help the animal to maintain an equitable relationship with the events that happen around it. Exteroceptive stimuli are especially useful in providing cues that are involved in reproduction.

Interoceptive stimuli are those that result from changes in the physicochemical conditions within the body. Ultimately they may reflect the external conditions: For instance, a lack of drinking water and a hot, dehydrating environment will lead to an increase in the osmotic concentration of the body fluids. Internal stimuli include changes in the concentration of salts, such as sodium, potassium, and calcium, in the body fluids, alteration of the hydrostatic pressure of the blood vasculaı system, oscillations in the levels of nutrients, like glucose, amino acids, and fatty acids, as well as changes in the body temperature. The physiological factors influencing release of hormones are summarized.

Hormones can exert tropic effects on other endocrine glands and the terminal secretions, once released, may travel back to the region where the tropic hormone originated and inhibit its further release. This last effect completes the cycle of events that closes the loop of a *negative-feedback system* that plays a vital role in regulating the endocrine system. Such a control system is well known to engineers and its action is illustrated. The release of hormones from the median eminence, which in turn controls the formation and discharge of the tropic hormones of the adenohypophysis, is regulated in this manner. The hormones that exert the inhibitory effects in the median eminence may alter the thresholds for stimulation of the neurosecretory cells. In addition, such a negative-feedback can also act directly on theadenohypophysis as seen with the action of thyroid hormones that inhibit the release of **thyro**

Table. 8.1. Principal Stimuli Influencing the Relase of Hormones

Hormone	*Releasing stimuli*
Aldosterone	Low plasma Na concentration, angiotensin II and III
Angiotensin	Renin
Calcitonin	Hypercalcemia
Cholecystokinin	Digestive products in the upper intestine
Cortisol and corticosterone	Corticotropin
Enterogastrone	Fats and oils in intestine
Enteroglucagon	Feeding
Epinephrine	Neural stimuli (mediated by acetylcholine)
Estrogens	FSH
FSH	External stimuli, such as light, low estrogen levels
Gastrin	Feeding (vagal reflex; local reflex from food in stomach)
GIP	Fats in intestine
Glucagons	Hypoglycemia, gastrin, CCK, high amino acid and low fatty acids in plasma, exercise
Growth hormone	Sleep, exercise, apprehension, hypoglycemia
Hypophysiotrophic hormones: CRH, TRH, etc.	Hypothalamic neuronal stimuli (dopamine and monoamine transmitters), inhibited by negative-feedback mechanisms carried by hormones and metabolites
Insulin	Hyperglycemia and amino acids in plasma, glucagons, growth hormone, in ruminants high levels of propionic and butyric acid, vagal stimulation, CCK and secretin. Inhibition by epinephrine
LH or ICSH	External stimuli, sexual excitement (male), estrogen "surge" (female), low progesterone or testosterone levels
MSH	Light on retina, low plasma corticosteroid levels, inhibition by neural stimuli
Melatonin	Darkness (adrenergic neural stimulation)
Oxytocin	Suckling, parturition
Parathyroid hormone	Hypocalcemia
Progesterone	LH, chorionic gonadotropin, prolactin
Prolactin	Diurnal rhythm (sleep), suckling, parturition, plasma osmotic concentrations) low in fish, high in mammals?), estrogens
Rennin	Low Na in plasma, hemorrhage, reduced renal blood flow, nerve stimulation (β-adrenergic), increased osmotic concentration in renal blood supply
Secretin	Acid in upper intestine
Testosterone	LH
Thyroid hormones	TSH
TSH	Low thyroxine, temperature reduction
Vasoperssin, ADH	Increased osmotic concentration of plasma
$1\alpha,25\text{-(OH)}_2$-vitamin D_3	Low Ca and phosphate levels in plasma; parathyroid hormone

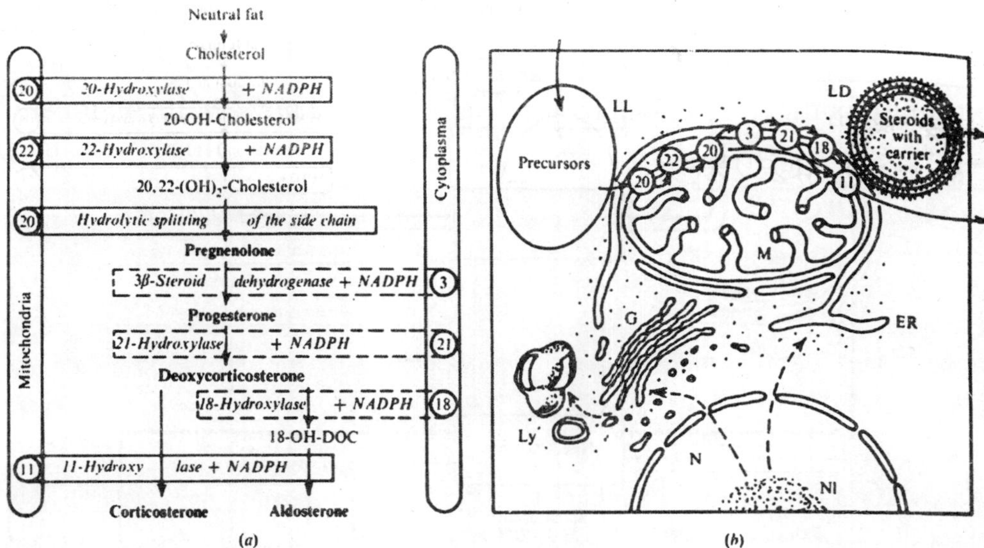

Fig. 8.7. The biosynthesis of corticosteroid hormones (*a*) in the adrenocortical tissue of a frog. This is illustrated with reference to the cell organelles (*b*). Corticosterone can also act as a precursor for aldosterone. In some species, 18-hydroxylase may be a mitochondrial rather than a microsomal enzyme. In vertebrates that form cortisol another microsomal enzyme, 17α-hydroxylase, active on progesterone and leading to 17a-OH-progesterone, deoxycortisol, and cortisol, is present. ER = endoplasmic reticulum; G = Golgi apparatus; N = nucleus; LD = electrondense lipid droplet; LL = electron-lucid lipid droplet; Ly = lysosomes; M = mitochondrion; NI = nucleolus. Solid arrows are pathways of steroid synthesis from precursors to steroid bound to a carrier; broken arrows indicate cellular responses activated by ACTH.

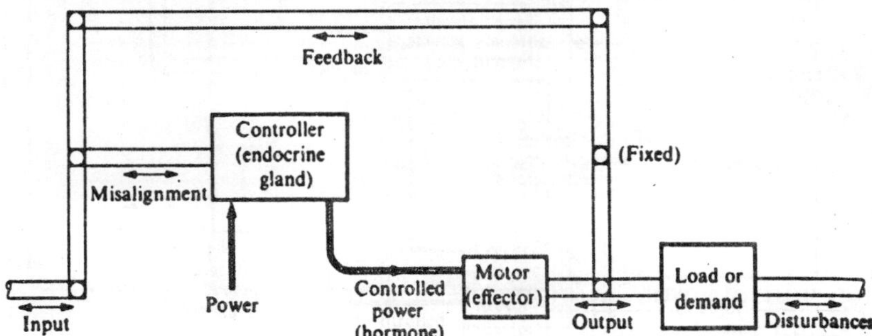

Fig. 8.8. Schematic diagram of a servosystem showing the analogies of the classic engineering control mechanism of physiological coordination by the endocrines. The double arrows indicate that the links can move to and fro. Stimuli (both input and feedback) are fed into the controller (endocrine gland in the physiological analogy) through a misalignment detector that is sensitive to changes from a certain set-point. It is conceivable that the latter may be reset in certain physiological conditions (such as hibernation). In response the controller varies its power output (or hormone secretion) to the motor or effector. The latter adjusts the physiological needs and tends to restore equilibrium (of metabolites, salts, water, other hormones, temperature, etc.), the degree of which is transmitted back to the controller via the feedback arc.

tropic hormone in this way. The interrelations of the hormones of the hypothalamus, adenohypophysis, and more peripheral endocrine glands.

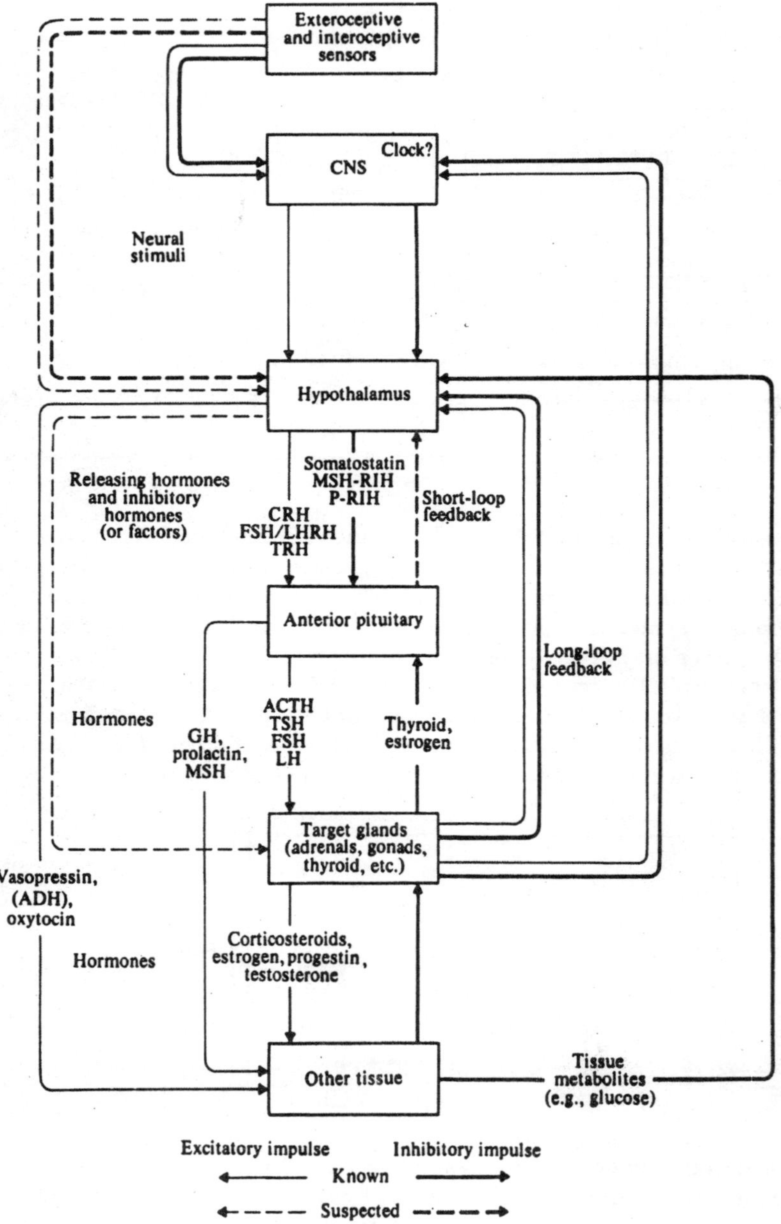

Fig. 8.9. Factors controlling the release of hormones from the anterior pituitary (adenohypophysis). Shown is the use of negative-feedback inhibition of secreted hormones and tissue metabolites in influencing further hormonal release. As can be seen, the hypothalamus (and its associated median eminence) plays a central role in this process. Stimulatory effects are shown by thin lines and inhibitory processes by thick lines.

The feedback mechanism involving the action of the peripheral endocrine secretions on the hypothalamus is called a *long-loop feedback.* There is also evidence suggesting that the adenohypophysial secretions may exert a similar action on the hypothalamus by what is termed a *short-loop feedback.* It should be noted that peripheral hormones do not necessarily initiate a negative-feedback inhibition on the hypothalamus. High estrogen levels can stimulate the release of LHRH, which initiates the events that result in ovulation. Such *a positive feedback* has also been shown in the hypothalamus of the goldfish where thyroxine stimulates the release of a thyrotropin-inhibiting hormone. The sporadic information available from experiments on nonmammals suggests that feedback control working through the hypothalamus and the adenohypophysis is widespread. More information is needed and there is some doubt as to the importance of such effects in cyclostomes.

A negative-feedback inhibition of hormone secretion also results from changes in the concentrations of the products of the hormone's actions. The retention of water or sodium, as a result of the actions of antidiuretic hormone and aldosterone, respectively, reduces the further release of these hormones. Comparable mechanisms exist involving glucose levels and the regulation of insulin, glucagon, and growth hormone, as well as calcium, parathyroid hormone, and calcitonin.

The endocrines, apart from influencing each other's release through tropic and feedback mechanisms, may also interact with each other and so modify their secretory activity. Epinephrine can inhibit the release of antidiuretic hormone from the neurohypophysis and that of insulin from the islets of Langerhans. This inhibition is an α-adrenergic effect that contrasts with the β-adrenergic effect that increases the release of insulin. Glucagon promotes the release of insulin and growth hormone directly; such effects do not depend on changes in blood glucose levels. Excesses of growth hormone are diabetogenic and inhibit the formation of insulin, though smaller amounts apparently stimulate its release.

Many less precise and less specific stimuli than those described in the foregoing can initiate the discharge of hormones from endocrine glands. Such stimuli may contribute to the homeostatic process, though they may also confuse it. No endocrine gland can exist and be uninfluenced by events outside what we may like to think of as its homeostatic area of influence. Nonspecific stimuli, especially if strong enough, can elicit a discharge of many hormones. This is sometimes referred to as "stress" and is particularly likely to occur in experimental situations that contribute to the confusion of the perpetrating scientists.

The tropic effects of hormones on the secretion of other endocrine glands can contribute to the processes of biological amplification in the body. An initial stimulus may only produce a change that involves a very small amount of energy while the energetic demands of the response may be relatively immense. The quantities of the excitants necessary to initiate an effect and the amount of the products formed (on a weight for weight basis) are illustrated. We have no information as to the quantity of neural transmitters necessary to trigger a release of corticotropin-releasing hormone from the median eminence, but quantitative changes have been estimated for the subsequent physiological events. The final release of cortisol promotes the formation of 5600 μg of glycogen. The initial amount of CRH required to do this is about 0.1 μg so that the final response represents an amplification of 56,000 times.

"Pulsatile" Release of Hormones

Hormones are not necessarily released in a continuous stream but also in discrete bursts. This process may take place in a single large "surge" or in a pulsatile manner, one "pulse" succeeding another at regular intervals. The latter is also referred to as "episodic" release. This phenomenon did not become apparent until sensitive and convenient radioimmunoassay methods were developed for assaying hormones in small, serially collected samples of plasma. The pulses of released hormones may take place as frequently as every few minutes to as little as three or four times a day.

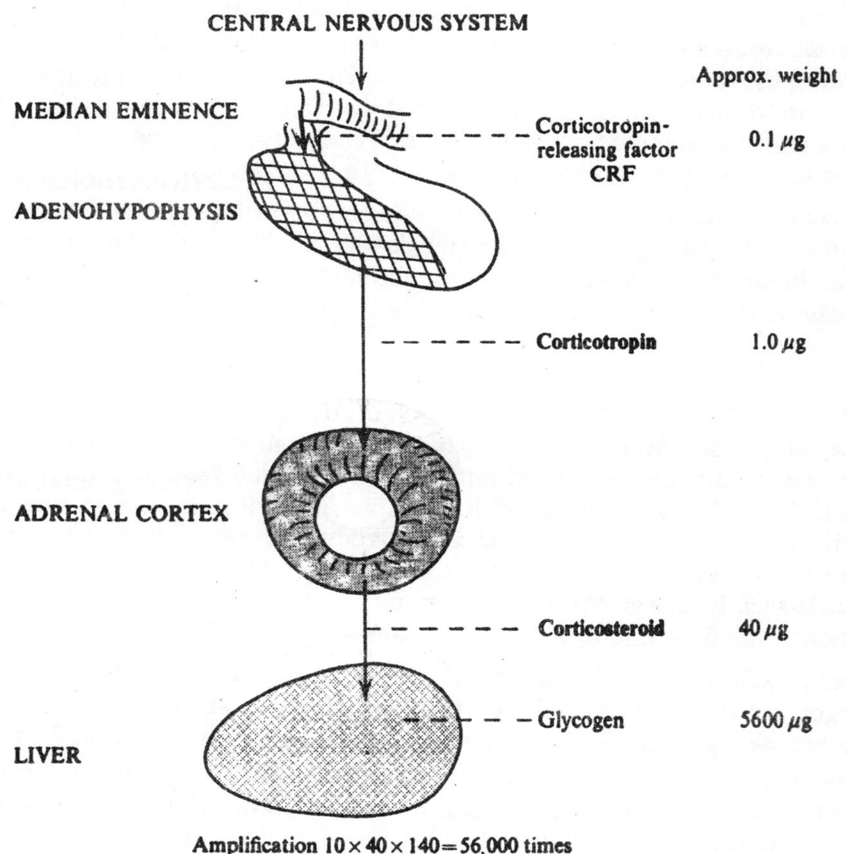

Fig. 8.10. An illustration of biological amplification as shown in the endocrine system. Corticotropin-releasing hormone initiates the release of corticotropin (ACTH), which leads to the deposition of glycogen in the liver.

The rhythmical, pulsatile release of hormones can result in their attainment of stable basal, or tonic, concentrations in the plasma. When a hormone is released in such a manner it will at first be diluted, as a result of its redistribution in the body fluids, and it will then be removed from the circulation. The rate of this process is called its "clearance time." This occurs as a result of its uptake by tissues, its metabolic breakdown, and its excretion in the urine and bile. Its levels will thus decline in the intervals between the pulses. The level to which they drop will depend on the factors that influence the hormone's clearance and the quantity that was originally released. If no further pulses occur, as in a surge of released hormone, the plasma concentrations will eventually be reduced to zero. A useful measure of this decline is the half-life ($t_{1/2}$) of the hormone in the circulation, which is the time it will take to reach 50% of its previous level. If a hormone is released more frequently than the time of its half-life, then it will tend to accumulate and its concentration in the plasma will rise. It will then reach a new equilibrium level after the equivalent of about four half-lives. The particular basal level attained will depend on the frequency of the pulses and the amount of hormone released on each such occasion. The basal level will appear graphically oscillating-type system with abrupt increases and declines centered around a steady median value. Such a mechanism provides a system whereby different tonic levels of hormones can be attained, and maintained, such as in daily rhythms of hormone concentrations in the plasma or in more long-term changes that accompany reproductive cycles. The factors that determine the frequency of the pulses and the amount of hormone released on each occasion are not well understood and often seek hypothetical refuge in engineering jargon, such as "pulse generator."

The testes of rams are small during the spring; these animals do not normally come into breeding condition until the autumn. When LHRH is injected into rams in the spring, in an episodic manner, for 60 seconds every 2 hours, the testes enlarge.

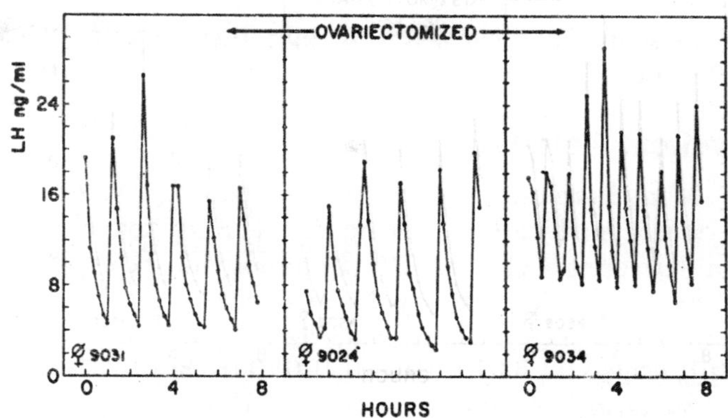

Fig. 8.11. The spontaneous pattern of release of luteinizing hormone (LH) in three ovariectomized ewes. Blood samples were collected every 12 minutes for 8 hours during the normal midanestrous period in May. It can be seen that the plasma concentrations of LH varied in a pulsatile manner from about 2 to 30 ng/ml. The time between the peak (or trough) levels reflects the times of release of the hormone from the pituitary gland and varies from about once every 50 to once every 100 minutes. The particular pattern was consistent in each sheep. The mean LH concentration in the plasma and the basal level, below which it did not drop, also varied, the highest being in the ewe that released LH most frequently.

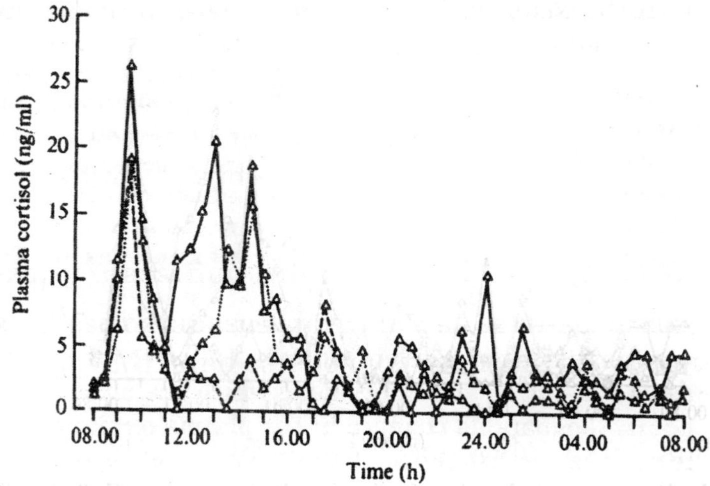

Fig. 8.12. Diurnal variation in the levels of plasma cortisol in three sheep. It can be seen that the highest concentrations were recorded during the early hours of daylight. The concentrations of cortisol in the plasma were not uniform but showed continual oscillations, suggesting that it is released in sudden "pulses".

These sheep then come into a reproductive condition similar to that which occurs normally in autumn.

Cyclical Release of Hormones

Many hormones are released periodically: at certain well-defined hours of the day (diurnal or circadian rhythms), during the reproductive cycle, or at certain seasons and times of the year (circannual rhythms). Such timing may be especially important in coordinating the events of the reproductive cycle and insuring that this occurs during the times of the year most appropriate to the survival of the young. A predictably functioning release mechanism also insures that adequate hormone levels, necessary for the animal's optimal daily activities, are available. Such release of hormones is usually controlled by centers in the brain (sometimes called "biological clocks") that are programmed by stimuli that include the length of the daily period of light, the external temperature, and changes in the seasons, as well as certain interoceptive stimuli.

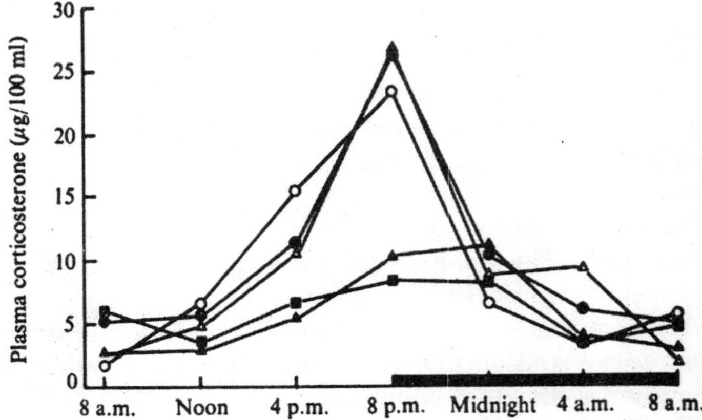

Fig. 8.13. The circadian pattern in the concentration of corticosteroids in the plasma of rats. The effects of the administration of exogenous corticosteroids to young, developing rats on the subsequent circadian periodicity in the endogenous corticosteroid levels are shown. It can be seen that the administration of dexamethasone or hydrocortisone (cortisol) on days 2 to 4 after birth suppressed the rhythmical release. However, when dexamethasone was given on days 12 to 14 after birth no effect was seen: •, control; ○, saline, 0.1 ml, days 2-4; ■, hydrocortisone acetate, 500 μg, day 3; ▲, dexamethasone PO_4, 1 μg, days 2-4; Δ, dexamethasone PO_4, 1 μg, days 12-14.

The levels of corticosteroids in the blood vary in a distinct pattern during the course of the day. In mammals, this is well-known in primates including man, and dogs, rats, and mice. Release is related to the incidence of light and the corticosteroids are lowest in concentration during the night and reach a distinct peak after various periods of daylight. In man and sheep this is seen in the morning hours, soon after dawn, though in laboratory rats, which are nocturnal, it is delayed until the early evening hours. Comparable changes in the plasma corticosteroid concentrations have been observed in some teleost fishes [the channel catfish, *Ictalurus punctatus*, and the gulf killifish, *Fundulus grandis* where peak concentrations occur about 8 hours after the onset of light. In another teleost (the eel *Anguilla rostrata),* diurnal variation of plasma cortisol levels does not seem to occur. Prolactin and growth hormone in man are released in greatest amounts during the period of sleep do not appear to be directly dependent on light.

The release of gonadotropins from the adenohypophysis may also occur at precise times of the day. In the Japanese quail *(Coturnix coturnix japonica),* the pituitary is depleted of gonadotropin over a period of 4 hours commencing about 16 hours after dawn. This release is seen in birds kept on a long-day cycle of light, 20 hours light/ 4 hours of dark. When they are exposed to a short day of 6 hours light/18 hours dark such changes are not observed. This long-day photoperiodic pattern is similar to that which quail experience in nature where it is associated with the onset of breeding. Comparable precise patterns in the daily release of pituitary gonadotropins have been observed in brook trout *(Salvelinusfontinalis)* and rainbow trout *(Salmo gairdneri)*, as well as in leopard frogs *(Rana pipiens)*.

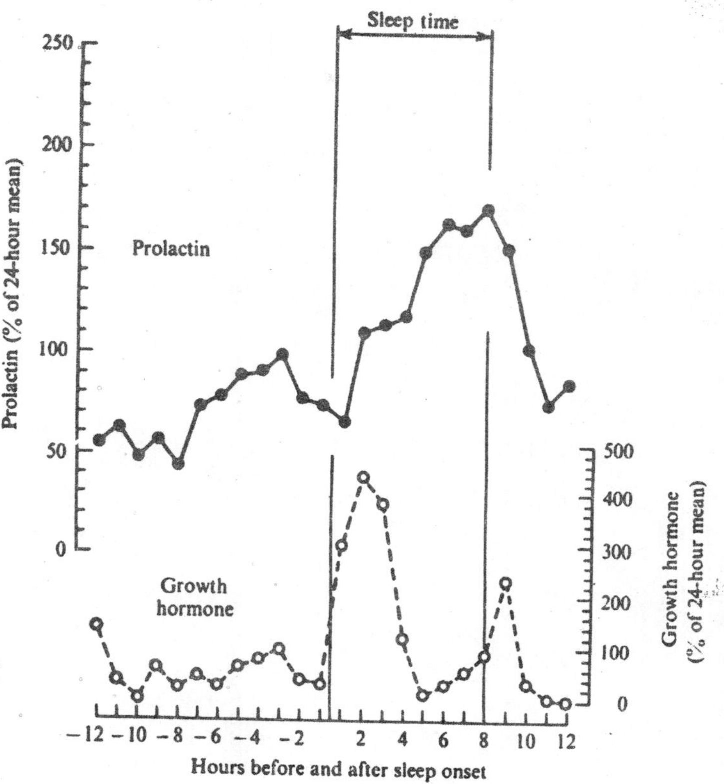

Fig. 8.14. The diurnal pattern in the release of prolactin and growth hormone in man. It can be seen that the release of both these hormones was accentuated during sleep.

The pineal gland of rats and chickens interesting cyclical pattern in activity during the day, which may be related to reproduction. The melatonin levels in the gland increase during the hours of darkness. The enzymes *N*-acetyl-transferase and hydroxyindole-*O*-methyltransferase (HIOMT) mediate the synthesis of melatonin, and the activity of the former enzyme increases at night. In rats, blinding, cutting the sympathetic nerve supply to the pineal, or placing the animals in continuous light prevents this effect. In chickens, however, extraretinal photoreceptors somewhere in the brain can mediate this rhythm.

The precise events involved in this cycle in the pineal of rats appear to be as follows. During the hours of darkness there is an increase in the release of the neurotransmitter norepinephrine and a rise in the sensitivity of its pineal receptor. These changes result in the synthesis of the enzyme serotonin-*N*-acetyltransferase. This response is a β-adrenergic one involving cyclic AMP. The *N*-acetylserotonin that is formed from serotonin is then converted to melatonin under the influence of HIOMT. This rhythm in pineal activity is thought to be controlled by a biological clock that is situated in the hypothalamus, near the suprachiasmatic nucleus, which can be inhibited as a result of the receipt of light by the retina.

In many vertebrates, melatonin inhibits growth of the gonads and this can be correlated with the inhibitory effects of darkness on this process. It is unknown how widespread this nocturnal rhythm in the activity of the pineal is. A diurnal rhythm in pineal or plasma melatonin, however, has been ob-served on reptiles, teleosts, and man. Whether such changes are always related to the receipt of light is uncertain, however.

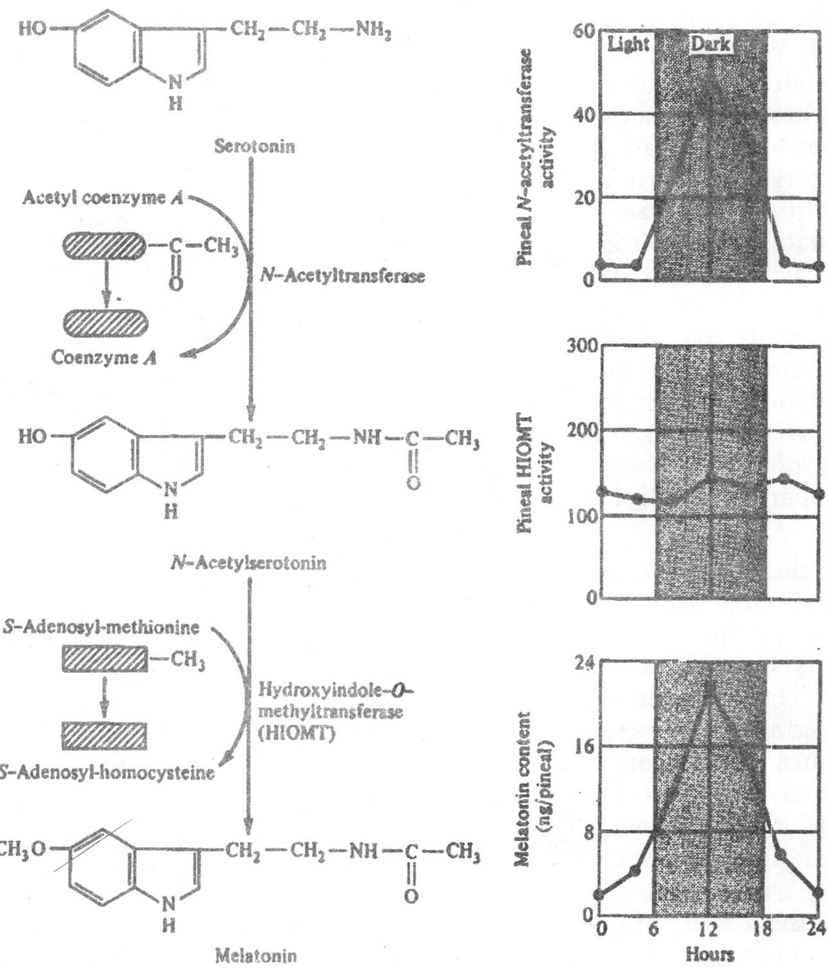

Fig. 8.15. Circadian changes in the synthesis of melatonin by the chicken pineal gland. The melatonin content of the pineal rises considerably during the hours of darkness (lower right). This change can be related to change in the biosynthetic process. The latter process is illustrated on the left side of the figure. Serotonin (5-hydroxytyptamine, 5-HT) under the influence of *N*-acetyltransferase gains an acetyl group. The activity of this enzyme increases during the hours of darkness (upper right). The *N*-acetylserotonin that is formed in this reaction is then *O*-methylated under the influence of HIOMT to melatonin. The latter enzyme does not appear to be rate limiting to the process, and its activity does not appear to change significantly over the period of the day (middle right).

It has been suggested that the pineal, by responding to a biological clock, functions like the hypothalamus as a neuroendocrine-transducer that modifies the actions of other endocrine glands. It thus may provide an important pathway whereby light provides cues to the endocrine system. The pineal may be utilized to control various rhythms such as those involved in reproduction and the release of prolactin.

The activity of the endocrine glands and the release of their hormones often show profound changes that are associated with the season of the year. The diurnal patterns of release just described may be initiated at such times or be superimposed on these gross changes in activity. Such changes have most often been observed in relation to the breeding season, which, especially in animals from nonequatorial regions, usually only occurs at certain times of the year. The relationship of the size of the testes and cloacal gland of Japanese quail to the length of the day. These glands reach their maximum size in May, which corresponds to the onset of the breeding season. The morphological changes are related to the rising levels of gonadotropins and testosterone. The rates of secretion of these hormones start to increase in March, when the length of daylight is about 12 hours. Reproduction may be dictated or influenced by predictably favourable seasons or in less favoured areas, like deserts, by the sudden appearance of rain. Apart from pituitary and gonadal sex hormones, cyclical changes in the activity of the thyroid gland and the adrenal cortex and medulla have been observed. The thyroid of the Japanese quail follows a pattern of activity that parallels that of the gonads. Thyroid activity increased in fishes undergoing seasonal migrations. In the African lungfish *(Protopterus annectens),* thyroid activity is lowest during estivation at the time of seasonal drought. In toads and frogs, epinephrine attains its highest concentration in the plasma during autumn and winter. Stores of growth hormone are highest in the pituitary of the perch *(Perca fluviatilis)* in June, a few weeks prior to the summer rapid-growth period.

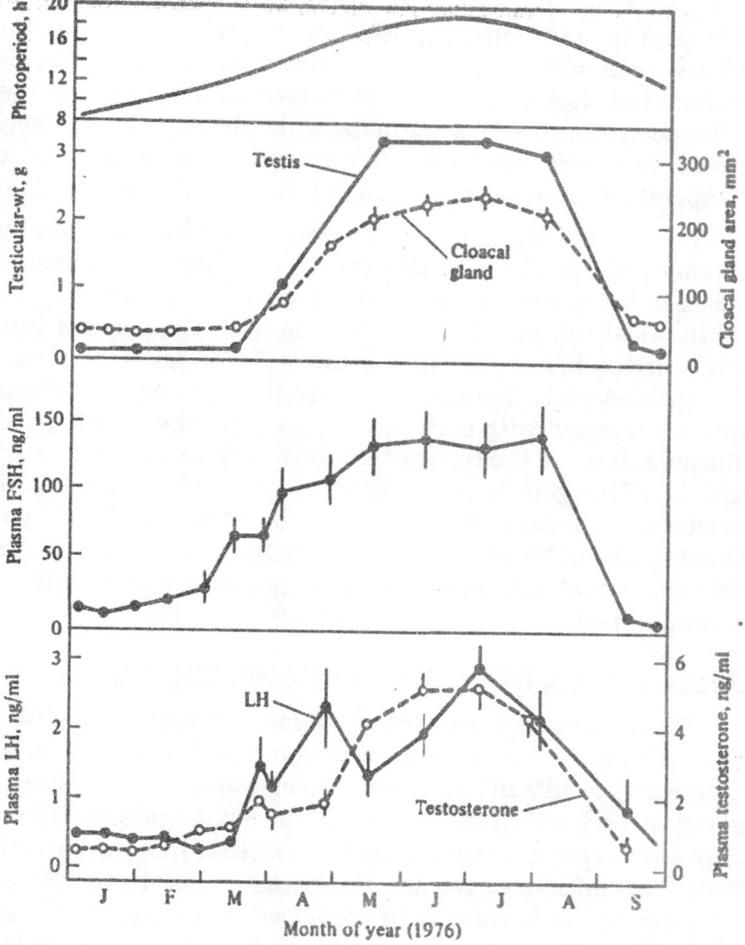

Fig. 8.16. The relationship of the growth of the testis and cloacal gland of quail to the hours of daylight (the photoperiod, upper panel) throughout the year. The lower two panels show the related changes in the plasma concentrations of luteinizing hormone (LH), follicle-stimulating hormone (FSH), and testosterone. Each point is the mean from 12 birds; the vertical lines are standard error of the mean.

The cyclical and episodic release of hormones is usually controlled through the brain. This is especially apparent when light provides the cues for these changes. External temperature may also be involved in controlling reproductive and possibly other cycles, especially in poikilotherms. It is likely that such heat stimuli act through the central nervous system, but they could also exert more direct effects on the pituitary and gonads and even influence the rates of a hormone's metabolism. The principal photoreceptors are the eyes, but there is evidence that in some vertebrates others may exist in other parts of the brain. Temperature impinges on receptors present in the skin but may also directly influence the brain.

The diurnal rhythm of activity of glands such as the pineal and the adrenal cortex is not usually present at birth. However, in embryonic chickens the diurnal rhythm of melatonin levels in the pineal is observed before hatching, after 17-19 days of incubation. In rats, the rhyth-mical release of corticosteroids from the adrenal cortex usually arises about 21 days after birth whereas in man in takes 2-8 years to appear. The adequate development of a biological clock is thought to require the presence of optimal levels of hormones. If newborn rats are injected with cortisol 2-4 days after they are born, the subsequent normal pattern of diurnal release of the corticosteroids is suppressed. Similarly, the normal pattern in the cyclical activity of the gonads is prevented when newborn female rats are injected with androgens. If newborn male rats are castrated soon after birth the hypothamalmus develops along the female pattern. When ovaries and a vagina are grafted into these animals they undergo cyclical changes typical of the female. Such endocrine manipulation can also contribute to behavioral abnormalities.

Systematic Differences in Releasing Stimuli

There are many instances among the vertebrates in which the physiological roles of analogous hormones exhibit systematic differences. Such a change in the use of a hormone necessarily results in an altered responsiveness to excit-atory stimuli that prompt the endocrine gland to discharge its secretion. When certain teleost fish move from fresh water to seawater they lose water and accumulate salt so that the concentration of sodium in their body fluids rises. This initiates a release of corticosteroids. Tetrapods, on the other hand, usually release such analogous hormones in response to declines in the sodium concentration of the body fluids. Prolactin and oxytocin are released during suckling in mammals. Other vertebrates lack mammary glands so that this represents a unique and phyletically novel stimulus. On the other hand, in certain teleost fish a prolactin-type hormone is released when the fish migrate from the sea into fresh water. The prolactin is then released in response to the lower calcium concentration in the fresh water. In homeotherms, thyroid hormones are discharged following a decline in the temperature of the blood flowing through certain areas in the hypothalamus but there is no indication that this happens in poikilotherms. Indeed, thyroid secretion usually increases in cold-blooded vertebrates exposed to elevated temperatures. Changes in glucose concentrations usually determine insulin release, but in ruminants (sheep and cattle) fatty acids are more important. In fishes, amino acids may be the major stimulant for secretion of insulin. Such phyletic differences in the propensity of endocrine glands to respond to certain stimuli are necessary for the evolution of a hormone's physiological role.

Conduction of Stimuli to the Endocrines

The conveyance of a stimulus to an endocrine gland may involve a complex series of events that take place along rather circuitous pathways. These are consistent with, and indeed are dictated by, the particular physiological requirements of the animal. The initial stimulus is usually translated into another form that may be a chemical compound or an electrical event or both. It travels to the endocrine gland, in such a modified form, by routes of varying complexity and may suffer further translation on the way. During this voyage, the stimulus may be modulated and interpolated with other information that is already available and other stimuli that also impinge on that particular communication pathway. This may take place in the brain and endocrine glands that are temporally proximal to the gland destined to receive, eventually, the final message. Such intermediary substations may involve neural areas in the brain, interconnecting endocrine glands, like the hypothalamus and the pineal, as well as the pituitary gland.

These events can be illustrated by summarily following the effects of light on reproduction. The stimulus is usually received by the eye (or in hypothalamic photoreceptors in many birds) where it is translated by the retinal receptors into electrical impulses that travel along nerves within the brain. These messages after further translation to other transmitter substances (such as acetylcholine, norepinephrine, and dopamine) eventually reach the hypothalamus and, probably also, in some species, pass through the pineal gland, which may release melatonin. The hypothalamus modulates its release of LHRH that crosses in the short portal blood vessels to the adenohypophysis. The receipt of this information will be further interpreted (in a process that probably involves the formation of cyclic AMP) in terms of an appropriate release of FSH and/or LH, which are carried in the blood to the ovaries or testes. Such an effect is illustrated by the levels of gonadotropin-releasing hormones and the gonadotropins in the hypothalamus and pituitary of the Japanese quail. There is a drop in the concentration of gonadotropin-releasing hormone that corresponds to the release of adenohypophysial gonadotropins. The gonadotroins may influence such events as ovulation and the formation and release of estrogens and progesterone as well as testosterone. These latter steroids, in turn, are carried back to the hypothalamus and pituitary gland where they provide information about the current hormone levels that will be used to modify stimuli that subsequently pass through this tissue. The process is, in detail, undoubtedly even more complex than that which has been described.

In other instances, the mechanism of the hormone's release may be simpler. The discharge of vasopressin (or antidiuretic hormone) from the neurohypophysis can occur in response to small increases in the osmotic concentration of plasma. This is thought to induce changes in osmoreceptors in the region of the supraoptic nucleus, possibly by releasing small amounts of acetylcholine, which initiates a wave of depolarization along the axons of the supraopticohypophysial tract. This results in the release of ADH from the storage granules at the terminus of the nerve. Neurohypophysial hormones may also be released in other circumstances. This release may involve non-specific stimuli, often termed "stress," that pass through higher centers in the brain to the nerve cells of the gland. Oxytocin is discharged in response to suckling in mammals; the initial receptor is in the nipple of the mammary gland from which it is transmitted along nerves to areas in the brain that initiate the release of the hormone from oxytocinergic neurons in the neurohypophysis.

Even simpler processes, not involving nerve pathways, may exist and determine the release of hormones in the body. The release of insulin can be demonstrated in isolated, perfused pieces of pancreatic tissues containing the islets of Langerhans. Elevated glucose and certain amino acid concentrations in the perfusate initiate a release of insulin. In a similar way, glucagon is discharged when the blood glucose concentration is depressed. The release of insulin from the B-cell is thought to occur as a result of the combination of glucose with a receptor (the glucoreceptor) in the cell wall. This event triggers a rise in the intracellular calcium concentration, possibly by increasing the permeability of the cell to this ion. The calcium initiates, and supports, the fusion of the insulin storage granules with the cell wall and the extrusion of their products into the extracellular fluid. A number of other excitants, including acetylcholine, glucagon, and some gut hormones, can also initiate a release of insulin, whereas others, such as prostaglandins and somatostatin, inhibit it. Separate receptors for these substances also appear to exist on the cell wall.

The Mechanism of Release of the Hormones

Upon receiving a stimulus an endocrine gland may release its hormones from their storage sites. Many hormones, such as catecholamines, neurohypophysial peptides, and insulin, are spewed from their storage granules; steroid hormones are released from lipid droplets; and thyroid hormones are detached from thyroglobulins. In some instances, this is preceded by the formation of cyclic AMP as in the thyroid gland, the adrenal cortex, the ovary, and sometimes the adenohypophysis and pancreatic B-cells.

The release of hormones from intracellular storage granules has been studied in detail in the instances of the secretion of catecholamines from the adrenal medulla and vasopressin and oxytocin from the neurohypophysis. Hormones that are stored in granules exist there as a nondiffusible complex with proteins and adenine nucleotides. Their release from the cell involves the process of emiocytosis (exocytosis or reverse pinocytosis) across the cell membranes. The nature of the events that result in such a release of hormones can be studied *in vitro* and is as follows:

(*i*) As a result of nerve stimulation the cell membrane is depolarized and ions enter the cells. *In vitro,* this can be performed by exposing the tissue to high concentrations of potassium or by stimulating it electrically.

(*ii*) An increase in the concentration of intracellular Ca^{2+} occurs, as a result of its uptake across the cell membrane and probably also its dissociation from binding within the cell. No release of hormone occurs *in vitro* following depolarization if the external media contain no Ca^{2+}, which thus appears to be vital for excitation-release coupling.

(*iii*) Excitation-release coupling involves a migration of the hormone storage granules toward the cell membrane, a process that may involve the cell microtubular system.

(*iv*) When contact between the cell membrane and the granules is made they fuse and Ca^{2+} may be important in the structural links. The entire contents of the granule, hormones, proteins, and adenine nucleotides are then extruded and pass into the capillaries. The empty granule may then be reconstituted and return to the cell cytoplasm.

Modifications of this process may occur in different endocrine glands. Insulin is released as a result of nerve stimulation, but glucose and amino acids have direct effects and these apparently also result in the admission of Ca^{2+} into the cell, or mobilization of intracellular calcium, which results in hormone release. The activation of the enzyme adenylate cyclase and the formation of cyclic AMP (as will be discussed) also appear to be involved in the release of insulin (as well as several other hormones) where it may promote an increase in intracellular calcium concentration or act more directly in the excitation-release coupling process.

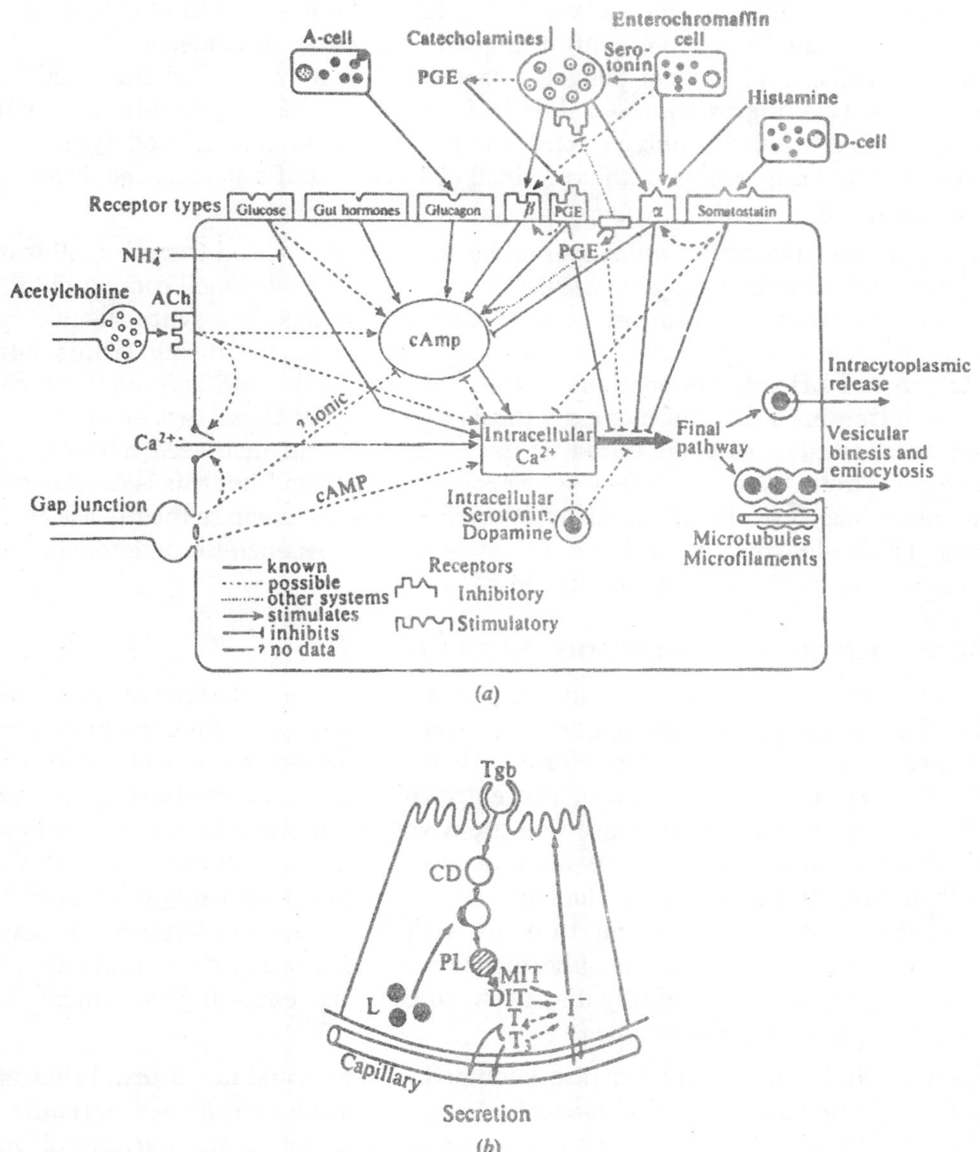

Fig. 8.17. (*a*) Hypothetical model describing the various interacting systems that influence the release of insulin from the B-cell. (*b*) The secretion of thyroid hormones. Thyroglobulin (Tgb) is transported by endocytosis from the follicles into the thyroid cells where it appears as the colloid droplets (CD) that fuse with lysosomes (L) to form phagolysosomes (PL) in which proteolytic digestion of the thyroglobulin occurs. The iodothyronines (MIT, DIT, T_4, and T_3) are released and the active hormones pass out of the cell. Iodide $(I)^-$ is reaccumulated.

Thyroid hormones are stored extracellularly, combined with thyroglobulin in the colloid of the thyroid gland follicles. Stimulation by TSH results in an activation of adenylate cyclase in the thyroid cells and the formation of cyclic AMP. In a manner that is not yet understood, this nucleotide is thought to stimulate pinocytosis during which fragments of follicular colloid are broken off and taken up by the thyroid cells; this process is thus called endocytosis. Lysosomes in the cell cytoplasm contain digestive protease enzymes combine with these ingested pieces of colloid to form phagolysosomes, which move toward the basal regions of the cell. During this migration, the colloid is digested and its components are separated so that thyroxine and triiodothyronine are freed to pass out of the cell. The iodotyrosines present are deiodinated and the iodine is either retained by the cell or, if also extruded, is subsequently reaccumulated.

The precise mechanism by which steroid hormones are released from the cell is not clear. The amounts stored in lipid droplets within the cell are relatively small compared to the large amounts of those hormones that are accumulated in granules. The synthesis and release of steroid hormones are promoted by various substances, including tropic hormones. Corticotropin and LH activate adenylate cyclase so that cyclic AMP is formed, which plays a role in the initiation of the hormone synthesis and release; these two processes have not, however, been satisfactorily separated. It is possible that an increased hormone synthesis inevitably leads to its release by a type of "over-flow mechanism," but this is usually considered to be an unsatisfactory explanation. Adenylate cyclase does not mediate the release of all steroid hormones. Thus, aldosterone secretion, which occurs in response to angiotensin II and III, appears to involve calcium but not cyclic AMP.

The Concentration of Hormones in the Blood

Hormones are normally present in the peripheral plasma at concentrations as low as 10^{-12}M and *in vitro* observations suggest that these levels may, in some instances, be similar to those necessary to stimulate the effector. However, for several reasons it is difficult to predict with certainty what the latter concentration will be. As we have seen, endocrines discharge their secretions in response to a great variety of stimuli; in some instances they may be sudden, or acute, occurrences whereas at other times they are of a more sustained, or chronic, nature. In the first case, the hormones may appear as a single surge or "spurt" of activity in the blood, such as seems to occur with the release of oxytocin in response to suckling. Thus, high concentrations of hormones may exist locally that briefly stimulate the effector and which are subsequently dissipated so that the remaining concentrations in the peripheral blood are not effective ones.

Differences in the patterns of release combined with seasonal and diurnal rhythms make it difficult to generalize as to the concentrations of hormones that are normally present in the blood. The antidiuretic hormone is normally present in the peripheral plasma of mammals at concentrations of 10^{-12} to 10^{-11} M. It is higher in small animals, such as the mouse, 2×10^{-11} M, than in large ones such as man, 4×10^{-12} M. In the laboratory rat, ADH concentration increases about 25-fold during dehydration: from 10^{-11} to 2.5×10^{-10} M. Some dehydrated amphibians (frogs and toads) have similar circulating levels of vasotocin: about 6×10^{-10} M. In others such as the mudpuppy and a teleost fish, the eel *Anguilla rostrata,* the hormone is not detectable (less than 10^{-11} M) under such conditions. Other polypeptide hormones like glucagon and insulin are normally present in the blood of mammals at a concentration of about 10^{-9} M.

Table 8.2. The concentrations of vasopressin or vasotocin in the plasma of vertebrates. The animals were normally hydrated except where indicated

	Hormone concentration (moles / liter)
Mammals	*Vasopressin*
Mouse	2×10^{-11}
Rat (*i*) normal hydration	1×10^{-11}
(*ii*) dehydrated	2.5×10^{-10}
Guinea pig	8×10^{-12}
Cat	2×10^{-12}
Rabbit	1.6×10^{-12}
Dog	2×10^{-12}
Man	4×10^{-12}
Bandicoot, wallaby, and possum (marsupials) dehydrated	1.8×10^{-10}
Amphibians	*Vasotocin*
(Anura) dehydrated	4 to 8×10^{-10}

Source: Robinson and MacFarlane, 1957; Heller and Stulc, 1960; Heller, 1966; Bentley, 1969.

The steroid hormones attain much higher concentrations in the blood than the polypeptides; thus, plasma cortisol and corticosterone levels in teleostean fish are about 10^{-7} M, though in the holostean and chondrostean fishes it is 10 times less than this. In chondrichthyeans 1α-hydroxycorticosterone usually has a concentration of 10^{-8} M in plasma. Aldosterone, compared to other corticosteroids, is present at much lower concentrations, about 10^{-10} M in mammals and 10^{-9} M in birds, though it is 10^{-8} M in amphibians. The steroid sex hormones, due to the cyclical nature of their release, are present in widely differing concentrations in the peripheral plasma: $10^{-7} - 10^{-10}$ M.

Hormones act when present in extremely dilute solution. The levels vary somewhat with different species. The reasons for variation appear to be related to the animal's size, metabolic rate, and the hormone-binding capacity of the plasma. Receptors may also have different requirements for hormones, to which they may be bound with different strengths (affinities). It seems likely that differences in concentration may also be influenced by the body temperatures at which the animals usually function, because they affect the rate of the hormones' reactions with the effectors. Differences in the potencies of homologous hormones, as reflected in their chemical structure, may also contribute to variations in plasma levels. A mutation that initiates the forma-tion of a less active hormone analogue could be compensated for physiologically by its release in greater quantity. An example of such an adaptation may be seen in mammals (Suiformes) that have lysine- instead of arginine-vasopressin.

Transport of Hormones in the Blood

Following their release from storage sites in the endocrine glands the hormones are carried, in the blood, to their various effector sites. This may be for very short distances, such as from the median eminence to the adenohypophysis, or involve much longer journeys, like from the neurohypophysis to the kidney. Some hormones have a relatively long life in the circulation and may recirculate many times before they are finally destroyed or excreted.

The blood plasma is an aqueous solution that contains high concentrations of proteins made up of several distinct components, or fractions. The hormone molecules may be dissolved in this solution or a substantial proportion can be bound to some of the proteins that are present. The binding of hormones in the blood is particularly important in the instances of the thyroid and steroid hormones.

The chemical nature of the binding of such hormones to plasma proteins is such that an equilibrium exists between the molecules that are bound and those that remain in solution. In other words, the binding is a reversible phenomenon and thus involves relatively weak chemical forces such as hydrogen and ionic bonds and Van der Waals' forces. The relative strength of these bonds, however, may vary considerably (high and low affinity) depending on the particular hormone and the nature of the binding protein. As we shall see, in some instances special proteins are present that appear to be able specifically to bind certain hormones.

There are several consequences of a hormone's binding to a protein in the plasma.

1. Hormones are usually assumed to be unable to initiate their effects when so bound. The receptor is envisaged as interacting with hormone molecules present in the aqueous phase. Their removal from solution shifts the equilibrium and may thus result in a dissociation into solution of some of the hormone that is bound to the proteins. It is implicit that the receptor has an even stronger ability than the plasma proteins to bind the hormone, and it is even possible that more direct exchanges may occur between binding proteins and the receptors.
2. The process of the hormone's inactivation, such as can occur in the liver, and its excretion, mainly in the urine and bile, is delayed while it is in the bound form.
3. The bound hormone may constitute a circulating pool of the excitant that can extend or moderate the hormone's action. For instance, in pregnant women and guinea pigs, a plasma protein is formed that can bind testosterone and so may protect the mother from the effects of this steroid.
4. The distribution of the hormone within the body can be influenced by its binding to plasma proteins. A hormone-protein complex may have a special propensity to pass through the capillaries in certain vascular beds. It may thus help specifically to determine where the hormone is going.
5. Protein binding may contribute to the specificity of a hormone's action in the cell. In the instance of the corticosteroids, two types of response can occur, mineralocorticoid and glucocorticoid (referring to their effects on electrolytes and intermediary metabolism), for which there are two types of receptors in the cell. Although the aldosterone receptors have a higher affinity for aldosterone than corticosterone, the normal excess of the latter hormone (in the plasma of rats) would be sufficient to negate this difference so that corticosterone would be expected, inappropriately, to occupy the aldosterone receptors. Most of the corticosterone in the plasma, unlike aldosterone, is bound to proteins in such a way that it does not obscure the effect of the aldosterone.

High proportions of hormones often exist in a bound form in the plasma. In the instance of testosterone, it is usually greater than 90% of the total present though it is less in other cases. Cortisol binding in the plasma of teleost fishes varies from 30 to 55% of the total. Thyroxine and triiodothyronine also are substantially associated with plasma proteins; thus, in the plasma of kangaroos, more than 95% of the thyroxine and 90% of the triiodothyronine is so bound.

The hormones may be bound to different protein components present in the blood plasma. In some instances, specialized proteins are present that have a high affinity (they are strongly bound) for the hormones. These include a globulin that binds cortisol (cortisol-binding globulin, or CBG), which is also called tran*scortin.* This protein is present in most vertebrates though quantitative differences in cortisol-binding capacities of the plasma exist, indicating that interspecific variations occur. The echidna (Monotremata) lacks CBG, a deficiency that may reflect its low plasma levels of corticosteroids. Cortisol binding is much less in the plasma of fishes than in most mammals, and it is unlikely that the proteins are identical in the different species. Transcortin also binds corticosterone and progesterone (but not aldosterone), and in mammals its levels increase during pregnancy as a result of the action of estrogen, which promotes its formation in the liver.

Cortisol-binding globulin may exert a protective role and guard animals against excessive toxic levels of corticosteroids in plasma. The males of a small species of shrew-like marsupials, *Antechinus stuartii,* all die about 2 to 3 weeks after the start of the breeding season. These little beasts become very aggressive at this time and their plasma corticosteroid levels rise sharply. This change is paralleled by a decline in the CBG concentration, a change which appears to be androgen dependent as it can be prevented by castration or increased by the injection of testosterone. The death of these marsupials is associated with gastrointestinal hemorrhage and infection, apparently reflecting the high levels of free or loosely bound corticosteroids in the plasma. The females do not experience comparable oscillations in steroids or CBG and carry on, mateless, to bear their young.

Plasma proteins that can bind sex steroids, androgens, estrogens, and progesterone with a high affinity have been identified in many vertebrates. They appear, however, to have undergone several evolutionary developments and the proteins involved are not identical. Cyclostome and elasmobranch fishes possess a plasma steroid-binding protein that binds not only androgens and estrogens but also corticosteroids, and so may serve a multiple role. Similar proteins have been identified in the plasma of teleost fish and amphibians. A *sex hormone-binding globulin* (SHBG) is found in women (where its levels increase during pregnancy) and a protein that exhibits crossimmunoreactivity with it has also been found in other primates. This globulin binds both androgens and estrogens. In nonprimate eutherians an analogous protein is present, but it only binds androgens and displays no cross-immunoreactivity to the primate protein. Marsupials possess a SHBG, binds androgens. The appearance of a SHBG that can bind both androgens and estrogens may thus be a specialization that, among mammals, is confined to the primates. The marsupials belonging to the order Polyprotodonta (which includes the opossums) lack, SHBG, it is also absent in rodents. The occurrence of SHBG thus appears to be quite sporadic among the mammals, possibly reflecting differences in their reproductive requirements for the sex steroids.

Many mammals possess an α-globulin that preferentially binds thyroxine, *thyroxine-binding globulin* or *TBG,* but which is absent in nonmammals. It also binds triiodothyronine but only about one-third as strongly as it does thyroxine. Although such plasma proteins strongly bind hormones, they are present in relatively small quantities so that the total amount of hormone that they can associate with is limited (low binding capacity). The plasma albumins, on the other hand, have a high binding capacity though their affinity is low. In species that lack such specialized hormone-binding proteins, or if the amount of hormone present exceeds their binding capacity, large amounts are bound to plasma albumin. In primates, and even their phyletically distant relatives, the marsupials, thyroxine is principally bound to a prealbumin (thyroxine-binding prealbumin, *TBPA)* whereas triiodothyronine combines with albumin. In some birds, reptiles, and fishes, thyroxine is bound principally to albumin and in others it is associated with both prealbumin and albumin.

Differences in thyroxine binding proteins exist between species. In some Australian marsupials (kangaroos), 75% of the thyroxine is bound to the prealbumin and only about 10% is associated with the postalbumin. On the other hand, in an American marsupial, the opossum, about 60% of the thyroxine is bound to the postalbumin. Even among these opossums differences were observed: The postalbumin exists in two different polymorphic forms, PtA_1 and PtA_2- Six out of 177 sera examined possessed PtA_2, which has different physicochemical properties from PtA_1; for example, PtA_1 binds triiodothyronine, but PtA_2 does not. In birds plasma lipoproteins may be the principal sites for the binding of thyroid hormones, but in mammals such a role for these substances is insignificant.

Peripheral Activation of Hormones

Some hormones are chemically altered, at sites peripheral to the endocrine gland from which they originated, in a manner that enhances their biological activity. This process is sometimes referred to as "activation" and may occur in the plasma or at tissue sites.

Angiotensin I is thus converted to its active forms, angiotensin II and III, by the action of enzymes. Not all the triiodothyronine present in the circulation is directly released from the thyroid gland. Some is formed from thyroxine in the liver and kidneys. An integral step in the action of testosterone involves its conversion at its target site by 5α-reductase, to 5α-dihydroxytestosterone. Cortisone can also be converted peripherally to the more active steroid cortisol. Cholecalciferol (vitamin D_3) undergoes several transformations before it can exert its effects. These changes involve the formation of 25-hydroxycholecalciferol in the liver and a further hydroxylation to 1α,25-dihydroxycholecalciferol in the kidney. It is also possible that some of the large protein hormones are fragmented peripherally into smaller pieces prior to their action. There is evidence that this may occur in the instance of parathyroid hormone while large amounts of proinsulin are. normally present in the plasma that, possibly, may also be converted into the active hormone.

Termination of the Actions of Hormones

The durations of action of hormones vary: They may persist for many hours and even days, or only have a short-lived, transitory effect. This is usually in keeping with the nature of the homeostatic processes they mediate. Even a prolonged effect, however, will necessitate the renewed release of hormones because of the metabolic and excretory processes that inevitably result in their inactivation and elimination from the body.

Hormones persist in the circulation for different periods of time. The half-life of vasopressin in man is about 15 minutes, that of cortisol is about 1 hour, and that of thyroxine is nearly a week. This reflects the speed of their degradation in the body, the rate at which they may be eliminated in the urine and bile, the protection afforded them as a result of binding to proteins, and the nature of their actions at the effector site. The effects of binding are well illustrated by comparing the rates of removal (clearance) of corticosteroid hormones from the blood. In man, about 1600 liters of plasma are normally completely cleared of aldosterone each day. In the instance of cortisol, how-ever, only about 180 liters are so purged in this time. The difference principally reflects the strong binding of cortisol to cortisol-binding globulin (transcortin) in the plasma. Considerable interspecific differences are apparent in the half-lives of hormones: arginine-vasopressin (ADH) has a half-life of about 1 minute in laboratory rats, compared to 15 to 20 minutes in man. This probably reflects the effects of size: Small animals destroy and eliminate hormones more rapidly than large ones. The precise chemical structure of the hormone may also be important; lysine-vasopressin has a half-life that is nearly twice as long in the rat as arginine-vasopressin. The considerable differences that have been observed in the potency of human, porcine, and salmon calcitonin substantially reflect the differences in their degradation rates. In rats, porcine calcitonin is destroyed by the liver whereas that from man and the salmon is degraded by the kidney. Salmon calcitonin is much more resistant to inactivation (in rats, dogs, and man) than the mammalian hormones. Body temperature will also be expected to have an effect on the rate of inactivation of hormones so that, in cold-blooded vertebrates, the inactivation and excretion process will be modified accordingly. In the toad *Bufo marinus,* at 26°C arginine-vasotocin has a half-life of 33 minutes while in the domestic fowl (at 43°C) it is only 18 minutes. The effects of differences in size, species, and temperature on hormone metabolism have not yet, however, been thoroughly evaluated. The action of a hormone may be terminated in several ways.

1. In order to act it must attain a certain critical concentration in the neighbourhood of its receptor site. If a hormone is released in a short burst as, for instance, usually occurs with oxytocin, the receptor wili respond to a local high concentration that is sequestered like a small packet in the plasma. The response will then be terminated simply as a result of the subsequent dilution and redistribution of the hormone in the body fluids. A hormone may also be removed if it is bound or has accumulated at tissue sites. Epinephrine is readily taken up by the adrenergic nerve terminals. Less specific binding to tissues, such as that of neurohypophysial peptides to skeletal muscle, may also contribute to the removal of hormones from the circulation.

2. Small amounts of hormones may be eliminated unchanged in the urine and bile, but this usually amounts to less than 5% of the total released. The activity of hormones is generally reduced or destroyed as a result of their metabolism by enzymes in the tissues, particularly in the liver and kidneys. They are transformed in various ways and the by-products are usually then excreted in the urine and bile. The hormone's chemical structure may be altered in several ways. The catecholamines can be methylated by catechol-*O*-methyltransferase (COMT) or, to a lesser extent, they may be deaminated as a result of the action of monoamine oxidase (MAO). The action of the thyroid hormones is largely destroyed by the removal of iodine from the molecule by a deiodinase enzyme. Protein hormones are broken up by proteolytic enzymes. Steroid hormones (and, to some extent, thyroxine) are combined chemically with glucuronic and sulfuric acids, in a process called "conjugation" that results in increased

water solubility, which enhances their chances for excretion in the urine and bile. Prior to such a conjugation considerable changes may be wrought in the chemical architecture of the steroids. Despite such chemical alterations, some hormones may still retain some of their biological activity. This is especially apparent in the gonadotropins that appear in large quantities in the urine of mammals during pregnancy. The activity of such urine in promoting ovulation and spermiation in various animals is used as a basis for pregnancy tests. Despite the retention of such biological effects, the products in the urine exhibit a different chemical behaviour from that of the gonadotropins present in the pituitary gland.

Mechanisms of Hormone Action

Hormones exert actions on every major group of tissues in the body. Their effects are numerous and include changes in the intermediary metabolism of fats, proteins, and carbohydrates, growth and development of the tissues, changes in the permeability of membranes, and the contraction or relaxation of muscles. The precise manner in which they effect such processes is only incompletely understood.

It is generally considered that a hormone, in order to act, must eventually influence an enzyme activity. It may do this by a process of activation, promoting enzyme formation (induction) or possibly acting as, or providing, a cofactor in the chemical reaction that it promotes. As described earlier, this process is thought to be initiated as a consequence of the hormone's combination with a precise chemical moiety in the cell that is called its receptor. The exact role of this hormone-receptor unit in triggering the response is uncertain. It may provide a carrier for the hormone and so facilitate its transfer to a vital site in the cell or, conversely, the hormone may facilitate transfer of the hormone receptor to an essential site. The receptor-hormone complex itself may act as an enzyme activator (or inhibitor) or cofactor. The receptor could exist as part of an enzyme complex so that an activating effect of a hormone could be direct, but it usually seems to comprise a separate subunit.

Hormonally mediated responses to hormones may take place in successive steps (a "cascade") involving numerous enzymes. Conceivably, by acting at a single rate-limiting stage in such a process and altering the supply of an essential metabolite, a hormone could influence a complex series of metabolic events in the cell. It is also possible that a hormone may directly influence more than a single process in such a chain of events.

Hormones often initiate responses in several different tissues in the body. In such instances, the basic effects may be similar or even differ from one another. For instance, aldosterone initiates a change in membrane permeability to sodium in the kidney, colon, salivary glands, sweat glands and, in amphibians, the skin and urinary bladder. Alternatively, epinephrine mobilizes fatty acids from fat cells but glucose from muscle cells. Despite the diverse tissue sites and the nature of the processes involved, the underlying initiating mechanism is generally the same. For instance, the effects of epinephrine on fat and muscle cells are both mediated by the action of cyclic AMP (see the following section).

At the present time the mechanisms of action of hormones are thought to be affected through either of two major groups of processes in the cell (others may also exist). The nucleotide adenosine-3′, 5′-monophosphate, or cyclic AMP, is a vital link in many endocrine responses and its formation is promoted, or inhibited, by numerous hormones. Other hormones, by regulating genetic transcription in the cell nucleus, can control the formation of proteins and enzymes that are essential for a response. Other processes, however, may also be involved, such as calcium and prostaglandins, which may modulate these responses or, possibly, even mediate them. Not all the actions of all the hormones are even partially understood. It is likely that some with multiple actions may utilize several different mechanisms. Some hormones, notably insulin, do not easily fit into either of these classifications. It is possible that some hormones, including insulin, may control translational processes on the ribosome.

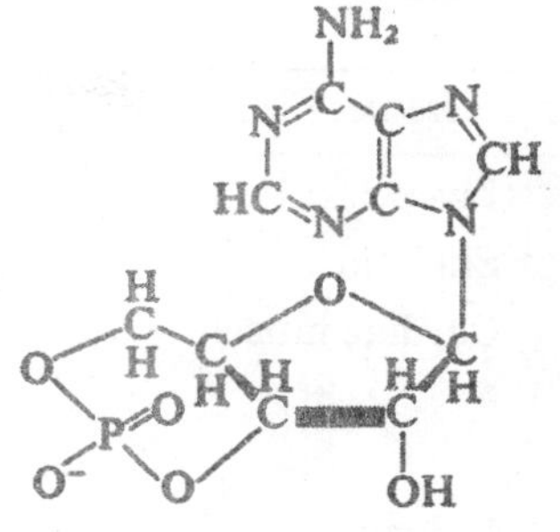

Fig. 8.18. The structural formula of adenosine-3′, 5′-monophosate (cyclic AMP).

The Role of Adenosine-3′, 5′-monophosphate (Cyclic AMP)

Our knowledge of the part played by adenosine-3′, 5′-monophosphate, or cyclic AMP, in the action of many hormones is primarily due to the work of Earl Sutherland. This adenine nucleotide is chemically related to ATP, from which it is formed. Its chemical structure.

The discovery of the endocrine role of cyclic AMP was made during an investigation of the mechanism of action by which epinephrine and glucagon convert glycogen to glucose in the liver. This reaction is dependent on several enzymes but the presence of a phosphorylase is rate limiting. This enzyme exists in two forms, a relatively inactive phosphorylase *b,* which on the incorporation of phosphate is converted to the much more active form, phosphorylase *a.* This activation takes place not only in intact cells but also in broken-cell preparations upon the addition of epinephrine or glucagon to them. The phosphorylase is a soluble enzyme that can be separated in the supernatant fraction of the broken cells. This enzyme, however, cannot be activated by the hormones when alone in this solution; the presence of the particulate cell material is a necessary condition. When the latter fraction is exposed to epinephrine or glucagon a substance can subsequently be washed from the cell particles that, when added to the supernatant, activates the phosphorylase. This activating chemical was found to be cyclic AMP and it is formed from ATP as a result of the action of an enzyme, *adenylate cyclase,* which is part of the cell membrane. In the liver, this enzyme is activated by epinephrine or glucagon and in skeletal muscle preparations only by epinephrine.

Adenylate cyclase has been found in many animals: mammals, birds, amphibians, and fishes, as well as many invertebrates. This enzyme is also present in bacteria, though apparently not in higher plants. Most tissues in the body contain adenylate cyclase. Apart from liver, they include muscle, kidney, heart, brain, adipose tissue, bone, and many endocrine glands.

The role of adenylate cyclase and cyclic AMP in mediating the effect of hormones was first described for the actions of glucagon or epinephrine on glycolysis in the liver or skeletal muscle. A description of this process can thus be used as a prototype for its role. It should, however, be remembered that although the initiating reactions may be similar for many tissues and various hormones, the responses of the final effector (or responding system) will often differ. As summarized, the initiating event is the interaction of the hormone and its receptor, in the cell membrane, which results in the activation (or sometimes the inhibition) of adenylate cyclase.

Table 8.3. Distribution and hormonal sensivity of mammalian adenylate cyclase

Tissue	*Hormone*
Liver	Glucagon and epinephrine
Skeletal muscle	Epinephrine
Cardiac muscle	Catecholamines
	Glucagon
	Triiodothyronine
Kidney	Vasopressin
	Parathyroid hormone
Bone	Parathyroid hormone
	Calcitonin
Brain	Catecholamines
Adrenal	ACTH
Corpus luteum	LH and prostaglandins
Ovary	LH
Testes	LH and FSH
Thyroid	TSH
	Porstaglandins
Parotid	Catecholamines
Pineal	Catecholamines
Lung	Epinephrine
Spleen	Epinephrine
Adipose	Epinephrine
Brown adiopose	Catecholamines
Platelets	Prostaglandins
Leucocytes	Catecholamines and prostaglandins
Erythrocytes	None demonstrated
Uterus	Catecholamines
Pancreas	None demonstrated
Anterior pituitary	Several
Vascular smooth muscle	None demonstrated

Source: Robison, Butcher, and Sutherland, 1971.

There is now a considerable amount of information available about the nature of the receptor-adenylate cyclase system. This complex is situated in the plasma membrane, the receptor facing outward, toward the extracellular space, the adenylate cyclase inward. These complexes have been isolated in fragments of plasma membranes and can be studied in this form. The receptors from several hormones have been separated from the enzyme and, with the aid of detergents, prepared in a soluble form. They are glycoproteins with molecular weights ranging from about 100,000 to 200,000. Such isolated receptors have been shown to be able to specifically bind hormones in a manner similar to that in the intact tissue. The peptide hormone receptors may normally form a relatively fixed unit with adenylate cyclase or they may be able to float laterally in the lipids of the plasma membrane. The latter receptors, when they interact with their hormones, can then combine with adenylate cyclase and so modulate its activity. Receptors for several different hormones can be present in a single plasma membrane, each of which, when combined with its specific hormone, can interact with a common adenylate cyclase moiety. This phenomenon is called the "mobile receptor theory". Such receptors have been separated and then grafted onto foreign cells where they can mediate a response that is foreign to the recipient, but similar to that of the donor. Thus, 3-adrenergic receptors from turkey erythrocytes have been transplanted onto Friend erythroleukemia cells. These "F"-cells possess adenylate cyclase but it does not normally respond to epinephrine. The trans-planted β-receptors can, however, mediate its response to this type of catecholamine hormone, so that cyclic AMP is then formed.

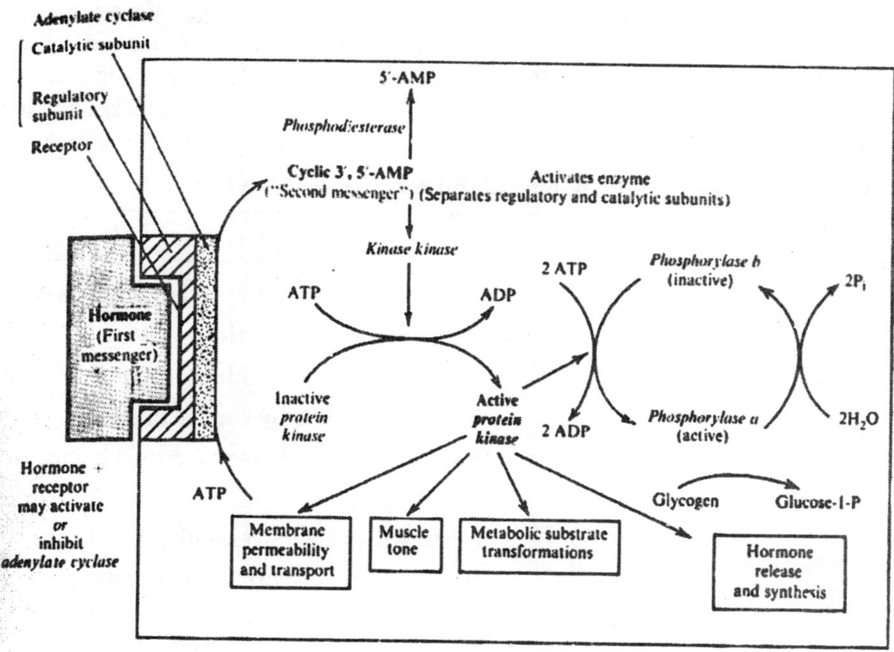

Fig. 8.19. A diagrammatic representation of the role of cyclic AMP as a "second messenger" in the mechanism of a hormone's action. A variety of protein kinases are thought to be present in different cells, which, when activated by cyclic AMP, can mediate the phosphorylation or dephosphorylation of certain proteins. This may result in diverse responses.

The receptor-adenylate cyclase system contains three important sites.

(*a*) A "catalytic site" (or unit) with which the substrate, ATP, can combine,

(*b*) The "receptor site," which modulates the enzyme's activity, and

(*c*) A "regulatory site," which combines with guanosine triphosphate (GTP), an event that is vital for the function of the enzyme. When the bound GTP is broken down the response is terminated.

All three of the units in the receptor-adenylate cyclase complex have been chemically separated from each other. The activity of the system can subsequently be restored by rejoining these pieces.

Cyclic AMP can mimic the actions of many hormones on a variety of tissues. The formation of this nucleotide, as a result of hormonal activation of adenylate cyclase, appears to mediate the actions of such hormones as (apart from epinephrine and glucagon) melanocyte-stimulating hormone, vasopressin, corticotropin, thyrotropic hormone, luteinizing hormone, and parathyroid hormone. The ultimate responses are very diverse and include glycolysis, lipolysis, changes in the permeability of membranes to water, sodium, and calcium, the dispersion of melanin in the melanophores, and the formation and release of many other hormones including thyroxine, corticosteroids, and sex steroids.

The nature of the effector response to hormones differs widely. It has been suggested, however, that in those processes that involve the action of cyclic AMP as a "second messenger," this is always the result of the phosphorylation, or dephosphorylation, of a protein and involves *a protein kinase.* Protein kinases, ubiquitous cellular enzymes that can transfer phosphoryl groups onto proteins, usually at the site of a serine residue. Such kinases normally exist in an inactive form, but when they combine with substrates such as calcium and cyclic nucleotides they become active. Cyclic AMP-dependent kinases, when combined with cyclic AMP, dissociate

Table. 8.4. Relations between the metabolic actions of some hormones and cyclic AMP

Hormone	*Actions shared by cyclic AMP*
Epinephrine	Glycolysis (liver and muscle), lipolysis (fat cell)
	Heart muscle contraction
ACTH	Steroid synthesis in adrenal cortex
LH	Steroid synthesis by corpus luteum
TSH	Production and release of thyroid hormones
Growth hormone	Lipolysis (?)
Vasopressin	Increased water movement in renal tubule
MSH	Melanophore, dispersion of melanin
Glucagons	Glycolysis (liver), lipolysis (fat cell)
Parathyroid hormone	Mobilization of bone calcium

into two part, a regulatory subunit, containing the nucleotide, and an active catalytic unit. The holoenzyme is inactive:

$$\underset{\text{inactive holoenzyme}}{\text{R.C}} + \text{Camp} \rightarrow \underset{\text{regulatory subunit}}{\text{R.cAMP}} + \underset{\text{active catalytic subunit}}{\text{C}}$$

Different types of proteins may be activated as a result of their phosphorylation by protein kinases. Metabolic enzymes can be activated; for instance, epinephrine can increase the conversion of glycogen to glucose and fats to glycerol and fatty acids, processes that depend, respectively, on a phosphorylase and lipase, both of which are activated by a cyclic AMP-dependent protein kinase. Changes in the configuration of proteins in the plasma membrane, and hence increased permeability to water, result from the activation by antidiuretic hormone of a protein kinase in the renal tubule.

Apart from initiating increases in levels of cyclic AMP in cells it is possible that some hormones may act by bringing about a decrease in the nucleotide's concentration. An understanding of the mechanism of action of insulin has been somewhat elusive but in some instances, especially when the basal levels of the nucleotide are elevated, it has been found to decrease the concentration of cyclic AMP. Decreases in cyclic AMP levels have also been observed in smooth muscles during contractions elicited by catecholamines. Such effects could be due to a reduction in the activity of the adenylate cyclase or an increase in the rate of its destruction.

The inactivation of cyclic 3′, 5′-AMP to 5′-AMP in cells is normally due to the action of the enzyme *phosphodiesterase*. This enzyme could provide a potential site for the action of hormones but such effects have not been described. Phosphodiesterase can be inhibited by certain drugs, notably the methylxanthines, which include caffeine.

The activation of adenylate cyclase can be accomplished by several hormones, in various tissues, and can result in diverse responses. The actions of hormones are, however, relatively specific, so how, when several utilize cyclic AMP, can a separation of their roles and effects be accomplished? In some instances more than one hormone is thought to act on a single adenylate cyclase. This seems to be so in the effects of epinephrine, glucagon, ACTH, and TSH on fat cells *in vitro*. The effects of the two latter hormones may not be physiological but reflect the relative, nonspecific reactivity of the receptors, and the adenylate cyclase, that are present. The adenylate cyclase that mediates the effects of ADH on the kidney is much more specific in its reaction with hormones. This selectivity is probably dictated by the nature of the hormone receptors. In some instances, a tissue exhibits more than one adenylate cyclasemediated response to a single hormone. Neurohypophysial hormones increase both water and sodium transfer across amphibian skin and urinary bladder, and these effects are mediated by different receptors and effectors. In the amphibian membranes, it seems likely that two adenylate cyclases and/or two "pools" of cyclic AMP exist in the tissue, each mediating a distinct response.

Differences in the ability of various hormones to activate adenylate cyclase from different tissues or species could reflect a polymorphism in the enzyme's structure; however, it appears that differences in the associated receptor mainly determine the particular response that is manifested. Mutations that result in changes in the protein linking receptors to the catalytic component of the adenylate cyclase system have been identified, however, in the mouse and man. Cell-free (membrane fragment) preparations of adenylate cyclase exhibit different quantitative responses to hormone analogues. Adenylate cyclase preparations from rat, mouse, rabbit, and ox kidney are more readily activated by arginine-vasopressin than by lysine-vasopressin (and much less by oxytocin). On the other hand, the enzyme complex prepared from the kidneys of pigs is most responsive to lysine-vasopressin. This relative sensitivity corresponds to that of the homologous hormones present in each species, whereas oxytocin does not normally act on the kidney, anyway. These differences, however, as discussed, appear to reflect the response of a receptor that need not necessarily be a permanent part of the enzyme itself. The properties of renal adenylate cyclase systems from a number of vertebrates have been compared. Mammal, bird, reptile, and amphibian enzymes were all strongly activated in the presence of fluoride. The mammalian enzyme systems were also activated by neurohypophysial hormones but this response was much less prominent, or even undetectable, in the nonmammals. This lack of response could, however, merely reflect the presence of

Table. 8.5. Stimulation of renal adenylate cyclase system in various vertebrate species

Species	*Neurohypophysial hormones*	*Parathyroid hormone*	*Fluoride*
Rat	++++	++++	++++
Mouse	++++	++++	++++
Pigeon	0	+++	++++
Alligator	0	++	++++
Toad	±	0	++++
Bullfrog	+	0	++++

+Indicates strength of response; 0, no response; ±, rudimentary response. Homologous neurohypophysial hormones were tested.
Source: Dousa *et al.*, 1972.

only small amounts of the appropriate enzyme in the kidneys of the nonmammals or its lack of importance in mediating the renal responses in those species. It is notable that the adenylate cyclase system obtained from amphibian urinary bladder is stimulated by neurohypophysial hormones. The renal adenylate cyclase system that is activated by parathyroid hormone (a distinct entity from that responding to the neurohypophysial hormones) was stimulated in enzyme preparations made from the kidneys of mammals, birds, and reptiles but not in those made from amphibian kidneys. This may reflect an absence of a renal effect for parathyroid hormone in the Amphibia. Except in bacteria, adenylate cyclase has not been prepared as a soluble enzyme so that the precise studies that may confirm its polymorphism are not yet feasible.

Prostaglandins are lipid molecules that exhibit many biological actions. Their distribution is ubiquitous in the tissues of the body. They are formed from some fatty acids, especially arachidonic acid. The latter is a component of phospholipids that are present in the plasma membrane, from which they are released by the action of phospholipase A_2. The prostaglandins are formed from their substrates under the influence of the prostaglandin synthetase (or cyclooxygenase) enzyme system. A large family of such substances has been identified, including prostaglandin E_1, prostaglandin E_2, and prostaglandin $F_{2\alpha}$. They have numerous effects such as the contraction or relaxation of smooth muscle, decrease in blood pressure, and changes in the activity of hormones (*e.g.*, an inhibition of the effects of antidiuretic hormone). Prosta-glandins can interact with the adenylate cyclase system to increase or decrease the formation of cyclic AMP. It has been suggested that they may normally contribute to the regulation and modulation of a variety of intracellular processes. Thus, they may interact with hormone-dependent processes, whereas hormones, in turn, can influence the levels of prostaglandins. In some instances, such as the action of prolactin on mammary glands, they may even help mediate a hormone's effect.

The Role of Calcium and Calmodulin

Calcium plays an important role in the mechanism of action of hormones, and it has sometimes, like cyclic AMP, also been called a "second messenger". It may be involved in several types of reactions in cells, including changes in the activity of enzymes such as adenylate cyclase and phosphodiesterase. Calcium also acts as a coupling agent and can link the primary actions of excitants to the ultimate expression of their response. This type of role includes the process of muscular contraction and the secretion of exocrine and endocrine glands. Changes in the levels of soluble "free" intracellular calcium may occur as a result of an increase in the influx of the ion across the plasma membrane into the cell or its mobilization from intracellular stores in the sarcoplasmic reticulum, mitochondria, or plasma membrane. Such an intermediary role of calcium generally appears to result from its interaction with an intracellular binding protein (or Ca-"receptor") called *calmodulin*. As a result of this binding the configuration of this protein changes, so that about 50% of the molecule assumes an α-helical configuration. When in this form it can activate a large number of enzymes, including phosphodiesterase, adenylate cyclase, phospholipase A_2, and Ca-activated ATPase, as well as some cell kinases.

Calmodulin is a protein, with a molecular weight of about 16,700, containing 148 amino acids. It appears to be present in all animal cells, which is consistent with its ubiquitous function, and it has been identified in species from both the animal and plant kingdoms. Its structure is remarkably conservative and displays few differences in amino acid sequences between such species as cattle and sea anemones. Antibodies to rat testis calmodulin can interact with calmodulin prepared from this coelenterate. "It is likely that calmodulin...will provide a link in our understanding of the interactions of calcium and cyclic nucleotides in the control of cellular metabolism".

Hormonal Effects Mediated by Changes in the Transcription of DNA in the cell Nucleus

The synthesis of specific proteins is primarily controlled by genes, which, through their coding patterns of DNA, provide a template for the formation of RNA. The latter is formed in a process that involves the action of RNA polymerase. This RNA (messenger RNA) can move to the ribosomes in the cytoplasm where it, in turn, acts as a template for the synthesis of a specific protein This basic mechanism of genetic transcription and translation was originally proposed as the result of the work of J. Monod and F. Jacob in bacteria. Although the details differ, it basically also applies to animal cells. In 1963 Karlson proposed that some hormones may act by initiating transcription of genetic material on the chromosome in a manner analogous to that of depressors in the genome of bacteria. These observations had their origins in invertebrate endocrinology. It was observed that the giant salivary gland chromosomes of various insects often displayed a loosening, or "puffs," in the regions of the bands of DNA. The number of such puffs could be increased following the injection of the insect metamorphosis hormone ecdysone. These puffs were identified as the sites of RNA synthesis. On the basis of these observations Karlson proposed that other hormones, including vertebrate steroid and thyroid hormones, may be acting at the site of the genes on the chromatin to initiate the synthesis of the protein for which it is encoded. This was a remarkably prophetic hypothesis.

The effects of the steroid hormones, as well as the thyroid hormones and growth hormone, are most obviously manifested as increases in growth, development, and differentiation of tissues. These changes are especially clear when one observes the effects of estradiol-17β on the uterus. Such growth is associated with a marked increase in the rate of protein synthesis.

More subtle changes may accompany other responses to such hormones. For instance, aldosterone, which increases the rate of sodium transfer across some membranes, may induce the formation of a permease, or some other enzyme, that increases active sodium transport. It is not always possible to identify the specific proteins formed, but progesterone is known to increase the formation of ovalbumin in the chick oviduct, whereas cortisol enhances the production of Na–K-activated ATPase in several tissues including the kidney and, in some fish, the gills and intestine.

Protein synthesis is a cytoplasmic process taking place in association with the ribosomes that follow the translation pattern provided by the messenger RNA derived from the chromosomal DNA. Although it is possible that hormones exert some direct effects on translation at the ribosome, this does not appear to occur usually. However, it has been suggested that insulin and growth hormone may have some action at this site. One of the earliest distinguishable effects of the actions of hormones that influence protein synthesis is the formation of messenger RNA by the nucleus. This process is mediated by the action of RNA polymerase II. It is now considered unlikely that this enzyme is induced or activated by hormones. Rather a perturbation of the structure of the DNA template occurs, which facilitates the action of the enzyme. The hormones could, for instance, be creating "initiation" binding sites for RNA polymerase on the chromatin. The protein synthesis by the ribosomes can be blocked by puromycin and cycloheximide, which act only on the cytoplasmic translational process and do not stop the early formation of the messenger RNA. A further indication as to the nuclear site of action of hormones is their autoradiographic identification in the chromatin.

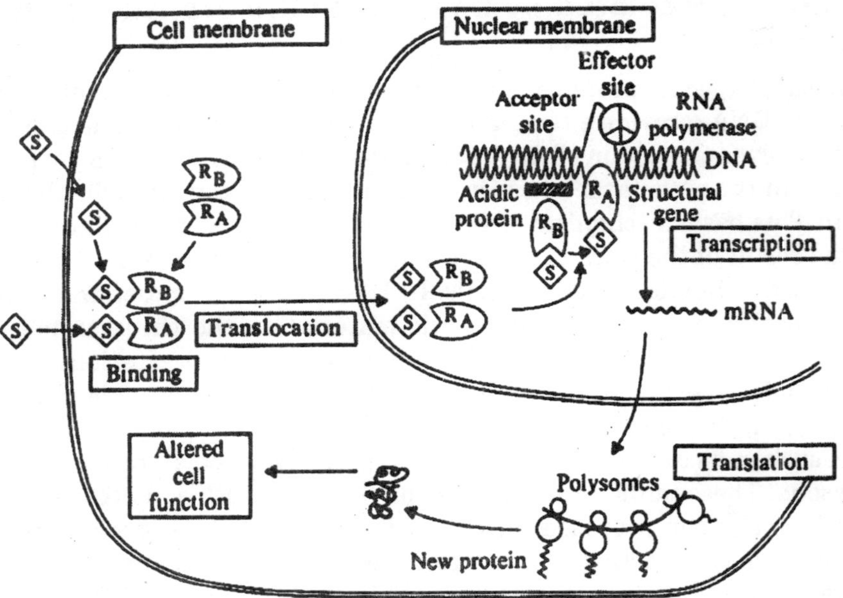

Fig. 8.20. General mechanism of action of sex steroid hormones as constructed by Chan and O'Malley. This scheme was based principally on information gathered from studies on the progesterone receptor in the chick oviduct. S = steroid hormone; R_A and R_B = subunits of the steroid hormone receptor.

The processes by which hormones act to initiate protein synthesis, via increases in nuclear transcription of the genome, have been most completely described for the action of estradiol-17β on the uterus progesterone and estrogen on the chicken's oviduct. These results are summarized. Comparable mechanisms also appear to exist for progesterone, the androgens, cortisol, aldosterone, and thyroid hormone. In the instance of the latter, however, differences do exist.

The estrogenic hormone molecule, upon crossing the plasma membrane and entering the cell's cytoplasm, is bound to a protein that has been called its cytoplasmic receptor. These proteinaceous units are also called *estrophiles.* The hormonereceptor combinations can be isolated *in vitro.* They pass intact through a gel filtration column and can be separated from other cell constituents by ultracentrifugation in a sucrose gradient where they have a sedimentation coefficient of about 8 S. In the presence of salt solutions this changes to 4 S, indicating that the receptor-hormone unit-can be broken into two subunits, A and B, each of which combines with a molecule of the hormone. There are bout 100,000 estrophiles in each uterine cell, which is a much higher number than is present in nontarget tissues. They have a molecular weight of about 200 000. The binding for a particular hormone is relatively specific and can be prevented by substances that block their effects *in vivo.* The binding sites are 50% saturated by the hormone when its concentration is 10^{-10} M (the K_D).

At a temperature of 2°C, 75% of the estrogen is present in the 8 S complex in the cytoplasm; but if the tissue is warmed to 37° C most of it moves across into the nucleus. The estrophilic complex may be acting as a carrier for the hormone or the hormone may facilitate the transfer of one of the subunits of 8 S that could initiate the formation of messenger RNA. The estradiol-17p in the nucleus is bound in another proteinaceous form that can be separated in KCl solution and which has a sedimentation coefficient of 5 S. This complex is clearly distinguishable from the 8 S one in the cytoplasm. The 5 S nuclear complex can be freed from the chromatin by the action of DNAase, indicating that it is closely associated with DNA. The 5 S nuclear receptor has not been found in the nucleus prior to its exposure to the 8 S complex, suggesting that the latter may first be converted to 5 S, which has an *acceptor site* in the nucleus. The free hormone does not readily bind to the chromatin. The site of the combination with the chromatin is not known, but it appears to be a nonhistone acidic protein.

In the instance of progesterone, the action of this steroid on the chicken oviduct is thought primarily to involve an increase in the frequency of the initiation of transcription on the DNA template. This process may involve a perturbation of the structure of the chromatin by the A-subunit of the hormone-receptor complex. This could direct the action of the RNA polymerase. The B-subunit may act as a "binding site specifier protein" that guides the hormone-receptor complex to the appropriate site on the chromatin.

At the present time not all the actions of hormones can be accounted for by these two types of mechanism. Hormones have multiple effects and not all the effects of a single hormone can always be accounted for by a single mechanism. Thyroxine, growth hormone, and prolactin can stimulate protein synthesis in the cell and, although these actions can be prevented by inhibition of RNA polymerase, just how primary, or universal, this effect is in these hormones' actions is not clear.

The action of prolactin on the mammary gland, for instance, may involve the activation of phospholipase A_2. This enzyme releases arachidonic acid from lipids in the plasma membrane that is a substrate for prostaglandins. The latter are thought to initiate an increase in protein synthesis.

Triiodothyronine binds to chromatin and can initiate the formation of messenger RNA. It can also, however, bind specifically to receptor sites on the inner membrane of mitochondria. Although it is bound to a component in the cell cytosol, this complex, in contrast to steroid hormones, is not necessary for the transfer of the hormone to its receptor sites in the nucleus, or the mitochondria. An important enzyme that can be induced by thyroid hormones, and which may mediate its effects on oxygen consumption, is Na-K-activated ATPase.

Some Speculations on the Evolution of the Actions of Hormones in cells

The two basic control mechanisms involving cyclic AMP and the regulation of genetic transcription exist in all animals as well as many microorganisms. We may thus suspect that these basic mechanisms, upon which hormones can impinge their actions, have always been present in animals, even in unicellular ones.

Cyclic AMP exists in bacteria, where it also appears to have a role in regulating cellular activities. Adenylate cyclase in bacteria is an intracellular enzyme, and it has been suggested that this represents the primitive condition. This internal site would appear to be most suitable for an enzyme that has to respond to changes in the intracellular nutrients and metabolites. In metazoan animals, adenylate cyclase appears to be confined to the cell membrane, a position that may be more apt for its interaction with metabolites and chemicals coming from other cells. These include the hormones. Such a membrane site for adenylate cyclase may thus be more opportune for intercellular cooperation and especially for interactions with molecules like the polypeptide hormones that cross cell membranes with some difficulty. It is unnecessary to postulate evolutionary changes in the structure of adenylate cyclase, only in the receptor which may be a subunit of it. The specificity of the receptor for a particular hormone has also been described. The simultaneous evolution of the complementary nature of both the hormone and its receptor is difficult to envisage. One wonders, when considering all the possible differences in structure, how they ever got together and how, at the same time, they acquired their complementary relationships.

The transcription of genetic material plays a basic role in the life of cells. In unicellular organisms, this need only be controlled by internal accumulations of nutrients and metabolites. These may act by combining with structural units, analogous to the subunits of the 8 S cytosol receptors. With the onset of need for intercellular communication these controlling subunits may have contracted an ability to combine with materials originating outside the cell. In other words, they may have acquired, or been transformed so as to incorporate, a hormonal receptor. Steroid hormones are nonpolar materials that readily gain access to cells and so would appear to be well suited to such an intracellular reaction.

Conclusions

In the following chapters we will be examining the roles of hormones in coordinating different physiological processes in the body. In the present chapter we have looked at the manner by which the endocrine system itself works. Although information about nonmammals is rather sparse it appears that the underlying mechanisms of a hormone's synthesis, release,

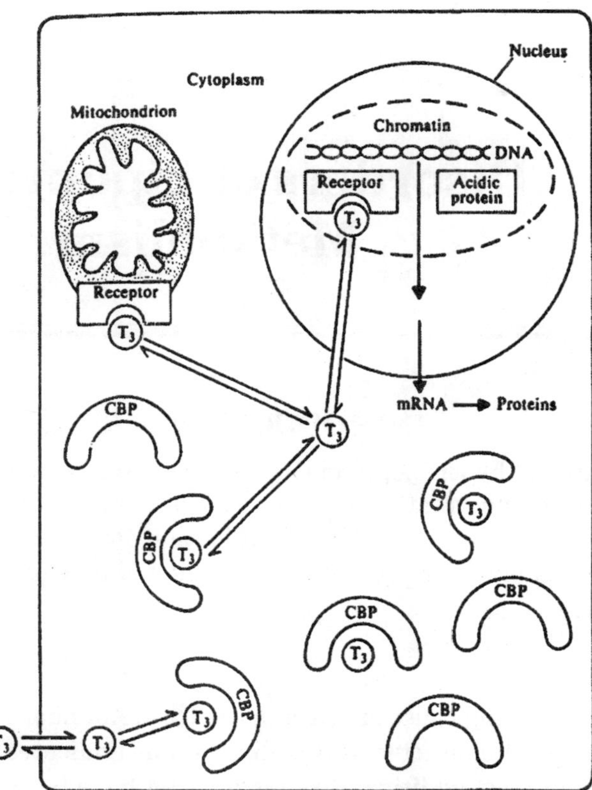

Fig. 8.21. Model showing the sequence of events that occurs when thyroid hormone enters its target cell. The unbound triiodothyronine (T_3) enters the cell by diffusion and binds to the cytosol-binding protein (CBP). The T_3-CBP complex remains in a reversible equilibrium with a very small amount of free T_3 in the cytoplasm. This unbound T_3 may interact with receptors in the nuclei or mitochondria, where it can trigger a response.

transport, mechanism of action, and its destruction are rather similar in all vertebrates. Even when different hormones are involved, the general underlying processes involved are often similar; but major differences are often apparent between the general types of hormones, especially those made from cholesterol (steroids) and those derived from amino acids. There are a number of interspecific differences in the "life history" of particular hormones in the body and these can be related to the animal's manner of life. The natures of the stimuli that initiate a particular hormone's release are especially variable among the vertebrates and are dictated by the different physiological roles that a hormone may have assumed. In addition, quantitative differences may arise with respect to a hormone's rates of synthesis, destruction, the quantities that are stored in the gland, and the concentrations that appear in the blood. Such differences can arise at distinct stages of the life cycle of an animal, but they are also observed between various species where they can be related to such characteristics as size, rates of metabolism, and environmental factors like temperature and the availability of different nutrients, salts, and water.

Disorders of Lipoprotein Metabolism

INTRODUCTION

The clinical importance of hyperlipoproteinemia derives chiefly from the role of lipoproteins in atherogenesis. However, the greatly increased risk of acute pancreatitis associated with severe hypertriglyceridemia is an additional indication for intervention. Detection and characterization of hyperlipoproteinemia is also important and may provide clues to underlying primary clinical disorders.

Arteriosclerosis

Arteriosclerosis is the leading cause of death in the USA. Abundant epidemiologic evidence establishes the multifactorial character of this disease and indicates that the effects of the multiple risk factors are at least additive. Risk factors that have been convincingly identified for atherosclerosis of the coronary arteries are hyperlipidemia, arterial hypertension, cigarette smoking, diabetes mellitus, physical inactivity, and decreased levels of high-density lipoproteins (HDL) in plasma. Coronary atheromas are complex lesions containing cellular elements, collagen, and lipids. It is clear, however, that the progression of the lesion is chiefly attributable to its content of unesterified cholesterol and cholesteryl esters. It is now firmly established that the cholesterol in the atheroma is delivered to the site by circulating lipoproteins. Epidemiologic evidence indicates that the atherogenic lipoproteins are the low-density (LDL), intermediate-density (IDL), and very low density (VLDL) species, all of which contain the B-100 apoprotein. Interaction between this apoprotein and a mucopolysaccharide (chondroitin sulfate B) of the artery wall may play an important role in the degradation of lipoproteins in situ. In animal models, hypertension is associated with increased access of lipoprotein to the subintimal region, probably resulting from endothelial damage. Smoking may accelerate the process of atherogenesis chiefly through its influence on blood platelets, though it is also associated with decreased levels of HDL in plasma. Increased platelet interaction at sites of damaged endothelium leads to release of a protein that stimulates migration of cells of smooth muscle origin into the lesion, where they proliferate. Cholesteryl esters appear in macrophages in the atheroma. They also appear in the extracellular matrix, where they induce the production of collagen by fibroblasts. HDL appear to inhibit the process of atherogenesis, probably by enhancing the removal of cholesterol from the atheroma.

Acronyms Used in this Chapter

ACAT	Acyl-CoA: cholesterol acyltransferase
Apo-	Apolipoprotein or apoprotein
FFA	Free fatty acids
HDL	High-density lipoprotein(s)
HMG-CoA	Hydroxymethylglutaryl-CoA
IDDM	Insulin-dependent diabetes mellitus
IDL	Intermediate-density lipoprotein(s)
LCAT	Lecithin: cholesterol acyltransferase
LDL	Low-density lipoprotein(s)
LPL	Lipoprotein lipase
NIDDM	Non-insulin-dependent diabetes mellitus
VLDL	Very low density lipoprotein(s)

Reversal of Atherosclerosis

A number of studies in animals, including higher primates, indicate that atherosclerotic coronary disease resulting from diet-induced hyperlipidemia is reversible when levels of lipoproteins in plasma are restored to normal. These findings support the concept that the atheroma is a dynamic lesion. The striking similarity of the lesions induced in primates to those occurring in human coronary arteries suggests that the progression of human lesions can also be retarded, or perhaps that regression can occur if hyperlipidemia is controlled. Because of the very limited changes in serum lipoproteins that have been attained in the large-scale intervention studies reported thus far, no real test of this hypothesis has been made. Until the results of definitive experiments in humans are available, intervention in coronary vascular disease by reducing elevated levels of atherogenic lipoproteins to normal would appear to be based upon a reasonable presumption. It should be noted that average levels of LDL in plasma in the United States population are considerably higher than in many other nations, where the levels appear to approach the biologic norm for humans. This probably accounts in large part for the markedly higher incidence of coronary vascular disease in industrialized Western nations.

OVERVIEW OF LIPID TRANSPORT

The Plasma Lipoproteins

Because all the lipids of plasma are relatively insoluble in water, they are transported in association with protein. The simplest complexes are those formed between unesterified, or free, fatty acids (FFA) and albumin, which serve to carry the FFA from peripheral adipocytes to other tissues. Similarly, lysolecithins (phospholipids formed when 1 mol of fatty acid is removed from lecithin) are bound to albumin.

The remainder of the plasma lipids are transported in lipoprotein complexes. All of the major lipoproteins of plasma are pseudomicellar in structure, *i.e.*, they are spherical, and each has a core region containing hydrophobic lipids. The principal core lipids are cholesteryl esters and triglycerides. Triglycerides predominate in the cores of the chylomicrons, which transport newly absorbed lipids from the intestine, and of the very low density lipoproteins, which originate in the liver. The relative content of cholesteryl ester is increased in remnants derived from these 2 classes of lipoproteins, and cholesteryl esters predominate in the cores of low-density and high-density lipoproteins. Surrounding the core in each type of lipoprotein is a stabilizing monolayer containing amphiphilic (detergentlike) lipids, chiefly phospholipids and unesterified cholesterol, frequently termed free cholesterol. Amphiphilic (or amphipathic) lipids are characterized by having a polar or charged region that associates readily with the aqueous environment surrounding the lipoprotein particle and a hydrophobic region that preferentially interacts with hydrophobic core lipids.

Proteins called apoproteins (also called apolipoproteins), which are noncovalently bound to the lipids, are mostly located in or on this surface monolayer. The preponderance of the bonding energy between these proteins and lipid appears to be hydrophobic; however, some ionic interaction between headgroups of phospholipids and charged amino acid side chains of apoproteins occurs, stabilizing helical regions.

B Apoproteins

Several of the lipoproteins contain proteins of very high molecular weight known as the B apoproteins, which behave like intrinsic proteins of cell membranes. Unlike the smaller apoproteins, the B apoproteins do not migrate from one lipoprotein particle to another. The B apoproteins of intestinal and hepatic origin are different. They are now described by numbers in a centile system based on the relative apparent molecular weights of the different species. Thus, VLDL contain the B-100 protein, which is retained in the formation of LDL from VLDL remnants by the liver. Two complementary B apoproteins, B-74 and B-26, which appear to be derived from B-100, are also found in LDL. The intestinal B protein, B-48, is found in chylomicrons and their remnant particles but is completely absent from LDL.

Other Apoproteins

In addition to the B apoproteins, the following apoproteins are present in lipoproteins. (The distribution of these proteins in the different lipoproteins).

(1) C apoproteins : These are low-molecular-weight (7000-10,000) proteins that equilibrate rapidly among the lipoproteins. There are 3 distinct species with unique amino acid compositions, designated C-I, C-II, and C-III. Apoproteins C-II and C-III each have several subspecies with slightly differing electrostatic charges. Apoprotein C-II has been identified as a requisite cofactor for lipoprotein lipase.

(2) E apoproteins : At least 2 normal isoforms (E-3 and E-4) of this MW-35,000 protein exist that appear to be the products of allelic genes. Normal individuals thus may have either or both isoforms, which share with B-100 protein the property of interacting with certain high-affinity receptors (B-100-E receptors) on cell membranes. Another isoform (E-2) which lacks this property contains 1 more mol of cysteine than E-3 and 2 more than E-4. About 15% of individuals are heterozygous for this isoform.

(3) Apoprotein A-I : This protein of MW 28,300 is the major apoprotein of HDL; it is also present in chylomicrons and is the most abundant of the apoproteins of human serum (about 125 mg/dL). It is a cofactor for lecithin:cholesterol acyltransferase (LCAT).

(4) Apoprotein A-II : This protein of MW 17,400 is an important constituent of HDL. It contains cysteine, which permits the formation of disulfide-bridged oligomers with apo-E.

Table. 9.1. Major lipoproteins of human serum.

Lipoprotein	*Electrophoretic Mobility in Agarose Gel*	*Density Interval g/cm³*	*Predominant Core Lipids*	*Diameter*	*Apolipoproteins in Order of Quantitative Importance*
High-density (HDL)	Alpha	1.21-1.063	Cholesteryl ester	7.5-10.5 nm	A-I, A-II C, E, D
Low-density (LDL)	Beta	1.063-1.019	Cholesteryl ester	21.5 nm	B-100, B-74, B-26
Intermediate-density (IDL)	Beta	1.019-1.006	Cholesteryl ester, triglyceride	25-30 mn	B-100, some C and E
Very low density (VLDL)	Prebeta; some "slow prebeta"	<1.006	Triglyceride	30–100 nm	B-100, C, E
chylomicrons	Remain of origin	< 1.006	Triglyceride	80–500 nm	B-48, C, E, A-I, A-II, A-IV, proline-rich apo-protein

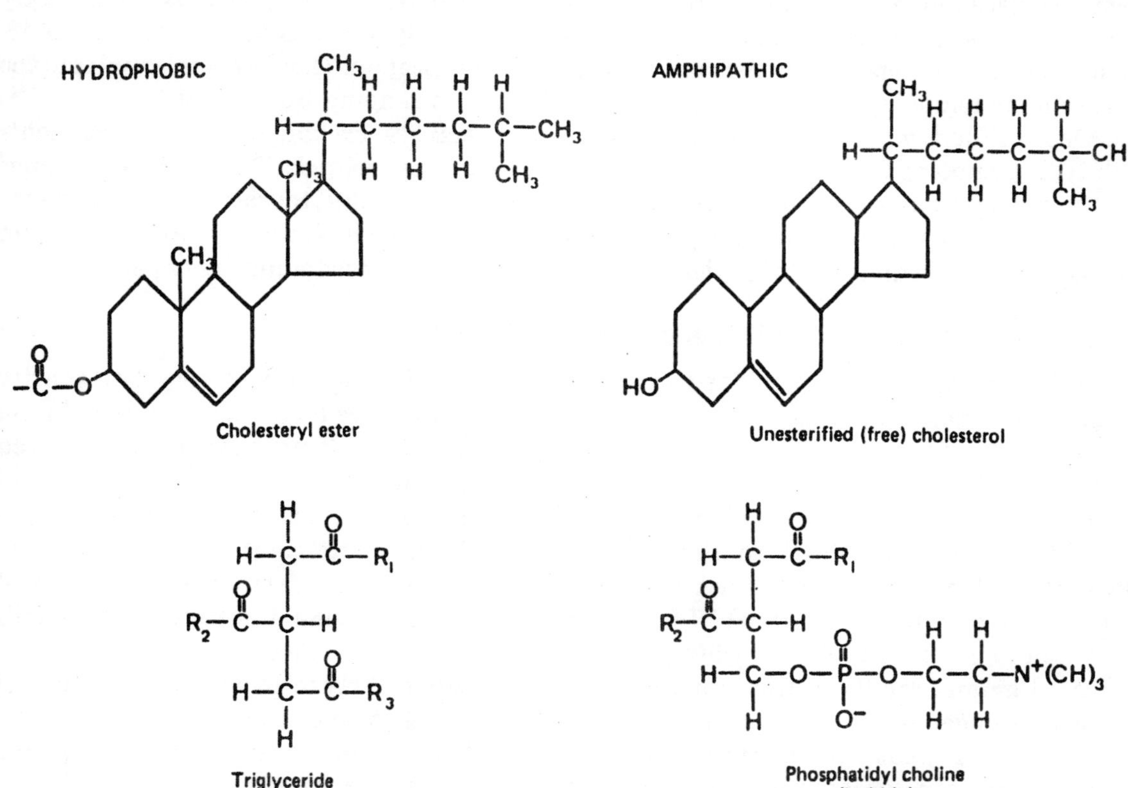

Fig. 9.1. Major lipids of plasma lipoproteins. R = hydrocarbon chain of a fatty acid.

(5) D apoprotein : This protein of MW 35,000 catalyzes the transfer of cholesteryl ester to VLDL and LDL from the lipoprotein subspecies of HDL that contain LCAT.

(6) Apoprotein A-IV : This protein of MW 46,000 is chiefly associated with chylomicrons.

(7) Proline-rich protein : This protein of MW 74,000 is associated with chylomicrons and occurs also as a multimer in plasma that is not associated with lipoproteins.

An additional apoprotein, the Lp(a) protein, is found in association with a minor species of lipoprotein, the Lp(a) lipoprotein that occurs in the density interval 1.030-1.090 g/cm^3.

Absorption of Dietary Fat; Secretion of Chylomicrons

Dietary triglycerides are largely hydrolyzed in the intestine to β-monoglyceride and fatty acids by pancreatic lipase. This enzyme requires activation by bile acids and a protein cofactor. The partial glycerides and fatty acids form micelles that are absorbed by intestinal epithelial cells. Within these cells, the fatty acids are reesterified with the monoglyceride to form triglycerides. Some dietary cholesterol absorbed with the micelles is esterified by the acyl-CoA:cholesterol acyl-transferase system (ACAT), and some appears as free cholesterol in the surface monolayers of the chylomicrons. Droplets of triglyceride containing small amounts of cholesteryl esters form in the vesicles of the Golgi apparatus. Phospholipids and free cholesterol form a surface monolayer. Some of the phospholipid originates in bile, some is from dietary sources, and some is synthesized by the intestine. Newly synthesized apoprotein B-48, apoprotein A-I, and apoprotein A-II are added, and the nascent chylomicron emerges into the extracellular lymph space. Chylomicrons have diameters ranging from about 80 nm to 500 nm. Once in the lymph spaces, the new chylomicron begins to exchange surface components with HDL, acquiring C apoproteins and apo-E and losing phospholipids. This process continues as the chylomicron is carried via the intestinal lymphatics to the thoracic duct and thence into the bloodstream. Increased triglyceride transport from the intestine results chiefly in an increase of particle diameter of chylomicrons rather than increased numbers of particles.

Formation of Very Low Density Lipoproteins

The liver exports triglycerides to peripheral tissues in the cores of very low density lipoproteins. These triglycerides are synthesized in liver from free fatty acids abstracted from plasma and from fatty acids synthesized de novo. Acetyl-CoA for de novo synthesis is derived from the catabolism of carbohydrate, certain amino acids, and acetate formed by oxidation of ethanol. Several major features distinguish VLDL from chylomicrons. The only B apoprotein of VLDL is B-100, and VLDL contain none of the major apoproteins of HDL. Whereas the intestine produces only very limited amounts of C apoproteins, the liver secretes the bulk of these proteins with newly formed VLDL. Upon reaching the plasma, the VLDL acquire still more C proteins from HDL and yield phospholipid in exchange. Release of VLDL by liver is increased by any condition that results in increased flux of FFA to liver in the absence of increased ketogenesis. Increased caloric intake, ingestion of ethanol, and the administration of estrogens all greatly stimulate release of VLDL from liver and are important causative factors in clinical disorders resulting in elevated levels of triglycerides in plasma.

Metabolism of Triglyceride-Rich Lipoproteins in Plasma

A. Hydrolysis by Lipoprotein Lipases : Fatty acids derived from the triglycerides of chylomicrons and VLDL are delivered to tissues predominantly through a common pathway

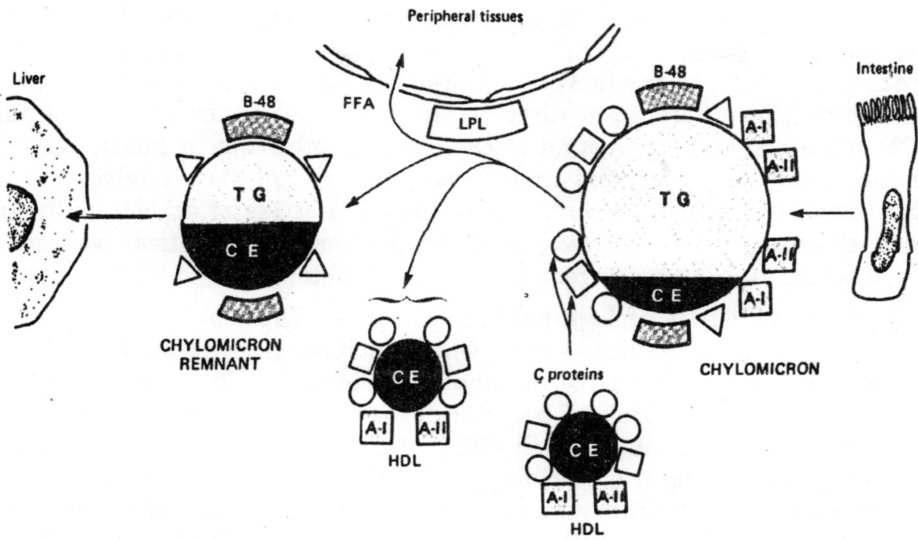

Fig. 9.2. Metabolism of chylomicrons.

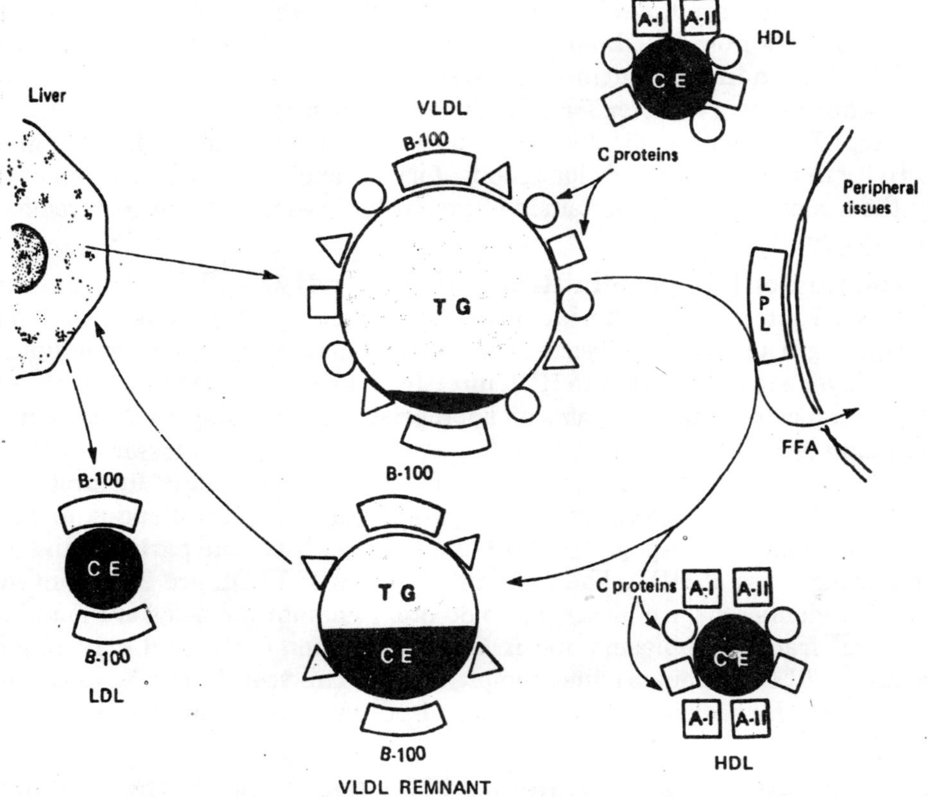

Fig. 9.3. Metabolism of VLDL.

involving hydrolysis by thelipoprotein lipase (LPL) system. Alternative pathways of low efficiency also degrade triglyceride-rich lipoproteins in peripheral tissues. These pathways can account for a larger than normal fraction of the degradaticn of VLDL and chylomicrons in patients with hypertriglyceridemia. Over three-fourths of the FFA derived during hydrolysis of triglycerides by LPL enters tissue directly. The remainder enters the pool of plasma FFA. Lipoprotein lipase enzymes are bound to capillary endothelium in heart, skeletal muscle, adipose tissue, mammary gland, and other tissues. The hydrolysis of triglycerides thus takes place within the vascular compartment. Removal of the first 2 fatty acids from the triglyceride appears to be mediated by LPL (triglyceridase and diglyceridase activities), whereas a separate monoglyceridase enzyme system is required for the last step.

B. Biologic Regulators of Lipoprotein Lipases : There are at least 2 forms of LPL with different apparent molecular weights. One, which has so far been identified in cardiac muscle, has a high affinity for triglyceride-rich lipoproteins and therefore yields fatty acids to the myocardium even when levels of triglycerides in plasma are low. In contrast, the form present in adipose tissue has a much lower affinity for its substrate. It operates most efficiently at higher levels of circulating triglycerides, yielding fatty acids for storage in adipose tissue. In addition to this fixed difference in affinity, there is a directional effect on the flux of triglyceride fatty acids based on nutritional status. When glucose levels in plasma are elevated and the release of insulin is stimulated, LPL activity in adipose tissue increases and fatty acids derived from triglycerides of circulating lipoproteins are stored. During prolonged fasting, however, LPL activity of adipose tissue falls to undetectable levels, completely preventing storage of fatty acids from VLDL and chylomicrons. The LPL activity of mammary gland is increased by the administration of estrogens and during lactation. Heparin, a mucopolysaccharide, is a cofactor for LPL. When heparin is given intravenously in doses of 0.1-0.2 mg/kg, LPL activity is displaced into plasma. This permits the in vitro measurement of LPL activity in plasma. An additional obligatory cofactor is one of the C apoproteins, C-II. Normally, the content of this apoprotein in plasma is far in excess of that required for activation of the LPL system.

C. Formation of Lipoprotein Remnants : Hydrolysis by LPL results in progressive depletion of the content of triglyceride in the hydrophobic core regions of chylomicrons and VLDL, producing a progressive decrease in particle diameter. Amphiphilic lipids from the surface monolayer are transported to HDL directly and also indirectly via membranes of cells, principally erythrocytes. This maintains an appropriate core:monolayer relationship for spheres of smaller diameters. During this process, C apoproteins—and, to a lesser extent, apo-E—are transferred to HDL. The product of this series of events is a "remnant" lipoprotein containing its original complement of apo-B, a portion of the original amount of apo-E, and little apo-C. In the case of chylomicrons, apoproteins A-I and A-II also leave the particles and become part of the circulating mass of HDL. The remnant particles of VLDL are 25-30 nm in diameter. Remnants of chylomicrons are larger, up to 80 nm. Remnant particles are found both in the $d < 1.006$ g/cm^3 fraction of plasma and in the IDL fraction ($1.006 < d < 1.019$ g/cm^3). They have lost about 70% of their original complement of triglyceride but have conserved their cholesteryl esters. In addition, they have acquired more cholesteryl esters, formed by the LCAT reaction.

D. Fate of Lipoprotein Remnants : The liver removes the remnants of both chylomicrons and VLDL from the circulation, but the fate of the 2 remnant lipoproteins differs significantly. Chylomicron remnants are disassembled in the liver, yielding their triglycerides, cholesteryl esters, and phospholipids to pools of these lipids in the hepatocyte. The B-48 protein appears to be degraded completely. The cholesterol derived from chylomicron remnants is the major mediator of feedback control of cholesterol biosynthesis in liver. In contrast with the metabolism of chylomicron remnants, in which no further lipoproteins are generated directly, the catabolism of VLDL remnants results in the formation and release into plasma of LDL. This process involves the removal of essentially all of the residual triglyceride and a portion of the cholesteryl esters and amphipathic lipids. The low-density lipoprotein particles released into plasma contain chiefly cholesteryl esters in their core regions. They retain virtually all of the original complement of the B-100 protein but only traces of other apoproteins. It is at this stage or afterward that the complementary B-74 and B-26 fragments appear in LDL. In normal individuals, a major fraction of the apoprotein B-100 of VLDL is converted to the apo-B of LDL, and all of the LDL apo-B comes from VLDL. In homozygous familial hypercholesterolemia, however, the liver also appears to secrete an LDL-like lipoprotein directly, without passing through VLDL as intermediary. In certain hypertriglyceridemic states, conversion of radiolabeled apo-B of VLDL origin to LDL apo-B has been reported to be less than in normals. However, because these data were obtained before it was recognized that the B-48 protein of chylomicrons does not enter LDL apo-B, it is likely that these findings could reflect contamination of "VLDL" by particles of intestinal origin.

The fact that LDL originates from the metabolism of VLDL remnants suggests that increases of LDL in plasma can arise from an increased rate of secretion of precursor VLDL as well as from decreased catabolism of LDL. The formation of LDL from VLDL also explains a clinical phenomenon referred to as the "beta shift." This is an increase of LDL (beta-lipoprotein) as hypertriglyceridemia is resolving. A classic example of the beta shift occurs following institution of adequate insulin treatment in ketoacidotic diabetes with lipemia. Insulin induces increased lipoprotein lipase activity, resulting in rapid conversion of VLDL to LDL. Because of its longer half-life, the LDL accumulates in plasma. Elevated levels of LDL may persist beyond the time when levels of triglyceride-rich lipoproteins have returned to normal.

E. Half-Lives of Lipoproteins : Normally, the half-life of chylomicron apo-B in plasma is 20-50 minutes; that of the apo-B in VLDL is 1-3 hours; and that of LDL about $2^1/_2$ days. At triglyceride levels of 800-1000 mg/dL, the lipoprotein lipase mechanism is at kinetic saturation; therefore, increases in the input of triglyceride-rich lipoproteins into plasma at those levels rapidly result in much higher levels.

F. Effect of Dietary Fat Restriction : Individuals consuming a typical American diet with about 45% of calories as fat transport 75-100 g or more of triglyceride per day into plasma in chylomicrons, whereas the liver exports 10-30 g of triglyceride in VLDL. Thus, the flux of triglyceride into plasma can be influenced most acutely by restriction of dietary fat. When the removal mechanisms involving lipoprotein lipase are saturated and plasma triglyceride levels are measured in thousands of milligrams per deciliter, acute restriction of dietary triglyceride intake will usually produce a significant reduction in triglyceride levels. This intervention is important in the lipemic patient with impending pancreatitis. If symptoms suggest that an attack of pancreatitis is imminent, all oral intake should be eliminated and the patient should be maintained on parenteral glucose until the symptoms subside and triglyceride levels decrease to less than 1000 mg/dL.

Catabolism of LDL

A. Endocytosis Via Specific High-Affinity Receptors : The catabolism of LDL appears to proceed by several mechanisms. The best-understood of these is a regulated adsorptive endocytosis mediated by high-affinity receptors on cell membranes. This mechanism has been demonstrated in cultured fibroblasts, arterial smooth muscle cells, lymphocytes, and endothelial cells. The receptors located in "coated regions" of the cell membrane bind the apo-B protein of LDL. Because they also bind apo-E, they are called B-100-E receptors. The coated region invaginates into the cell, forming an endocytotic vesicle that fuses with a lysosome. The apo-B of the LDL is degraded to amino acids, and the receptor returns to the cell membrane. The cholesteryl esters of the LDL core are hydrolyzed by a lysosomal hydrolase to yield free cholesterol, which passes to the cytosol, where it is utilized in the production of cell membrane bilayers. The free cholesterol suppresses the activity of hydroxymethylgultaryl-CoA (HMG-CoA) reductase, a rate-limiting enzyme in the biosynthetic pathway for cholesterol. Thus, the intake of cholesterol by this pathway decreases the formation of new cholesterol by this pathway decreases the formation of new cholesterol in the cell. Cholesterol entering the cell by this pathway in excess of need for membrane synthesis is esterified by the acyl-CoA:cholesterol acyltransferase (ACAT) system, chiefly forming cholesteryl oleate, which is tored in the cell. In addition to suppression of cholesterol biosynthesis, the entry of cholesterol via the LDL pathway leads to down-regulation of LDL receptors expressed on the cell surface, resulting in decreased uptake and catabolism of LDL.

B. Other Pathways: In addition to the high-affinity receptor-mediated pathway of degradation, LDL appears to be catabolized by at least 2 additional pathways. Macrophages take up chemically altered LDL by a high-capcity mechanism that is apparently not subject to feedback control. This mechanism resembles the uptake of other altered serum proteins. The biologic significance of this pathway remains unclear, because the chemical alterations of LDL that may occur in vivo have not yet been determined. LDL are also taken up by many types of cells by a low-affinity process, presumably involving bulk fluid endocytosis, which can contribute materially to LDL degradation. Certain observations indicate that the liver may also take up LDL from plasma. Increased production of bile acids by liver resulting from ingestion of bile acid-sequestering resins is associated with decreased levels of LDL in human serum, suggesting that some of the requirement for cholesterol as a precursor of bile acids is being satisfied by LDL cholesterol. Furthermore, hepatic receptors for LDL have been demonstrated in several species of experimental animals and in humans. The preponderant mass of LDL, however, appears to be removed from plasma by peripheral tissues.

Another type of receptor interaction occurs that does not lead to degradation of LDL. This involves special receptors on T and B lymphocytes that mediate the suppression of the immune response by lipoproteins. Although LDL is apparently the most efective inhibitor, IDL and VLDL share this property.

Metablosm of High-Density Lipoproteins

High-density lipoproteins, like the other major lipoproteins, are pseudomicellar in structure with cholesteryl ester-filled core regions. On the basis of their flotation properties, there appear to be 2 major subclasses of HDL: HDL_3, of density 1.21 -1.125, are about 7.5-10.5 nm in diameter; and HDL_2, density 1.125-1.063 g/mL, are about 9.5-10.5 nm in diameter.

HDL_3 predominate in human serum. Males and females have nearly equal amounts of HDL_3 in their serum (about 300 mg/dL of total lipid and protein mass). Women, however, have greater amounts of HDL_3 than do men (about 80 mg/dL versus about 36 mg/dL). There is epidemiologic evidence that the inverse relationship between coronary risk and HDL levels is chiefly dependent upon the content of HDL_2 in plasma. Other evidence indicates that a number of subspecies of HDL exist which may have individual roles in lipid metabolism.

A. Sources of HDL : HDL arise from 2 sources. The liver secretes a form of nascent HDL that is a bilayer disk composed of phospholipid and unesterified cholesterol. Apoproteins E, A-I, and possibly A-II are associated with this nascent particle. It has a bilayer structure, because it contains essentially no hydrophobic lipid. It is rapidly transformed into a spherical HDL particle by the action of LCAT. This enzyme transfers 1 mol of fatty acid from a lecithin molecule to the hydroxyl group of unesterified cholesterol, forming cholesteryl ester, which then enters the hydrophobic region between the lamellae of the bilayer. Lysolecithin, formed by transfer of the fatty acid from lecithin, leaves the lipoprotein complex, binds to albumin, and is transported to various tissues, where it is reacylated to form lecithin. The process of transesterification by LCAT rapidly forms sufficient cholesteryl ester to fill the hydrophobic core region to a spherical shape. LCAT enzyme is secreted by liver. In severe hepatic parenchymal disease, levels of this enzyme in plasma are low and esterification of cholesterol is impeded, leading to the accumulation of free cholesterol in lipoproteins and in membranes of erythrocytes and other cells. Excess free cholesterol in the membranes of erythrocytes transforms them into the target cells classically associated with hepatic parenchymal disease.

HDL also originate from the catabolism of chylomicrons. Nascent chylomicrons contain apoproteins A-I and A-II, the principal apoproteins of plasma HDL. As the triglycerides of chylomicrons are removed by hydrolysis by LPL in plasma, lipids of the surface monolayer and apoproteins A-I and A-II dissociate from the complex to form new HDL molecules. VLDL do not contain the major apoproteins found in HDL; however, phospholipids and cholesterol derived from their monolayers, released during intravascular hydrolysis of these lipoproteins, contribute to the lipid moiety of HDL.

B. Metabolic Roles for HDL : Several metabolic roles for HDL are now recognized. These lipoproteins serve as carriers for the C apoproteins, transferring them to nascent VLDL and chylomicrons. Whereas one of these, apoprotein C-II, serves as a requisite cofactor in the process of intravascular lipolysis, roles for the other C proteins remain undetermined. The relative concentrations of these apoproteins in triglyceride-rich lipoproteins and in HDL are apparently determined by their individual association constants. Apoprotein C-I, for example, comprises a much larger fraction of the C proteins in HDL than in VLDL. HDL as well as LDL deliver cholesterol to the adrenal cortex in support of steroidogenesis. A major role of HDL is its part in the centripetal transport of cholesterol—*ie*, the transport of surplus cholesterol away from peripheral tissues. Whereas this process is not fully understood, it appears that several subspecies of HDL are involved. Unesterified cholesterol is acquired from the membranes of peripheral tissues and is esterified by LCAT. The resulting cholesteryl esters, predominantly cholesteryl linoleate, are then transferred to LDL and to triglyceride-rich lipoproteins. Remnants of VLDL and chylomicrons are taken up by liver very quickly, providing for rapid transport of the newly formed cholesteryl esters to the liver. Recently, it has been discovered that the processes of esterification and transfer are both mediated by a species of lipoprotein that comprises but a small fraction of total HDL mass. In addition to lipids, this particle contains apoprotein A-I, LCAT, and apo-D. The apoprotein A-I is a cofactor for LCAT, and apo-D is required for transfer of the cholesteryl esters to recipient lipoproteins. At least one subspecies of HDL is involved in transferring free cholesterol from tissues to the esterification transfer particle.

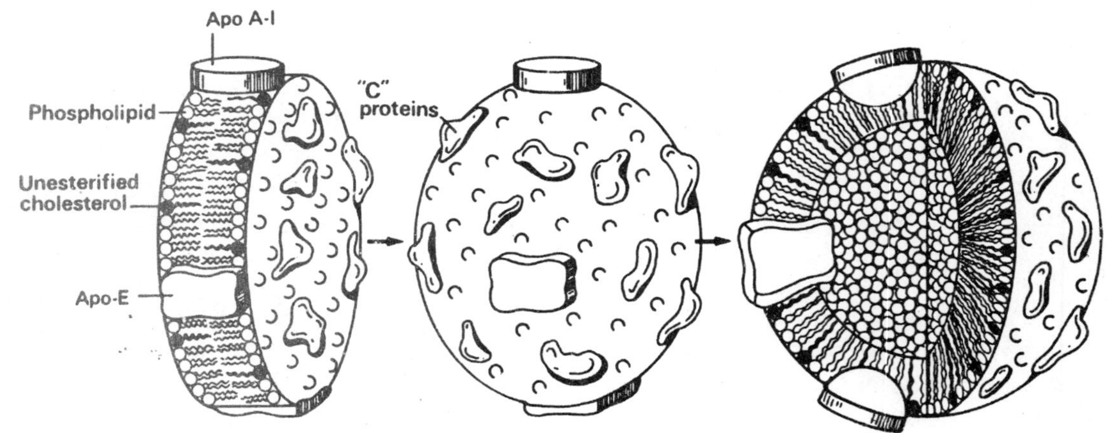

Fig. 9.4. Schematic diagram depicting conversion of hepatic discoidal HDL to spherical plasma HDL.

HDL appear to play a major role in the degradation of bacteria] endotoxins in plasma. They also apparently exert a potent inhibitory effect against certain species of trypanosomes.

C. Catabolism of HDL : The pathways of catabolism of HDL are not yet known. Radiochemical studies indicate that apoproteins A-I and A-II are removed from plasma synchronously and that a portion of the degradation occurs in liver. Because degradation by perfused rat livers is slower than for the intact animal, it must be presumed that peripheral tissues contribute substantially to the catabolism of HDL.

The Cholesterol Economy

Cholesterol is an essential constituent of the plasma membranes of mammalian cells and of the myelin sheath. It is also required for adrenal and gonadal steroidogenesis and for the production of bile acids by the liver. Virtually all nucleated mammalian cells can synthesize cholesterol, commencing with acetyl-CoA. Formation of HMG-CoA is the initial step. The first committed step, mediated by HMG-CoA reductase, is the formation of mevalonic acid from HMG-CoA. Mevalonic acid is then metabolized via a series of isoprenoid intermediaries to squalene, which cyclizes to form a series of sterols leading to cholesterol. A small amount of the mevalonate is converted eventually to the important isoprenoid substances ubiquinone and dolichol. The portion of the pathway leading to cholesterol synthesis is tightly regulated by cholesterol, or some metabolite of cholesterol, which suppresses the activity of HMG-CoA reductase. Thus, the cells have the ability to produce cholesterol to the extent that their requirements are not met by that derived from circulating lipoproteins. Cholesteryl esters stored in cells serve as an immediate reserve of cholesterol. Large requirements for cholesterol result from rapid proliferation of cells with attendant need for elaboration of cell membranes.

Hepatocytes and intestinal epithelial cells require large amounts of cholesterol for secretion of lipoproteins. In addition to these requirements, cells appear to be constantly transferring cholesterol to circulating lipoproteins, chiefly HDL. A relatively small amount of cholesterol is lost from the body in desquamated skin and through the loss of intestinal epithelial cells in the stool. The net daily loss of cholesterol from the gut, derived both from biliary cholesterol and from desquamated epithelial cells, is about 500 mg. In addition, about 2% of the mass of bile acids secreted into the intestine are lost in the stool, equivalent to about 250 mg of cholesterol.

Cholesterol is acquired from the diet as well as from endogenous synthesis. Humans, unlike many other species, do not absorb dietary chölesterol quantitatively; however, at normal levels of intake, about one-third of the amount ingested reaches the blood-stream. Most of this cholesterol is transported to liver in chylomicron remnants, leading to suppression of hepatic cholesterogenesis. Recent evidence suggests that individuals may differ substantially in the effect of dietary cholesterol on serum Jipoproteins, reflecting in part differences in the efficiency of absorption of cholesterol. The range of cholesterol content in the usual American diet is 0.4-2 g/d or more. In many parts of the world, the average daily intake of cholesterol is below 0.2 g.

About 1.5 g of cholesterol per day pass through the low-density lipoproteins, of which about one-fourth is unesterified. The remainder largely represents cholesterol esterified by the LCAT mechanism in plasma.

DIFFERENTIATION OF DISORDERS OF LIPOPROTEIN METABOLISM

Laboratory Analyses of Lipids & Lipoproteins

Because chylomicrons normally may be present in plasma up to 14 hours after a meal, they contribute triglyceride to the total measured during that period, raising the triglyceride concentration to as much as 600 mg/dL. For this reason, it is conventional to measure serum lipids and lipoproteins after a 14-hour fast. If blood glucose is not to be measured, patients may have fruit juice and black coffee with sugar (which provide no triglyceride) for breakfast.

A. Inspection : Much useful information is gained from inspection of the serum, especially before and after overnight refrigeration. As a screening technique, this will identify sera in which triglycerides need to be measured. Opalescence is due to light scattering by large triglyceride-rich lipoproteins. Serum begins to appear hazy when the level of triglycerides reaches 200 mg/dL, about the point at which clinical significance begins. The presence of chylomicrons is readily detected, because they form a white supernatant layer overnight. A marked increase in opalescence with chilling suggests familial dysbetalipoproteinemia, in which there is an increased tendency of the abnormal lipoproteins to aggregate in the cold. A further observation made on serum, which is critical to the detection of the uncommon cases in which binding of immunoglobulins to lipoproteins takes place, is the formation of a curdlike lipoprotein aggregate or a snowy precipitate as serum cools. If one of these disorders is suspected, blood should be drawn, allowed to clot, and serum separated at 37 °C, because the critical temperatures for precipitation of the cryoglobulin complex may be so high that precipitation will have occurred at room temperature before the serum and clot are separated.

B. Laboratory Techniques : The cholesterol and triglyceride contents of serum can be measured by several chemical techniques that provide reliable results. These measurements are an essential minimum for differentiation of disorders of lipoprotein metabolism. The usual methods determine unesterified and esterified cholesterol together, so that the reported value is the total content of cholesterol in serum. In most cases, electrophoresis of lipoproteins contributes little additional information; thus, this technique is not a necessary part of the routine laboratory examination. The exception is the identification of lipoproteins found in familial dysbetalipoproteinemia. The best electrophoretic separation of lipoproteins is made in agarose gel. The most complete characterization of a patient's lipoproteins is achieved by measurement of the cholesterol and triglyceride contents and electrophoretic behaviour of individual lipoprotein fractions, separated by sequential preparative ultracentrifugation, a technique usually available only in research laboratories. However, the content of high-density lipoproteins can be measured in many laboratories using a technique in which HDL are the only lipoproteins that remain in solution after treatment of the serum with heparin and manganese. The cholesterol content of the solution is taken as HDL cholesterol. Albeit rapid, the results of this technique tend to be unacceptably variable unless rigid quality control is exercised. The prognostic implications of small changes in HDL cholesterol make such controls necessary.

An important determinant of the content of cholesteryl esters in HDL is the amount of triglyceride-rich lipoproteins to which the HDL is exposed in plasma. Triglycerides from these lipoproteins exchange into the core regions of HDL, displacing cholesteryl esters and leading to an inverse logarithmic dependence of HDL cholesterol level upon the plasma triglycerides. Thus, the HDL cholesterol value cannot be interpreted without knowledge of the level of serum triglycerides. For example, a level of HDL that would normally contain 45 mg/dL of cholesterol would contain 37 mg/dL when the triglycerides were 200 mg/dL and 30 mg/dL when they reach 500 mg/dL. In certain forms of hypertriglyceridemia, there may also be a moderate decrease in the protein-phospholipid vehicle of HDL, but the principal effect on the measurement of HDL cholesterol levels remains the exchange of triglyceride for cholesteryl esters.

More sophisticated tests of composition of isolated lipoprotein fractions are of use in certain instances. The most important of these is analysis of the ratio of cholesterol to triglycerides by chemical techniques and of the apoproteins of VLDL by isoelectric focusing. The latter allows the identification of deficiency of the normal isoforms of apo-E, the apparent underlying molecular defect in familial dysbeta-lipoproteinemia. In this disorder, there is an unusually high content of cholesterol in the VLDL. Quantitative measurement of the apoproteins of VLDL can be made by electrophoresis in agarose gel, following delipidation by tetramethylurea.

Clinical Differentiation of Abnormal Patterns of Plasma Lipoproteins

A. Preliminary Screening : The first step in the diagnosis of hyperlipidemia is the determination that levels of lipids in serum are abnormal. Serum cholesterol and triglyceride levels are both continuously distributed in the population; therefore, some arbitrary levels must be established to define significant hyperlipidemia. In general, the greater the departure from these norms, the greater will be the clinical importance of the hyperlipidemia. Selection of these arbitrary values should not imply that the levels of circulating lipids prevalent in most of the population in Western countries are without risk. As indicated in the introductory paragraphs, average levels of LDL in these populations appear to be well above the biologic ideal and may be a major etiologic factor underlying the increased incidence of atherosclerotic vascular disease in Western societies. Epidemiologic studies in Europe and the USA have shown that there is a progressive increase in risk of coronary artery disease as levels of serum cholesterol increase above 180 mg/dL. Present evidence would indicate that physicians should at least encourage these patients to eat diets low in saturated fats and cholesterol to minimize the burden of LDL in plasma.

In general, levels of serum cholesterol above 270 mg/dL and serum triglycerides above 190 mg/dL merit investigation. One abnormality associated with increased risk of coronary artery disease that will not be detected if only cases with hyperlipidemia are studied is hypoalphalipoproteinemia, or deficiency of HDL. Many of these individuals have normal levels of both cholesterol and triglycerides in serum and no clinical stigmas to draw the attention of the physician. The argument has been offered that detection of these individuals is unavailing, because there is little that can currently be done to modify levels of HDL in serum. Appreciation that hypoalphalipoproteinemia is present, however, is important if for no other reason than to underscore the importance of controlling other risk factors and perhaps the avoidance of drugs that reduce HDL levels in serum. The recent demonstrations that cigarette smoking is associated with lower levels of HDL and that heavy exercise, at least in normal subjects, may be associated with increased levels suggest further that some modulation of HDL levels may be possible. Thus, measurement of HDL cholesterol appears to be a reasonable component of a screening evaluation of serum lipoproteins.

B. Identification of Abnormal Patterns : The second step in investigation of hyperlipidemia is determination of the species of lipoproteins that account for the increased content of lipids in serum. In some cases, multiple species may be involved; in others, qualitative properties of the lipoproteins are of diagnostic importance. After identifying the pattern of the lipoprotein abnormality, the physician must search for underlying disorders that cause secondary hyperlipidemias of similar pattern. Such disorders may be the sole cause of the lipoprotein abnormality or may aggravate primary disorders of lipoprotein metabolism. The differentiation of specific primary disorders usually requires additional clinical and genetic information.

The following diagnostic protocol, based upon initial measurement of cholesterol and triglycerides in serum after a 14-hour fast, supplemented by observation of serum and by additional laboratory measurements where essential, will serve as a practical guide in identifying abnormal patterns of lipoprotein distribution. The term "hyperlipidemia" denotes high levels of any class of lipoprotein; "hyperlipemia" denotes high levels of any of the triglyceride-rich lipoproteins.

Case 1. Serum Cholesterol Levels Increased; Triglycerides Normal

(*a*) If the serum cholesterol level is only modestly elevated (up to 290 mg/dL), the HDL cholesterol level should be measured. Hyperalphalipoproteinemia (elevated levels of HDL in serum) may account for the observed increase in serum cholesterol level. Hyperalphalipoproteinemia is not associated with disease processes. The LDL cholesterol level in serum may be estimated by subtracting the HDL cholesterol level and the estimated cholesterol contribution of VLDL from the total serum cholesterol level. The VLDL cholesterol is approximated as one-fifth of the serum triglyceride level.

LDL cholesterol = Total Cholesterol – (TG/5 + HDL Cholesterol)

Calculated values of LDL cholesterol over 200 mg/dL are clinically significant.

(*b*) Because HDL almost never contribute more than 120 mg/dL of cholesterol, serum cholesterol levels in excess of 290 mg/dL always represent significant hyperlipidemia. Unless the patient has obstructive hepatic disease, the abnormality may be assumed to be due to an increase in low-density lipoproteins. The abnormal lipoprotein of cholestasis, like LDL, is selectively rich in cholesterol. It can be differentiated because it has gamma mobility on electrophoresis in agarose gel and because it stains metachromatically with Sudan black in the gels.

Case 2. Predominant Increase of Triglycerides in Serum; Moderate Increase in Cholesterol Level May Be Present : Here it is apparent that the primary abnormality is an increase in the triglyceride-rich VLDL (hyperprebetalipoproteinemia) or chylomicrons (chylomicronemia), or both (mixed lipemia). Because both VLDL and chylomicrons contain free cholesterol in their surface monolayers and a small amount of cholesteryl ester in their cores, the total cholesterol level in serum may be increased, though to a much smaller extent than is the serum triglyceride level. The contribution of cholesterol of these lipoproteins to the total in serum is about 8-15% of the triglyceride content. Low levels of LDL cholesterol often seen in hypertriglyceridemia may largely offset the increase in cholesterol due to the triglyceride-rich lipoproteins, especially in primary chylomicronemia. A white supernatant layer in serum refrigerated over-night reveals the presence of chylomicrons. Because VLDL and chylomicrons compete as substrates in a common removal pathway, chylomicrons will invari-ably be present when triglyceride levels exceed 1,000 mg/dL.

Case 3. Cholesterol and Triglyceride Levels in Serum Both Elevated : This pattern can be the result of either of 2 abnormal lipoprotein distributions. One of these is a combined increase of VLDL (prebetalipoproteins), which provide most of the increase in triglycerides, and LDL (beta-lipoproteins), which account for the bulk of the increase in cholesterol in serum. This pattern is termed combined hyperlipidemia and is one of the 3 phenotypic patterns encountered in kindreds with the disorder termed familial multiple lipoprotein-type hyperlipidemia. The second distribution is an increase of remnant lipoproteins derived from VLDL and chylomicrons. These lipoprotein particles have been partially depleted of triglyceride by lipoprotein lipase and have been enriched with cholesteryl esters by the LCAT system, such that the total content of cholesterol in serum is similar to that of triglycerides. This pattern is almost always an expression of familial dysbetalipoproteinemia. Differentiation of these 2 patterns requires application of additional diagnostic tests. Presumptive differentiation can be made with high-quality agarose gel electrophoresis. In combined hyperlipidemia, the prebeta- and betalipoprotein bands are both increased in staining intensity, but each has its typical electrophoretic mobility, and they are well resolved from one another. Preparative ultracentrifugation of serum in this disorder shows elevated levels of both VLDL and LDL. In contrast, the remnant particles in dysbetalipoproteinemia are distributed in a "broad beta" pattern that obscures the resolution of beta- and prebeta-lipoprotein bands. A substantial portion of the triglyceride-rich Jipoproteins shows beta-electrophoretic mobility after separation from serum in the ultracentrifuge (at a density of 1.006 g/cm^3), and they have an increased content of cholesterol. Dysbetalipoproteinemia is confirmed by the absence or severe deficiency of the E-3 and E-4 isoforms of apoprotein E when the triglyceride-rich fraction of serum is analyzed by isoelectric focusing.

CLINICAL DESCRIPTIONS OF PRIMARY & SECONDARY DISORDERS OF LIPOPROTEIN METABOLISM

THE HYPERTRI GLYCERIDEMIAS

Atherogenicity

In addition to LDL, certain triglyceride-rich lipoproteins appear to be atherogenic. There is ample clinical evidence that the remnant lipoproteins of dysbetalipoproteinemia are associated with accelerated atherosclerosis in coronary and peripheral vessels. The atherogenicity of other triglyceride-rich lipoproteins is probably dependent upon particle diameter and perhaps on other properties as well. For example, patients with primary chylomicronemia do not appear to have accelerated atherosclerosis despite extremely high levels of triglycerides in serum, whereas there is some clinical evidence in support of the atherogenicity of VLDL of small to moderate particle diameter. Impaired capacity of some individuals to accept cholesteryl esters from the LCAT reaction may also contribute to atherogenesis by impeding centripetal transport of cholesterol.

Cause of Pancreatitis

Very high levels of triglycerides in plasma are associated with risk of acute pancreatitis, probably from the local release of amphipathic free fatty acids and perhaps lysolecithin from lipoprotein substrates in the capillary bed of the pancreas. If the concentrations of these lipids exceed the binding capacity of albumin, they could lyse membranes of parenchymal cells, initiating a chemical pancreatitis. Patients who have had previous attacks of pancreatitis appear to be at higher risk. Many patients with lipemia have intermittent episodes of epigastric pain during which serum amylase does not reach levels commonly considered diagnostic for pancreatitis. This is especially true in patients who have had previous attacks. The observation that these episodes frequently evolve into classic pancreatitis suggests that they represent incipient pancreatic inflammation. The progression of pancreatitis can be prevented by rapid reduction of triglyceride levels in serum, which can usually be accomplished by rigorous restriction of dietary fat and institution of other corrective measures. In more threatening cases, parenteral feeding may be required for a few days. The clinical course of pancreatitis in patients with lipemia is typical of the general experience with this disease. Fatal hemorrhagic pancreatitis occurs in a few; many develop pseudocysts; and some progress to pancreatic exocrine insufficiency or compromised insulinogenic capacity.

Clinical Signs

When triglyceride levels in serum exceed 3,000-4,000 mg/dL, light scattering by these particles in the blood lends a whitish cast to the venous vascular bed of the retina, a sign known as **lipemia retinalis.** Markedly elevated levels of VLDL (and perhaps small chylomicrons) in plasma may be associated with the appearance of eruptive cutaneous xanthomas. These lesions, filled with foam cells, appear as yellow morbilliform eruptions 2-5 mm in diameter, often with erythematous areolae. They usually occur in clusters on extensor surfaces such as the elbows, knees, and buttocks. They are transient and usually disappear within a few weeks after triglyceride levels are reduced below 2,000-3,000 mg/dL.

Effects of Hypertriglyceridemia on Laboratory Measurements

Very high levels of triglyceride-rich lipoproteins may introduce several important errors in clinical laboratory measurements. Light scattering from these large particles can cause erroneous results in most chemical determinations involving photometric measurements in spite of corrections for blank values. Amylase activity in serum may be inhibited by triglyceride-rich lipo-proteins; hence, lipemic specimens should be diluted for measurement of this enzyme. Because the lipoproteins are not permeable to ionic or polar small molecules, their hydrophobic regions constitute a second phase in plasma. When the volume of this phase becomes appreciable, electrolytes (measured by flame photometry) and other hydrophilic species in serum will be underestimated with respect to their true concentration in plasma water. A practical rule for correcting these values is as follows: for each 1000 mg/dL of triglyceride in serum, the concentrations of all hydrophilic molecules and ions should be increased by 1%.

PRIMARY HYPERTRIGLYCERIDEMIA

Deficiency of Lipoprotein Lipase or Its Cofactor

A. Clinical Findings : Because the clinical expressions of these defects are identical, they will be considered together. Both appear to be transmitted as autosomal recessive traits. On a typical American diet, lipemia is usually severe (2,000-25,000 mg/dL tri-glyceride in serum). Lipemia retinalis is apparent if triglyceride levels exceed 3,000-4,000 mg/dL. Hepatomegaly and splenomegaly are frequently present. Foam cells laden with lipid are found in liver, spleen, and bone marrow. Splenic infarct has been described and may be a source of abdominal pain. Hypersplenism with anemia, granulocytopenia, and thrombocytopenia can occur. Recurrent epigastric pain and overt pancreatitis are frequently encountered. Patients may have eruptive xanthomas, though many do not despite very high levels of triglycerides in serum. These disorders are present from birth and may be recognized in early infancy or may go unnoticed until an attack of acute pancreatitis occurs or lipemic serum is noted on blood sampling. Thus, diagnosis may be delayed until as late as middle age. Patients with these disorders are classically not obese and have normal carbohydrate metabolism, unless pancreatitis impairs insulinogenic capacity. Estrogens intensify the lipemia by stimulating production of VLDL by liver. Therefore, in pregnancy and lactation or during the administration of estrogenic steroids, triglyceride levels in serum rise and the risk of pancreatitis increases. There is no apparent increase in the risk of atherosclerosis, which suggests that chylomicrons are not an atherogenic lipoprotein species.

B. Laboratory Findings : These patients have a preponderance of chylomicrons over VLDL in serum such that the infranatant layer of serum refrigerated overnight may be nearly clear. Many have a moderate increase in VLDL, however, and in pregnant women or those receiving estrogens, a pattern of mixed lipemia is usually present. Levels of low-density lipoproteins in serum are decreased, probably representing the predominant catabolism of VLDL by pathways that do not involve the production of LDL. Levels of HDL are also decreased. A presumptive diagnosis of these disorders can be made by restricting the oral intake of fat to 10-15 g/d for 3-5 days. Because the quantity of exogenous triglyceride is so much greater than that secreted in VLDL, the triglyceride level of plasma drops precipitously, reaching 200-600 mg/dL within 3-4 days. Confirmation of deficiency of lipoprotein lipase is obtained by measurement in vitro of the lipolytic activity of plasma prepared from blood drawn 10 minutes after intravenous injection of 0.2 mg/kg of heparin. Analysis of lipolysis is carried out with and without 0.5 M sodium chloride, which inhibits lipoprotein lipase but does not suppress the activity of other plasma lipases, including a hepatic lipase. Classically, the lipolytic activity of plasma is very low and is similar in the salineinhibited and saline-uninhibited incubates. Recent findings suggest that several forms of LPL deficiency can be differentiated by prolonged infusions of heparin. In some patients, the LPL activity begins at low levels and rises with time, whereas in others it may be normal at first and decline to abnormally low levels during the infusion. These findings are compatible either with selective deficiencies of the LPL species in different tissues or with the existence of a form of the disorder in which binding to endothelium is abnormal. The latter is supported by heterogeneity in levels of another heparin-releasable enzyme, histaminase, in sera of patients with LPL deficiency. Absence of the cofactor protein of LPL, apoprotein C-II, is the counterpart of deficiency of LPL and can be demonstrated most readily by electrophoresis or isoelectric focusing of the proteins of VLDL.

C. Treatment : Treatment of primary chylomicronemia is entirely dietary. Intake of fat should be reduced to 10% or less of total calories. In an adult, this represents 15-30 g/d. Because the defect involves lipolysis, both saturated and unsaturated fats must be curtailed. Medium-chain triglycerides may be added to increase palatability, since their fatty acids are transported to liver in portal vein blood, bypassing transport in chylomicrons. The diet should contain at least 5 g of polyunsaturated fat as a source of essential fatty acids, and an ample supply of fat-soluble vitamins must be provided. Careful adherence to this diet will invariably maintain serum triglyceride levels below 1,000 mg/dL in the absence of pregnancy, lactation, or the administration of exogenous estrogens. Because this is below the level at which pancreatitis usually occurs, compliant patients with these disorders have unimpaired lives.

Endogenous & Mixed Lipemias

A. Etiology and Pathogenesis : Endogenous lipemia (primary hyperprebetalipoproteinemia) and mixed lipemia probably both result from several genetically determined disorders. The occurrence of multiple cases in a kindred is the basis for considering them primary. Thus, a number of "sporadic cases" may be similar, only lacking evidence of familial occurrence. Because VLDL and chylomicrons are competing substrates in the intravascular lipolytic pathway, saturating levels of VLDL will cause an impedance in the removal of chylomicrons. Therefore, as the severity of endogenous lipemia increases, a pattern of mixed lipemia may supervene. In other cases, the pattern of mixed lipemia appears to be present continuously. Though specific pathophysiologic mechanisms remain obscure, certain familial patterns are known. In all forms, however, factors that increase the rate of secretion of VLDL from liver aggravate the hypertriglyceridemia—*i.e.*, obesity with insulin resistance, or the appearance of fully developed maturity-onset diabetes; ethanol ingestion; and the use of exogenous estrogens. Studies of VLDL turnover indicate that either increased production or impaired removal of VLDL may be operative in different individuals. It appears that a substantial number of patients with mixed lipemia have partial defects in catabolism of triglyceride-rich lipoproteins. Increases in production rates of VLDL secondary to excess caloric intake, ethanol, or estrogens tend not to be accompanied by increased removal, as in normal individuals, but rather result in increased levels of circulating triglycerides. Some patients with mixed lipemia have decreased levels of lipoprotein lipase in plasma after a heparin stimulus, which may be of importance in this regard. Most patients with significant endogenous or mixed lipemia have the hypertrophic form of obesity, in which there is a reduced population of insulin receptors on cell membranes associated with impaired effectiveness of insulin. Mobilization of free fatty acids is maintained at a higher than normal rate, providing an increased flux of fatty acids to the liver, in turn increasing the secretion of triglyceriderich VLDL into plasma. Recent studies also suggest that the removal of triglycerides from circulating chylomicrons is decreased in obesity. This factor would be expected to be important in obese subjects with hypertriglyceridemia.

B. Clinical Findings : Clinical features of these forms of hypertriglyceridemia depend upon severity and include eruptive xanthomas, lipemia retinalis, recurrent epigastric pain, and acute pancreatitis. One constellation of clinical features that may have an apparent monogenic transmission is endogenous lipemia with obesity, insulin resistance, elevated baseline levels of insulin, hyperglycemia, and hyperuricemia. There is also a tendency toward the development of hypertension in patients with this syndrome. In some kindreds in which this constellation of symptoms is present, some individuals will be more severely affected than others and will have a mixed lipemia, as though they had a double dose of the involved gene or a compound genetic state, including an aggravating trait. Many of the clinical features described above are associated with endogenous lipemia that does not present a clearly monogenic pattern of inheritance, suggesting that several mecha-nisms may be operative and that endogenous lipemia and mixed lipemia may also be polygenie.

C. Treatment : The primary mode of treatment is dietary. In the short term, severe restriction of total fat intake will usually result in a rapid decline of serum triglyceride levels to 1,000-3,000 mg/dL, averting pancreatitis. The objective of long-term dietary management is reduction to ideal body weight. Because ethanol causes significant augmentation of VLDL production, abstinence is important.

If weight loss is achieved, the serum triglycerides almost always show a marked response, often approaching normal values. In cases of endogenous lipemia, where the fall in triglyceride levels is not satisfactory, clofibrate or nicotinic acid will usually produce further reductions. In mixed lipemia, clofibrate is less likely to be of value, but nicotinic acid is often capable of effecting substantial reductions in serum triglyceride levels. (See Treatment of Hyperlipidemia, below.) Certain synthetic androgenic steroids that increase the activity of heparin-releasable hepatic triglyceridase may be very effective in reduc-ing triglyceride levels in severe mixed lipemia.

Multiple Lipoprotein-Type Hyperlipidemia; Combined Hyperlipidemia

A. Etiology : Epidemiologic studies of the kindreds of survivors of myocardial infarction revealed this common heredofamilial disorder. Some of the affected individuals have increased levels of both VLDL and LDL in serum (combined hyperlipidemia); some have increased levels of VLDL only; and others have increased levels of LDL. Without family studies, the latter 2 patterns would not be identified as belonging with this syndrome. There has been sufficient experience with this syndrome to determine that the patterns in individual patients' serum may change with time. It is known, however, that a mating of an individual having any one of the 3 phenotypic patterns with a normal individual can result in the appearance of one of the other patterns of lipoprotein distribution. The hypertriglyceridemia and hypercholesterolemia are both relatively mild. Obese children in these kindreds may have hyperlipemia, but hypercholesterolemia tends not to become apparent until adulthood. The underlying process in this disorder is overproduction of VLDL.

B. Clinical Findings : Neither tendinous nor cutaneous xanthomas occur. Available data suggest that this disorder is inherited as a mendelian dominant trait. It appears that the factors that increase the severity of hypertriglyceridemia in other disorders aggravate the lipemia in this syndrome as well.

C. Treatment : The risk of coronary vascular disease is increased in these patients. Therefore, pending the results of intervention studies, treatment with diet and, usually, drugs is indicated. Clinical experience indicates that reduction of body weight and restriction of saturated fat and cholesterol are only partially effective. Patients with lipemia respond to either clofibrate or nicotinic acid, whereas those with elevated levels of LDL may require bile acid-binding resins.

Familial Dysbetalipoproteinemia (Broad Beta Disease, Type III Hyperlipoproteinemia)

A. Etiology and Pathogenesis : A permissive genetic constitution for this form of familial hyper-lipoproteinemia occurs commonly, but expression of hyperlipemia apparently requires additional genetic or environmental determinants. In its fully expressed form, the lipoprotein pattern is dominated by the accumulation of remnants of VLDL and chylomicrons. These particles are normally removed very rapidly from serum, principally by the liver. In familial dysbetalipoproteinemia, 2 populations of VLDL are usually present: normal prebeta-lipoproteins and remnants with beta-electrophoretic mobility. Remnant particles of intermediate density are also present. Characteristically, levels of LDL in serum are decreased, probably reflecting interruption of the normal transformation of VLDL remnants to LDL. Thus, the primary defect appears to involve the uptake of remnants of triglyceride-rich lipoproteins from plasma. Chylomicron remnants are frequently present in serum obtained after a 14-hour fast even when total serum triglycerides are only 300-600 mg/dL. All the remnant particles are enriched in cholesteryl esters such that the level of cholesterol in serum is often as high as the level of triglycerides. The "broad beta" electrophoretic pattern of VLDL is highly suggestive of familial dysbetalipoproteinemia. However, this pattern is seen also in hypothyroidism, resolving lipemias of other origins, or certain disorders involving immunoglobulin lipoprotein complexes. Virtual absence of the E-3 and E-4 isoforms of apo-E, detected on isoelectric focusing of VLDL proteins, confirms the diagnosis of genetic dysbetalipoproteinemia. The total content of apo-E is much higher in the VLDL from the serum of patients with dysbetalipoproteinemia than in normals. Although the mode of inheritance of the apo-E isoforms is not completely understood, it appears that a single set of alleles governs this trait. All children of affected individuals have no more than half-normal levels of apo-E-3 or apo-E-4, indicating that the affected parent is homozygous for the deletion and that their children are heterozygous. Heterozygosity for apo-E-2, which may be associated with intensification of hyperlipernia from other causes, occurs in about 15% of the white population. Whereas homozygosity is present in about 1% of the population, the incidence of clinical hyperlipidemia among these patients is much smaller.

B. Clinical Findings : Hyperlipidemia and clinical stigmas are not usually evident before age 20. In younger patients with hyperlipidemia, hypothyroidism or obesity is likely to be present. Adults with dysbetalipoproteinemia frequently have tuberous or tuberoeruptive xanthomas. Both tend to occur on extensor surfaces, especially elbows and knees. Tuberoeruptive xanthomas are pink or yellowish skin nodules 3-8 mm in diameter that often become confluent. Tuberous xanthomas—reddish or orange, often shiny nodules up to 3 cm in diameter—are usually movable and nontender. Another type, planar xanthomas of the palmar creases, strongly suggests dysbetalipoproteinemia. The skin creases assume an orange color from deposition of carotenoids in addition to other lipids. They occasionally are raised above the level of adjacent skin and are not tender. (Planar xanthomas can also be observed in cholestatic disease.) Xanthelasma—yellowish plaques on the canthus—may also occur in dysbetalipoproteinemia.

Some patients have impaired glucose tolerance, which is usually associated with higher levels of blood lipids. Obesity is commonly present and tends to aggravate the lipemia. Patients with the genetic constitution for dysbetalipoproteinemia often develop severe hyperlipidemia if they are hypothyroid.

Atherosclerotic vascular disease of the coronary and peripheral vessels occurs with increased frequency in hyperlipidemic subjects with dysbetalipoproteinemia. The prevalence of atherosclerotic disease of the iliac and femoral vessels appears to be especially high in this disorder.

C. Treatment : Management should begin with institution of a weight reduction diet providing a reduced intake of cholesterol. The use of alcohol should be minimized. Patients who achieve ideal body weight seldom require drug treatment. When the hyperlipidemia does not respond satisfactorily, clofibrate, even in doses as low as 0.5-1 g/d, is usually effective. Similar results can be obtained with nicotinic acid in moderate doses (1-2 g/d).

SECONDARY HYPERTRIGLYCERIDEMIA

Diabetes Mellitus

In patients with insulin-dependent diabetes mellitus (IDDM), levels of VLDL in plasma are frequently elevated despite the regular use of insulin, reflecting the difficulty of control of glucose levels in this disorder.

A. Clinical Findings : Lipemia may be very severe, with elevated levels of both VLDL and chylomicrons when control is poor. Lipemic patients usually have ketoacidosis, but lipemia can occur in its absence. Patients with IDDM who have been chronically undertreated with insulin may have mobilized most of the triglyceride from peripheral adipose tissue, so that they no longer have sufficient substrate for significant ketogenesis. These emaciated individuals may have severe lipemia and striking hepatomegaly.

B. Pathogenesis : The severe lipemia associated with absence or marked insufficiency of insulin is attributable to deficiency of LPL activity, reflecting the fact that this enzyme is induced by insulin. The administration of insulin in such cases usually restores triglyceride levels to normal within a few days. However, if massive fatty liver is present, weeks may be required for the VLDL levels to return to normal while the liver secretes its triglyceride into plasma. Conversion of massive amounts of VLDL to LDL, as the impedance of VLDL catabolism is relieved, leads to marked accumulation of LDL that may persist for weeks. This "beta shift" phenomenon may lead to a spurious diagnosis of hypercholesterolemia.

The moderately high levels of VLDL seen in diabetes under average control probably reflect chiefly an increased flux of FFA to liver that stimulates production of triglycerides and their secretion in VLDL. In addition to VLDL, LDL levels are also moderately high in insulin-dependent diabetics under poor control, probably accounting in part for their increased risk of coronary heart disease. Mild increases in VLDL and in FFA occur in many individuals with maturity-onset, non-insulin-dependent diabetes (NIDDM). Some have much higher levels of VLDL, suggesting that an additional genetic factor predisposing to lipemia is present. Still another cause of lipemic diabetes is the compromised insulinogenic capacity that can result from acute pancreatitis in individuals with severe primary lipemias. The deficiency may be severe enough to require exogenous insulin, often only in small doses. In diabetics who develop nephrosis as a consequence of their microvascular disease, the secondary lipemia of nephrosis compounds their hypertriglyceridemia.

C. Treatment : The rigid control of blood glucose levels, which can be attained with continuous subcutaneous insulin infusion (*eg*, with an insulin pump), is associated with sustained normalization of levels of both LDL and VLDL. HDL cholesterol levels usually rise to the normal range over a period of several months on this treatment.

The lipemia of IDDM responds well to control of the underlying disorder of carbohydrate metabolism. In obese, insulin-resistant individuals, weight loss is the key to treatment. Diets containing a large fraction of calories as carbohydrates are actually well tolerated, allowing a decrease in the burden of chylomicron triglycerides in plasma.

Uremia

Uremia is associated with modest isolated increases in VLDL. The most important underlying mechanisms are probably insulin resistance and impairment of catabolism of VLDL. Many uremic patients are also nephrotic. The additional effects of nephrosis upon lipoprotein metabolism may produce a combined hyperlipoproteinemia. Patients who have had renal transplants may be receiving glucocorticoids, which also induce a combined hyperlipidemia.

Corticosteroid Excess

In endogenous Gushing's syndrome, there is insulin resistance, and levels of both LDL and VLDL are increased. It appears that the combined hyperlipidemia is primarily due to increased secretion of VLDL, which is then catabolized to LDL. More severe lipemia ensues when steroidogenic diabetes appears. When this occurs, catabolism of triglyceride-rich lipoproteins via the LPL pathway is reduced.

Estrogens

When estrogens are administered to normal pre-menopausal women, triglyceride levels may increase by as much as 15%. This is believed to reflect chiefly increased hepatic production of VLDL, though other mechanisms may be operative. Paradoxically, estrogens increase the efficiency of catabolism of triglyceride-rich lipoproteins. Whereas estrogens tend to induce insulin resistance, it is not clear that this is an important mechanism, because certain nortestosterone derivatives decrease plasma triglyceride levels despite the induction of appreciable insulin resistance.

Certain individuals, usually with preexisting mild lipemia, show marked hypertriglyceridemia when receiving estrogens even in the relatively small doses employed in contraception. Thus, the triglyceride level of serum should be measured in any woman receiving exogenous estrogens. Contraceptive combinations with predominant progestational effects produce less hypertriglyceridemia than purely estrogenic compounds.

Certain synthetic androgenic steroids that increase the activity of heparin-releasable hepatic triglyceridase may be very effective in reducing triglyceride levels in severe mixed lipemia.

Alcohol Ingestion

Ingestion of appreciable amounts of alcohol may not necessarily result in significantly elevated levels of triglycerides in serum, but many alcoholics are lipemic. Furthermore, alcohol profoundly increases triglyceride levels in patients with primary and secondary hyperlipemias. In Zieve's syndrome, the lipemia is associated with hemolytic anemia and hyperbilirubinemia. Because LCAT originates in liver, severe hepatic parenchymal dysfunction may lead to deficiency in the activity of this enzyme. A resultant accumulation of unesterified cholesterol in erythrocyte membranes may account for the hemolysis seen in Zieve's syndrome.

Ethanol is converted to acetate, exerting a sparing effect on the oxidation of fatty acids. The fatty acids are incorporated into triglyceride in liver, resulting in hepatomegaly due to fatty infiltration and in marked enhancement of secretion of VLDL. In many individuals, there is sufficient adaptive increase in the removal capacity for triglycerides from plasma that tri-glyceride levels tend to return toward normal if alcohol intake is continued over a period of weeks. In individuals in whom the adaptive response is impaired, marked lipemia may ensue. Increased levels of HDL in plasma associated with ethanol ingestion may be coupled to the increased secretion of VLDL.

Nephrosis

The hyperlipidemia of nephrosis is biphasic. Before serum albumin levels fall below 2 g/dL, levels of LDL increase selectively. This is probably a result of increased secretion of albumin by liver to compensate for that lost in the urine. The synthesis of VLDL in the Golgi apparatus of liver appears to be coupled to that of albumin. The increased flux of VLDL from liver would be expected to increase the production of its daughter particles, LDL; however, altered surface properties of LDL, functional hypothyroidism, and even changes in capillary perfusion may contribute to increased levels of LDL. As albumin levels fall below 1-2 g/dL, lipemia ensues. Impaired hydrolysis of tri-glycerides by LPL may be due in large part to lack of albumin as an FFA receptor. Free fatty acids, which normally circulate complexed to albumin, bind to lipoproteins when albumin levels are low. The ability of these altered lipoproteins to undergo hydrolysis is thus impaired. Lysolecithin and bile acids that normally bind to albumin also associate with lipoproteins. In nephrosis, VLDL contain abundant cholesteryl esters, probably reflecting an increased rate of synthesis of these lipids.

Because coronary vascular disease is quite prevalent in patients with long-standing nephrotic syndrome, treatment of the hyperlipidemia appears to be indicated, though no studies of the effect of treatment have been reported. The hyperlipidemia is relatively resistant to diet. Clofibrate may precipitate myopathy even in relatively small doses. Bile acid-binding resins and nicotinic acid appear to be the drugs of choice. Nephrotic patients may be deficient in tryptophan, and oral administration of this amino acid has been reported to ameliorate the hypertriglyceridemia.

Glycogen Storage Disease

In type I glycogenosis, insulin secretion is decreased, leading to an increased flux of FFA to liver, where a substantial fraction is converted to triglycerides, leading to increased secretion of VLDL. The low levels of insulin in plasma also are the probable cause of reduced activity of LPL, which may cause impaired removal of triglycerides from serum. The fatty liver in these patients tends to progress to cirrhosis.

Frequent small feedings help to maintain blood glucose levels and ameliorate the lipemia. A program of nocturnal nasogastric drip feeding may be of considerable benefit in this disease. In less tractable cases, portacaval shunting may be of considerable benefit, both in maintaining blood glucose levels and in reducing the hypertriglyceridemia. Other forms of hepatic glycogen storage disease may be associated with elevated levels of VLDL and even LDL in serum.

Hypopituitarism & Acromegaly

Part of the hyperlipidemia of hypopituitarism is attributable to secondary hypothyroidism, but hypertriglyceridemia persists even in the face of thyroxine replacement therapy. Dwarfism due to isolated deficiency of growth hormone is associated with higher than normal levels of both LDL and VLDL. Decreased insulin levels may be the major underlying defect; however, deficiency of growth hormone may impair the disposal of FFA by oxidation and ketogenesis in liver, favoring synthesis of triglycerides. Mild hypertriglyceridemia is often associated with acromegaly, probably resulting from insulin resistance. Though growth hormone acutely stimulates lipolysis in adipose tissue, FFA levels are normal in acromegaly.

Hypothyroidism

Whereas significant hypothyroidism tends to produce elevated levels of LDL in serum in nearly all individuals, only a fraction will develop hypertriglyc-eridemia. The increase in LDL levels results at least in part from decreased numbers of B-100-E receptors on cell membranes, although decreased conversion of cholesterol to bile acids may also contribute. Lipemia is usually mild, though serum triglyceride levels in excess of 3000 mg/dL can occur. The underlying mechanisms are not fully understood, though it is probable that impaired removal of triglycerides from blood is involved. Increased content of cholesteryl esters and apo-E in the triglyceride-rich lipoproteins suggests that accumulation of remnant particles occurs. Hypothyroidism, even of very mild degree, causes expression of hyperlipidemia in otherwise latent carriers of familial dysbetalipoproteinemia.

Immunoglobulin-Lipoprotein Complex Disorders

Both polyclonal and monoclonal hypergamma-globulinemias may cause hypertriglyceridemia. IgG, IgM, and IgA have each been involved. Of the underlying monoclonal disorders causing hypertriglyceridemia, myeloma and macroglobulinemia are the most important, but lymphomas and lymphocytic leukemias have also been implicated. Lupus erythematosus and other collagen vascular disorders have been associated with the polyclonal type. Binding of heparin by immunoglobulin, with resulting inhibition of LPL, can cause severe mixed lipemia. More commonly, the triglyceride-rich lipoproteins have an abnormally high density, probably as a result of bound immunoglobulin, though some may be remnantlike particles. These complexes, which bind lipophilic stains, usually have gamma mobility on electrophoresis in agarose gel.

Xanthomatosis associated with immunoglobulin complex disease includes tuberous or eruptive xanthomas and xanthelasma. With the exception of cholestasis, planar xanthomas of large areas of skin are associated with immunoglobulin-lipoprotein complexes. Deposits of lipid-rich hyaline material can occur in the lamina propria of the intestine, causing malabsorption and protein-losing enteropathy. Circulating immunoglobulin-lipoprotein complexes can fix complement, leading to hypooomplementemia.

Treatment is directed at the underlying disorder. Because the critical temperature of cryoprecipitation of some of these complexes is close to body temperature, plasmapheresis should be done at a temperature above the critical temperature measured in the serum of individual patients.

THE HYPERCHOLESTEROLEMIAS

1. PRIMARY HYPERCHOLESTEROLEMIA

Familial Hypercholesterolemia

A. Etiology and Pathogenesis : This disorder, which in its heterozygous form occurs in approximately one in 500 individuals in the USA, is transmitted as a mendelian dominant trait with very high penetrance. Because nearly half of first-degree relatives are affected, including children, all members of an affected individual's family should be screened for, this disorder. Hypercholesterolemia, representing a selective increase in LDL, exists from birth. Levels of LDL tend to increase during childhood and adolescence such that average levels of serum cholesterol in adult heterozygotes are about 425 mg/dL. VLDL levels are usually normal, though some individuals, especially those in kindreds in which hypertriglyceridemia is present, may have higher than normal levels of both VLDL and LDL. Aside from an increase in content of cholesteryl esters, the LDL is normal in structure.

The underlying defect appears to be a deficiency of normal high-affinity receptor sites for LDL on cell membranes. Thus, heterozygotes have approximately half of the normal population of such receptors, and homozygotes have none. Three types of receptor defects are now recognized: receptors with no binding capacity; those with impaired binding capacity; and, rarely, a defect of internalization of bound LDL into cells. Clinically and biochemically, it is apparent that some individuals may have combined heterozygosity; *i.e.*, they may have a typical gene for absent receptor activity coupled with an allele for kinetically defective binding, or perhaps some other allele affecting LDL metabolism. Such individuals are more severely affected than heterozygotes, with serum cholesterol levels in the range of 500-800 mg/dL. They usually have aggressive coronary disease. Production rates for LDL appear to be nearly normal in heterozygotes but are greatly increased in the homozygous state. In the heterozygote, a greater fraction of LDL is removed by non-receptor-dependent mechanisms than in normal subjects. In homozygotes, all removal of LDL proceeds through such pathways.

B. Clinical Findings : One of the most striking clinical features is tendinous xanthomatosis, which usually appears in early adulthood. These xanthomas cause a broadening or fusiform mass in the tendon. They can occur in almost any tendon but are most readily detected in the Achilles and patellar tendons and in the extensor tendons of the hands. Affected adolescents or older patients who are physically active may complain of achillodynia. Arcus corneae, occur as early as the third decade. Xanthelasma present. Both arcus and xanthelasma are seen in individuals who do not have hyperlipidemia, however. Coronary atherosclerosis tends to occur prematurely in heterozygotes. The median age at onset of overt coronary disease is 42 years for men and 48 for women. Eighty-five percent of the men will have had a myocardial infarction before age 60, compared with 15% of unaffected men. Coronary artery disease is particularly prominent in individuals who are relatively deficient in HDL. It is probable that this represents a coincident inheritance of both traits. The homozygous form of familial hypercholesterolemia is catastrophic. Levels of cholesterol in serum may exceed 1000 mg/dL, and xanthomatosis progresses rapidly. Patients may have tuberous xanthomas elevated plaquelike xanthomas of the extremities, buttocks, and especially the interdigital webs. Many homozygotes have overt coronary disease in the first decade of life, and many do not survive through the second decade.

Homozygous familial hypercholesterolemia is easily diagnosed. Differentiation of the heterozygous disorder from other forms of primary hypercholesterolemia is more difficult; however, the presence of a serum cholesterol level in excess of 400 mg/dL in the absence of significant hypertriglyceridemia makes the diagnosis of heterozygous familial hypercholesterolemia likely. The presence of affected first-degree relatives is supportive of this diagnosis, especially if no other types of hyperlipidemia are present in the family that would suggest combined or multiple lipoprotein-type disease. The presence of affected children is typical of familial hypercholesterolemia but is not associated with multiple lipoprotein-type disease. The presence of tendon xanthomas is nearly pathognomonic—betasitosterolemia and cerebrotendinous xanthomatosis (cholestanolosis) excepted. Although the cholesterol content of serum from umbilical cord blood is usually elevated in patients with this disorder, the diagnosis is most easily established by measuring serum cholesterol levels after the first year of life.

C. Treatment : Treatment with various single drug regimens is of moderate benefit in decreasing LDL levels in serum. However, complete normalization of LDL levels can be achieved in compliant heterozygotes with a combination of a bile acid-binding resin and nicotinic acid, when they are eating a diet low in saturated fat and cholesterol. Treatment of homozygotes is extremely difficult. Perhaps the most effective means have been portacaval shunt or repeated plasmapheresis in conjunction with combined drug regimens.

Multiple Lipoprotein-Type Hyperlipidemia (Combined Hyperlipidemia)

In some individuals in kindreds in which this disorder is present (see Primary Hypertriglyceridemia, above), LDL will be the only lipoprotein that is elevated. This pattern may vary within an individual over time, and elevated VLDL alone or combined elevations of LDL and VLDL may be observed in the patient or the patient's relatives. In contrast to most cases of familial hypercholesterolemia, the serum cholesterol level will usually be lower than 350 mg/dL, and neither tendinous nor tuberous xanthomas occur. Few children in these kindreds show hyperlipidemia, but, if present, it is usually VLDL levels that are increased. Studies of kindreds suggest a mendelian dominant mechanism of transmission. Coronary atherosclerosis appears to be accelerated in this disorder, which is sufficiently prevalent to be observed in about 10% of survivors of myocardial infarction. The underlying biochemical mechanism appears to involve increased synthesis of apo-B-100.

Treatment of the hypercholesterolemia should begin with diet and a bile acid-binding resin; however, it may be necessary to add nicotinic acid to control rising levels of triglyceride or to normalize levels of LDL.

Lp(a) Hyperlipidemia

A lipoprotein that normally comprises a very minor fraction of circulating lipoproteins, Lp(a), is present in high concentrations in some individuals. Upon ultracentrifugation, this lipoprotein ranges on both sides of the density that discriminates LDL from HDL. It contains apo-B-100 and a specific glycoprotein. It is identified in LDL or HDL fractions obtained by ultracentrifugation by its prebeta-electrophoretic mobility. It appears to be an atherogenic lipoprotein. Little is known about its response to diet or drug therapy.

Poorly Defined Types of Hypercholesterolemia

Kindreds have been found in which there are several individuals with higher than normal levels of LDL who respond dramatically to diets low in cholesterol and saturated fats. There are also kindreds in which multiple genetic factors that lead to higher levels of LDL appear to aggregate in individuals (polygenic hypercholesterolemia). The incidence of hyperlipidemia is smaller in the latter kindreds than in those with monogenic disorders, and the LDL increase is not great. There is also a group of recessive types of hypercholesterolemia that show marked responsiveness to dietary restriction of saturated fat and cholesterol. Tuberous or raised planar xanthomas may be present, and in some individuals serum cholesterol levels may reach 600 mg/dL. Retrospective examination in some of these patients has revealed the presence of beta-sitosterol in the LDL; therefore, at least some of these disorders may be variants of betasitosterolemia.

2. SECONDARY HYPERCHOLESTEROLEMIA

Hypothyroidism

The most consistent disorder of lipoproteins associated with hypothyroidism is high LDL and IDL concentrations. Increased content of apo-E in the VLDL and IDL is consistent with an increase in remnant particles in plasma. In addition to elevated LDL, some patients may have lipemia as described in the section on secondary hyperlipemia. The hyperlipidemia of hypothyroidism may occur in individuals with no overt signs or symptoms of decreased thyroid function. Biliary excretion of cholesterol and bile acids is depressed; however, cholesterol biosynthesis is also decreased. Absorption of cholesterol from the intestine is unchanged. Cholesterol stores in tissues appear to be increased, although the number of B-100-E receptors on cells appears to be decreased. Atherogenesis is accelerated by myxedema. The hyperlipidemia responds dramatically to treatment with thyroxine.

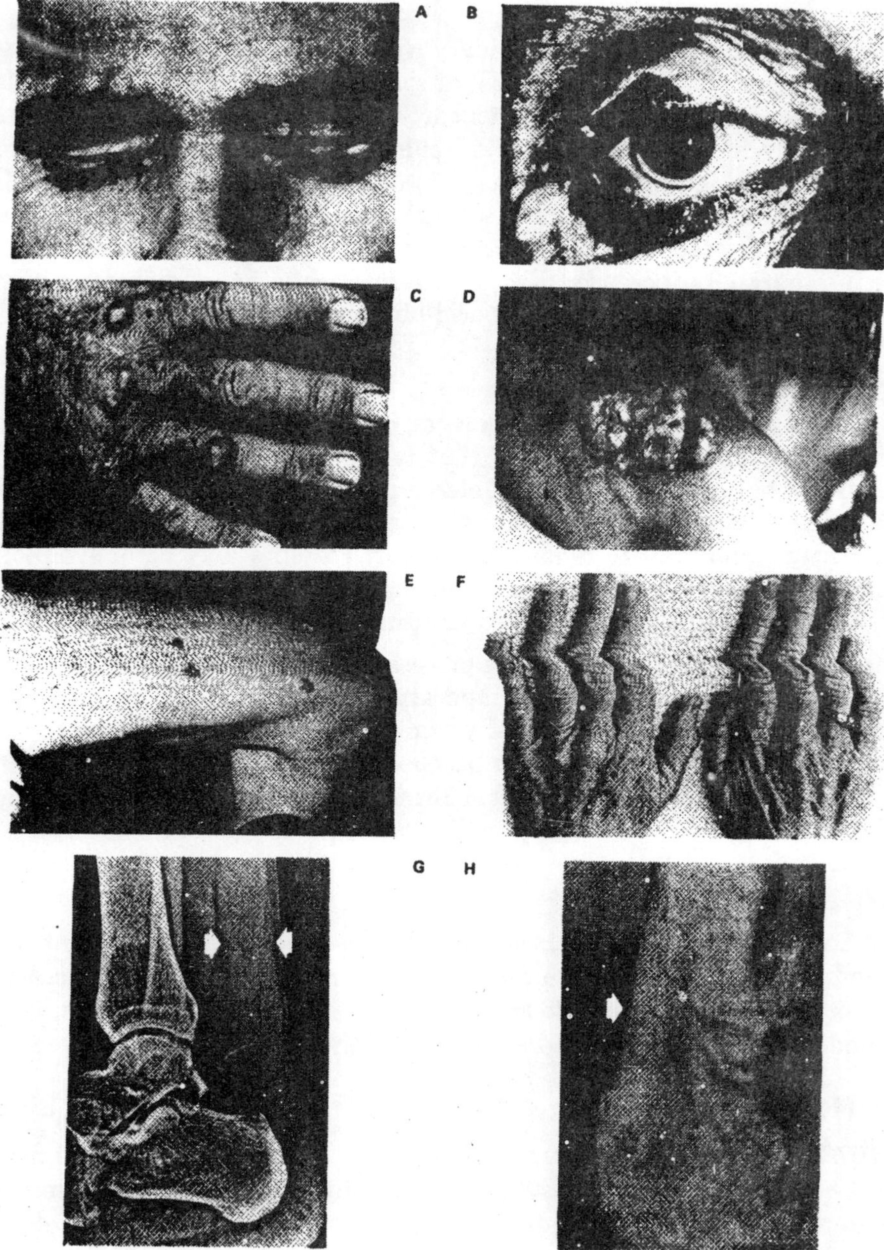

Fig. 9.5. Clinical manifestations of hyperlipidemias. *A:* Xanthelasma involving medial and lateral canthi. *B:* Severe xanthelasma and corneal arcus. *C:* Tuberous xanthomas. *D:* Large tuberous xanthoma of elbow. *E:* Eruptive xanthomas, singly and in rosettes. *F:* Xanthomas of extensor tendons of the hands. G; Xeroradiogram of Achilles tendon xanthoma. *H:* Xanthoma of Achilles tendon. (Normal Achilles tendons do not exceed 7 mm in diameter in the region between the calcaneus and the point at which the tendon fibers begin to radiate toward their origins.)

Nephrosis

As described in the section on secondary hypertriglyceridemias, nephrosis produces a biphasic hyperlipoproteinemia. The earliest alteration of lipoproteins in nephrosis is elevated LDL. The underlying mechanisms are not clearly identified, but increased secretion of VLDL by liver is probably involved. Because the lipids of the lipoprotein surfaces are altered by enrichment with sphingomyelin, lysolecithin, and FFA, the catabolism of LDL could be impaired. Perhaps the low metabolic rate in these patients introduces metabolic changes similar to those associated with hypothyroidism. The hyperlipidemia may be an important element in the markedly increased risk of atherosclerotic heart disease in these patients. The treatment of choice appears to be bile acid-binding resins with nicotinic acid.

Immunoglobulin Disorders

One of the lipoprotein abnormalities that can be associated with monoclonal gammopathy is elevated LDL. A "gamma lipoprotein" that is a stable complex of immunoglobulin and lipoprotein may be observed in agarose gel electrophoretograms of the serum of some patients. Cryoprecipitation, often in the temperature range encountered in peripheral tissues when the environmental temperature is low, may occur. Patients may have symptoms from the vascular effect of complement fixation resulting from complex formation and may have hyperviscosity syndrome from the elevated immunoglobulins per se.

Treatment is directed at the underlying process. Plasmapheresis is often effective; however, if cryoprecipitation occurs at critical temperatures near or above room temperature, the procedure must be carried out in a special warm environment. Transfusion of whole blood or serum may be dangerous in these patients because of rapid production of anaphylatoxins from complement in the serum, resulting from interaction with circulating antibody-antigen complexes.

Porphyria

Isolated increases in LDL levels have been described in intermittent acute hepatic porphyria and mixed porphyria, both during latency and during active disease. Cholesterol levels in serum are usually between 280 and 400 mg/dL in affected individuals. The mechanism is obscure, and little is known about the response of the hyperlipidemia to treatment.

Anorexia Nervosa

About 40% of patients with anorexia nervosa have elevated LDL in serum, and levels of cholesterol in serum may reach 400-600 mg/dL. The hyperlipidemia occurs in the absence of demonstrable abnormality of thyroid function and is probably a result of decreased fecal excretion of bile acids and cholesterol.

Serum lipoproteins return to normal when proper nutrition is restored.

Cholestasis

The hyperlipidemia associated with the obstruction of biliary flow is complex. It occurs either with extrahepatic or intrahepatic obstruction, though it tends to be more severe with the former. Levels of cholesterol in serum exceeding 400 mg/dL usually are associated with extrahepatic obstruction or with intrahepatic tumor. At least 3 types of abnormal lipoproteins are present in plasma. The most abundant, termed LP-X, is a bilayer vesicle about 40 nm in diameter. It is composed of unesterified cholesterol and lecithin, with associated apoproteins but not apo-B. Certain other plasma proteins are also associated with this particle, some of which may be trapped within the vesicle. LP-X is apparent on electrophoresis of lipoproteins in agarose gel as a band of zero to gamma mobility which shows metachromatic staining with Sudan black. It is these vesicular particles that cause the serum phospholipid content to be extremely high. In addition to LP-X, there is another abnormal species called LP-Y, which is also about 40 nm in diameter. It contains appreciable amounts of triglycerides and carries apo-B. It may represent a remnant particle derived from chylomicrons. The LDL in cholestasis also contain an unusually high amount of triglyceride.

Patients with cholestasis may have planar xanthomas of the skin, especially at sites of minor trauma, and may have xanthomas of the palmar creases. Occasionally, eruptive xanthomas are present. Xanthomatous involvement of nerves may lead to symptoms of peripheral neuropathy, and the abnormal lipoproteins may be atherogenic. Whereas bilirubin levels in serum may be nearly normal in some patients with chronic cholestasis, all have elevated serum alkaline phosphatase activity.

Neuropathy is the chief indication for treatment of the hyperlipidemia. Bile acid-binding resins may be of some value, whereas clofibrate causes an increase in serum cholesterol levels. Plasmapheresis is the most effective treatment.

THE HYPOLIPIDEMIAS

Although the clinician is confronted infrequently by the problem of a striking deficiency in plasma lipids, it is important to recognize the primary and secondary hypolipidemias. A serum cholesterol level less than 110 mg/dL in an adult patient is noteworthy. Since levels of triglycerides in normal fasting serum may be as low as 25 mg/dL, significance is limited to cases in which triglycerides are virtually absent.

1. PRIMARY HYPOLIPIDEMIA

Deficiency of High-Density Lipoproteins (Tangier Disease)

A. Etiology and Pathogenesis : Severe deficiency of HDL occurs in the primary disorder known as Tangier disease. Heterozygotes lack clinical signs but have about half the normal complement of HDL and of apoprotein A-I in plasma. Homozygotes lack normal plasma HDL, and apoproteins A-I and A-II are present at extremely low levels. Serum cholesterol levels are usually below 120 mg/dL and may be half that value. Mild hypertriglyceridemia is usually present. The genetic defect probably involves alteration of catabolism of apoprotein A-I.

Several biochemical defects in the metabolism of triglyceride-rich lipoproteins are probably the result of the deficiency of HDL. The chylomicrons and VLDL are deficient in C proteins, which may account for the mild hypertriglyceridemia. Abnormal oblate (oval) lipoproteins in the HDL density interval that contain cholesteryl esters are probably remnant particles derived largely from chylomicrons. The LDL also are abnormal; they are greatly enriched in triglycerides at the expense of cholesteryl esters.

The mechanism that accounts for accumulation of lipid in reticuloendothelial cells is incompletely understood; however, it appears likely that they phagocytose abnormal lipoproteins, probably chiefly the abnormal chylomicron remnants, though impaired centripetal transport of cholesterol may be involved.

B. Clinical Findings : The clinical features of this rare autosomal recessive disease include large, orange-colored, lipid-filled tonsils, accumulation of cholesteryl esters in the reticuloendothelial system, and an episodic and recurrent peripheral neuropathy with predominant motor weakness in the later stages. The course of the disease is benign in early childhood, but the neuropathy may *appear* as early as age 8. Cholesteryl ester accumulates most prominently in peripheral nerve sheaths. Carotenoid coloration may be apparent in pharyngeal and rectal mucous membranes. Splenomegaly and corneal infiltration may also be present.

C. Treatment : Because much of the lamellar lipoprotein material in plasma is believed to originate in chylomicrons, restriction of dietary fats should be of benefit. Restriction of dietary cholesterol is also indicated.

Familial Hypoalphalipoproteinemia

A. Etiology and Pathogenesis : This recently recognized pattern involves partial deficiency of HDL in serum and may involve heterogeneous mechanisms. These presumed constitutional disorders must be differentiated from the moderately low levels of HDL cholesterol seen in individuals consuming a diet very low in fat, perhaps reflecting decreased generation of HDL from chylomicrons. For example, white and Asian men on such diets usually have HDL cholesterol levels by ultracentrifugal analysis of 38-44 mg/dL, in contrast with a median value of 49 mg/dL when consuming a typical American diet. Such levels are commonplace, for example, in Asiatic populations and among vegetarians, where risk of coronary disease is small. The physician must further interpret HDL cholesterol levels in the light of the amount of triglyceride-rich lipoproteins in plasma. Because triglyceride is progressively substituted for cholesteryl esters in the core of HDL as the plasma triglyceride level rises, the HDL cholesterol will decrease as an inverse logarithmic function of the triglyceride level, causing an apparent decrease in HDL levels, since cholesterol is the component of HDL that is commonly measured. In severe forms of hypertriglyceridemia, there is an additional absolute decrease in levels of the HDL vehicle in the serum of many individuals.

B. Etiologic Factor in Coronary Disease : It appears that familial hypoalphalipoproteinemia is fairly common and an important etiologic factor in coronary vascular disease. Clinical experience already indicates that this abnormality may be the only apparent risk factor in many cases of premature coronary atherosclerosis. Furthermore, it may accelerate the appearance of coronary disease in patients with familial hypercholesterolemia. Several investigators have reported low levels of HDL cholesterol or total HDL mass in individuals with heterozygous familial hypercholesterolemia; however, because such kindreds are frequently brought to the attention of physicians by reason of having premature coronary disease, the associated hypoalphalipoproteinemia may merely reflect this preselection bias. Hypoalphalipoproteinemia shows a strong familial incidence. Although several mechanisms and modes of transmission may be involved, many kindreds show distributions consistent with mendelian dominance. The trait does not appear to cosegregate with the gene for familial hypercholesterolemia.

C. Treatment : Preliminary observations suggest that factors known to increase levels of HDL in normal individuals, such as regular exercise exceeding 7.5 kcal/min and moderate use of alcohol, may induce substantial increases of HDL cholesterol in plasma of some individuals with hypoalphalipoproteinemia. However, others show marked resistance to these interventions. Because cigarette smoking is statistically associated with lower levels of HDL cholesterol in the general population, it should be strongly interdicted in patients with hypoalphalipoproteinemia. Reports of substantial decreases of HDL cholesterol in plasma of patients treated with the drug probucol would appear to be a general contraindication to its use. Modest lowering of HDL cholesterol levels by propranolol must be weighed against the need for its cardiovascular effects in individual cases. Because nicotinic acid increases HDL cholesterol levels in serum of hyperlipidemic patients, it is potentially of value in treatment of hypoalphalipoproteinemia.

Deficiency of LCAT

A third genetic disorder associated with low serum levels of HDL is LCAT deficiency. This rare autosomal recessive disorder is not expressed in clinical or biochemical form in the heterozygote. In the homozygote, clinical characteristics are variable. The diagnosis is usually made in adult life, although corneal opacities may begin in childhood. Proteinuria may be an early sign. Deposits of unesterified cholesterol and phospholipid in the renal microvasculature, leading to progressive loss of nephrons and ultimate renal failure, have been a frequent cause of death. Many patients have mild to moderate normochromic anemia with target cells. Hyperbilirubinemia or peripheral neuropathy may be present. Red blood cell lipid composition is abnormal, with increased content of unesterified cholesterol and lecithin. Most have elevated plasma triglycerides (200-1,000 mg/dL), and levels of serum cholesterol vary from low normal to 500 mg/dL, only a small fraction of which is esterified. The large triglyceride-rich lipoproteins, presumably derived from VLDL and chylomicrons, are unusually rich in unesterified cholesterol and appear to have abnormal surface monolayers. The LDL are rich in triglycerides, and abnormal vesicular lipoproteins are present in the LDL density interval. Two abnormal HDL species are present: bilayer disks and small spherical particles. Restriction of dietary fat results in a decrease of VLDL-like particles and lamellar LDL in plasma.

Deficiency of Low-Density Lipoproteins; Recessive Abetalipoproteinemia & Familial Hypobetalipoproteinemia

A. Etiology and Pathogenesis : Previously recognized forms of genetic hypolipidemia involving deficiency of LDL are recessive abetalipoproteinemia and familial hypobetalipoproteinemia. The latter is inherited as an autosomal dominant. Affected heterozygotes with familial hypobetalipoproteinemia are usually asymptomatic but have moderately low levels of total plasma cholesterol (55-146 mg/dL), LDL cholesterol levels about 50% of normal (50-90 mg/dL), and normal HDL cholesterol levels. Individuals homozygous for this disorder share features in common with those who have recessive abetalipoproteinemia. In serum of patients homozygous for either disorder, all forms of apoprotein B are absent. No chylomicrons, VLDL, or LDL are found in plasma, leaving only

HDL. Plasma triglyceride levels are usually less than 10 mg/dL and fail to rise after a fat load. The plasma cholesterol, essentially all of which is found in HDL, is usually less than 90 mg/dL. There is a defect in the incorporation of newly synthesized triglycerides into chylomicron particles. However, about 80% of the ingested triglycerides are absorbed, probably by direct absorption of fatty acids via the portal vein.

B. Clinical Findings : Clinical features include a paucity of adipose tissue, associated with malabsorption of long-chain fatty acids due to failure of the intestine to secrete chylomicrons; red blood cells that may be acanthocytic, with a high cholesterol: phospholipid ratio; progressive degeneration of the central nervous system, including cerebellar degeneration and posterior and lateral spinal tract disease; retinal degeneration, which may be severe; and, usually, very low levels of fat-soluble vitamins in plasma. The neurologic defects may be related to deficiency of vitamin E (normally transported largely in LDL). Patients are apparently normal at birth and develop steatorrhea with impaired growth in infancy. The neuromuscular disorder often appears in late childhood with ataxia, night blindness, decreased visual acuity, and nystagmus. Cardiomyopathy with arrhythmias has been reported and may be a cause of death.

C. Treatment : Treatment includes administration of fat-soluble vitamins. Very large doses of tocopherols (1,000-5,000 1U/d) may limit the progressive central nervous system degeneration. Although vitamin A seems to correct the night blindness, it does not alter the course of the retinitis pigmentosa. Vitamins D and K may also be indicated. Restriction of long-chain triglycerides and use of medium-chain triglycerides, whose fatty acids are absorbed via the portal vein, help to increase needed caloric intake and minimize steatorrhea.

Normotriglyceridemic Abetalipoproteinemia

A newly recognized form of primary hypolipoproteinemia is ***normotriglyceridemic abetalipoproteinemia***, a disorder in which LDL is absent from plasma but fat absorption proceeds normally. The underlying defect appears to be deletion of the B-100 apoprotein, which is essential for formation of VLDL and its daughter particle LDL, whereas secretion of the B apoprotein of chylomicrons, B-48, is intact. Clinical features may include ataxia, minimal stomatocytosis of erythrocytes, and profound deficiency of tocopherols. Tocopherol levels in plasma can be significantly increased by supplementation with this vitamin.

2. SECONDARY HYPOLIPIDEMIA

Hypolipidemia may be secondary to a number of diseases characterized by chronic cachexia, *eg*, advanced cancer. A wide variety of conditions leading to intestinal malabsorption produce hypolipidemia. In both of these situations, levels of chylomicrons, VLDL, and LDL in serum are low, but these lipoproteins are never absent. Because most of the lipoprotein mass of fasting serum is of hepatic origin, massive parenchymal liver failure can cause severe hypolipidemia. In Reye's syndrome, there appears to be a contravention of the normal production of VLDL by liver, leading to near absence of both VLDL and LDL from serum. Reye's syndrome is a disorder usually associated with influenza B or varicella virus infection in which there is some anatomic disturbance of the mitochondrion attributable to the virus. There is often interference with the Krebs urea cycle resulting in hyperammonemia.

The hypolipidemias associated with immunoglobulin disorders result from diverse mechanisms. These patients usually have myeloma or macroglobulinemia, but disturbances of lipoprotein metabolism have also been described with some lymphomas and lymphocytic leukemia. Any of the major classes of immunoglobulins may be involved. In many cases the immunoglobulins are cryoprecipitins; thus, the diagnosis may be missed if blood is not drawn and serum prepared at 37 °C and observed for cryoprecipitation as it cools. Complexes between immunoglobulins and lipoproteins may circulate in plasma and fix complement, leading to hypocomplementemia and the risk of anaphylaxis on transfusion with whole blood or plasma. The complexes may precipitate in various tissues. When this occurs in the lamina propria of the intestine, a syndrome of malabsorption and proteinlosing enteropathy may result. Monoclonal IgA in myeloma may precipitate with lipoproteins, causing xanthomas of the gingiva and cervix. Lesions in the skin are usually planar and xanthomatous and may involve intracutaneous hemorrhage, producing a classic purple xanthoma. Planar xanthomas occurring in cholestasis may be confused with this condition, because the abnormal lipoprotein of cholestasis (LP-X), like the circulating lipoprotein complex of immunoglobulin and lipoprotein, has gamma mobility on electrophoresis.

OTHER DISORDERS OF LIPOPROTEIN METABOLISM

The Lipodystrophies

Classification

Current classification of these disorders is based on their familial or acquired origin and the regional or generalized nature of the fat loss. Among the associated metabolic abnormalities, insulin resistance is the common finding. Two of these disorders are inherited.

A. Familial generalized lipodystrophy (Seip-Berardinelli syndrome), a rare recessive trait, may be diagnosed at birth and is associated with macrosomia. Genital hypertrophy, hypertrichosis, acanthosis nigricans, hepatomegaly, insulin resistance, hypertriglyceridemia, and glucose intolerance are regularly observed.

B. Familial lipodystrophy of limbs and trunk (Köbberling-Dunningan syndrome) appears to be transmitted as a dominant gene, affects women predominantly, and is not evident until puberty. The face, neck, and upper trunk are usually spared. Growth is normal but otherwise this syndrome shares features of the generalized form noted above. It is frequently associated with Stein-Leventhal syndrome.

C. Acquired forms of lipodystrophy, generalized (Lawrence's syndrome) and partial (Barraquer-Simmons syndrome), usually begin in childhood, affect females predominantly, and often follow an acute febrile illness. The generalized type commonly shares the features described above, invariably involving the trunk and extremities but sometimes sparing the face. A sclerosing panniculitis, as seen in Weber-Christian syndrome, may appear at the outset. The partial type usually begins in the face and then involves the neck, upper limbs, and trunk. In this disorder, reduced levels of C3 complement are frequently encountered. This appears to be due to the presence of C3-splitting factor. Most have proteinuria, and some develop overt vascular nephritis.

Because a number of patients with disorders resembling both familial and acquired types of lipodystrophy have tumors or other lesions of the hypothalamus, appropriate neurologic evaluation should be obtained. Similarly, the physician should be alert to the association of collagen vascular disorders, including scleroderma and dermatomyositis, with some cases of acquired lipodystrophy.

RARE DISORDERS

Werner's Syndrome, Progeria, Infantile Hypercalcemia, & Sphingolipidoses

These rare disorders may be associated with hypercholesterolemia, but levels of triglycerides are usually normal. Some patients with Niemann-Pick disease have hypercholesterolemia, but most have hypertriglyceridemia, as do many patients with Gaucher's disease.

Wolman's Disease & Cholesteryl Ester Storage Disease

These rare, recessive lipid storage disorders involve the absence and partial deficiency, respectively, of lysosomal acid lipase, resulting in abnormal cholesteryl ester and triglyceride stores in the liver, spleen, adrenal glands, the small intestine, and the bone marrow. Most patients have elevated levels of both LDL and VLDL in plasma. Wolman's disease is fatal in infancy.

Cerebrotendinous Xanthomatosis

In this rare disorder, cholesterol and cholestanol, which are synthesized at increased rates, accumulate in body tissues, but plasma levels of cholesterol and triglyceride are normal. Cataracts, tendon xanthomas, progressive neurologic dysfunction, and premature coronary atherosclerosis are hallmarks of this disease. Its central nervous system effects include dementia, spasticity, and ataxia. Death usually ensues before age 50 from neurologic degeneration or coronary disease. There is no known treatment.

Betasitosterolemia

This rare disorder is distinguished by normal or elevated plasma cholesterol levels; high concentrations of plant sterols in serum, adipose tissue, and skin; and prominent xanthomas of both the tendinous and tuberous type. Individuals with this disorder absorb a substantially larger fraction of phytosterols from the intestine than do normal individuals. Restricting intake of foods rich in plant sterols is effective treatment. A more severe form apparently exists in which serum cholesterol levels may be as high as 700 mg/dL, reflecting an increase in LDL that contain sitosterol esters in addition to cholesteryl esters.

TREATMENT OF HYPERLIPIDEMIA

The first therapeutic measure in all forms of hyperlipidemia is institution of an appropriate diet. In most forms of hyperlipidemia, a single "universal" diet is indicated. In many subjects with lipemia or with hypercholesterolemia of mild to moderate severity, compliance with this diet will be sufficient to control lipoprotein levels. However, many patients with severe hypercholesterolemia or lipemia will require drug therapy. In all of these individuals, the prescribed diet must be continued to achieve the full potential of drug treatment.

Caution Regarding Drug Therapy

There are insufficient data to evaluate the effects on the fetus of drugs used in treatment of hyperlipoproteinemia. Therefore, women of child-bearing age should be advised of the potential risk and should be given these agents only if pregnancy is being actively avoided. If a contraceptive strategy is elected, estrogens should not be used in patients with hypertriglyceridemia. In children, hyperlipidemias other than familial hypercholesterolemia rarely require drug treatment. Children with heterozygous familial hypercholesterolemia should probably not be treated with drugs until after myelination of the central nervous system is complete (about 6 years of age), because data are lacking on the effects of lowering serum cholesterol levels on the myelination process. After 6 years, children with serum cholesterol levels exceeding 300 mg/dL might reasonably be treated with bile acid-binding resins.

DIETARY FACTORS IN THE MANAGEMENT OF LIPOPROTEIN DISORDERS

Restriction of Caloric Intake

The secretion of VLDL by liver is greatly stimulated by caloric intake in excess of requirements for physical activity and basal metabolism. Therefore, the total caloric content of the diet is of greater importance than its specific composition in treating endogenous hyperlipemia. There is a positive correlation between serum levels of VLDL triglyceride and various measures of obesity, but many obese patients have normal serum lipids. On the other hand, most patients with hypertriglyceridemia are obese. This association is more consistently observed in persons whose weight gain occurred in later childhood or adulthood and who have adipocyte hypertrophy with relative insulin resistance. As obese patients lose weight, plasma VLDL stabilizes at lower levels. In theory, reduction of output of VLDL by the liver should reduce the production rate of its daughter particle, LDL. In fact, there is a modest correlation of LDL levels with body weight in the general population.

Restriction of Fat Intake

In primary chylomicronemia, saturated and polyunsaturated fats both must be restricted rigidly, because the underlying defect is in lipolysis. Similarly, in the acute management of mixed lipemia with impending pancreatitis, elimination of dietary fat leads to a rapid decrease in chylomicronborne triglycerides in plasma.

The cholesterol-lowering effect of a significant reduction in total fat content of the diet is well known. It has also been shown that a 15-20% fall in serum cholesterol levels is achieved when individuals who have been consuming a typical American diet restrict their intake of saturated fats to 8% of total calories. The mechanism for these effects remains unclear, but increased bile acid and perhaps also neutral sterol excretion are known to occur at least transiently until a new steady state with lower levels of LDL and VLDL is established. This is accompanied by a decrease in the flux of the apoprotein of LDL, suggesting that production of the precursor lipoprotein, VLDL, is decreased. Substitution of polyunsaturated fats for saturates also reduces LDL cholesterol levels. The safety of diets containing large amounts of polyunsaturates has not been established, however. Diets very rich in these fatty acids may lower HDL levels and might lead to increased generation of free radicals as a result of hydroperoxidation. In obese subjects, increased intake of polyunsaturates may induce lithogenicity of bile.

Reduction of Cholesterol Intake

The amount of cholesterol in the diet affects serum cholesterol levels, but individual responses vary. Restriction of dietary cholesterol to less than 200 mg/d in normal individuals usually results in a decrease of 10-15% in serum cholesterol. This apparently in part reflects the fact that increased cholesterol intake in humans is not completely balanced by reduced cholesterogenesis in the liver. The ingestion of cholesterol is generally reflected in increased LDL content of serum. Some individuals whose serum levels increase very little with increased dietary intake may have greatly increased tissue stores of cholesterol. Excessive intake of dietary cholesterol can also lead to the production of HDLc lipoproteins, which are large HDL complexes rich in cholesteryl esters that contain apo-E. Cholesteryl esterladen VLDL of beta-electrophoretic mobility also appear. The dietary cholesterol content and saturated fat content have independent effects on levels of serum cholesterol. A decrease in the amount of saturated fat in the diet has also been shown to lower levels of triglyceride in serum, reflecting a reduction in VLDL.

Role of Carbohydrate in Diet

The role of dietary carbohydrate in lipid metabolism is still being investigated, but certain effects seem to have been uniformly observed. There is, however, great individual variation in these responses. When a high-carbohydrate diet is fed, serum triglyceride levels fall and remain lower for about 10 hours. If this diet is continued, hypertriglyceridemia then develops within 48-72 hours, and levels of triglyceride in serum rise to a maximum in 1-5 weeks. Persons with higher basal triglyceride levels and those consuming hypercaloric diets show the greatest effect. After 1-8 months on a high-carbohydrate diet, triglycerides fall to basal levels, with only an occasional subject showing sustained hypertriglyceridemia. Similar induction of lipemia by carbohydrate is seen in patients with endogenous and mixed lipemia, but the increases in serum triglycerides are not proportionately greater than in normal subjects. Thus, there is apparently no type of hyperlipemia that is particularly "carbohydrate-sensitive." It must be emphasized, however, that both of these patterns of lipemia are extremely sensitive to excess total caloric intake. Levels of HDL in serum are lower on a high-carbohydrate intake, but the observed differences are small.

Alcohol Ingestion

Ingestion of alcohol is a common cause of secondary hypertriglyceridemia, probably due to over-production of VLDL. Some individuals with endogenous hypertriglyceridemia on a familial basis are particularly sensitive to the effects of alcohol on triglyceride levels in serum, and abstinence may normalize their triglyceride levels. Chronic alcohol intake may also be associated with hypercholesterolemia. Increased cholesterol synthesis and decreased conversion to bile acids have been observed. Alcohol ingestion may account for alimentary lipemia persisting beyond 12-14 hours. This possibility should be excluded by the history or a repeat lipid analysis before further diagnostic studies are undertaken. In population studies, a positive correlation has been found between alcohol intake and HDL cholesterol levels; however, increased HDL levels are not observed in all individuals when alcohol is added to the diet. When it is present, much of the effect is obtained at moderate levels of intake (2 oz daily).

Fiber in Diet

Although much attention has been devoted to the possible role of fiber in the development of coronary heart disease and levels of triglyceride and cholesterol in serum, there is little evidence that plasma lipids can be significantly affected by fiber intake.

Other Dietary Substances

Several other nutrients have been studied in relation to atherosclerotic heart disease, including calcium, magnesium, trace elements, vitamins D, E, and C, and pyridoxine. The results of these studies are generally equivocal, and the observed effects on serum lipid levels are small. Caffeine and sucrose have negligible effects on serum lipids, and their statistical relationship to coronary heart disease is unimpressive when data are corrected for cigarette smoking.

The "Universal Diet"

Dietary treatment is an important aspect of the management of all forms of lipoprotein disorders and may in some cases be all that is required. Knowledge of the dietary factors reviewed above allows the physician to select appropriate modifications for an individual patient. However, a basic diet, termed the universal diet, is useful in the treatment of most patients with hyperlipoproteinemia. The elements of this diet are as follows:

(1) Ideal body weight should be achieved and maintained.

(2) Total fat calories should be reduced by substantial reduction in saturated fat. The latter should be less than 8% of total calories.

(3) Cholesterol should be reduced to less than 250 mg/d.

(4) Caloric difference should be made up with carbohydrate.

(5) Alcohol should be avoided in any patient with hypertriglyceridemia.

Caloric restriction and reduction of adipose tissue mass are particularly important for patients with increased levels of VLDL and IDL. Diets of 800-1,200 kcal are usually appropriate for weight reduction in adults.

Patients with familial hypercholesterolemia are usually of normal weight. When such an individual is obese, losing weight may achieve modest reduction in serum cholesterol levels, but effects are variable. All patients who have hypercholesterolemia should eat only foods low in cholesterol and saturated fat. Serum cholesterol concentration usually decreases only about 10% on this fat-modified diet. Patients with the highest initial levels of cholesterol in serum usually have the poorest response. Since some hypertriglyceridemic patients on the fat-modified diet also achieve some lipid-lowering effect, the diet is generally employed after weight reduction in all hyperlipidemic subjects. The exception to this is the patient with primary chylomicronemia due to deficiency of LPL or its cofactor apoprotein C-II whose diet must contain extremely low levels of total fat, usually 10-20 g/d. At least 5 g of the total fat intake in these patients should be a vegetable oil, such as safflower, rich in essential fatty acids.

Patients on low-fat diets should be given fat-soluble vitamin supplementation.

Sources of Dietetic Information

Dietary management must be started early, as soon as the diagnosis of hyperlipidemia is made. Referral to a local American Heart Association office or other source of dietetic consultation is often helpful in ensuring compliance. Several recipe books have been published, and food manufacturers are developing products that add variety and palatability to the restricted diet. Since the prudent diet, as proposed by the American Heart Association and others, is generally recommended, it is helpful to urge the entire family of a hyperlipidemic patient to eat a modified diet, providing greater ease of preparation and an added measure of psychologic support.

DRUGS USED IN TREATMENT OF HYPERLIPOPROTEINEMIA

BILE ACID SEQUESTRANTS

Mechanism of Action

Cholestyramine and colestipol are cationic resins that bind bile acids electrostatically in the intestinal lumen. The resin particles are not absorbed by the bowel and therefore increase the excretion of bile acids in the stool. Normally, 98% of the bile acid secreted in bile is reabsorbed. However, excretion of bile acids can be increased up to 10-fold when bile acid-binding resins are given. There is a large attendant increase in the synthesis of cholesterol by the liver, but the pool of cholesterol in circulating LDL is also tapped for bile acid synthesis by increased expression of high-affinity receptors on cell membranes of the liver and other tissues. These agents are useful only in disorders involving elevated LDL. In fact, patients who have increased levels of VLDL may have further increases in serum triglyceride levels during treatment with resins. Thus, in combined hyperlipidemia, where the resins may be given because of high LDL levels, a second agent such as nicotinic acid may be required to control the hypertriglyceridemia. Levels of LDL will fall approximately 20% in compliant patients with heterozygous familial hypercholesterolemia who are receiving maximal doses of the resins. Larger decrements of LDL cholesterol may be seen in patients with other, less severe forms of hyperbetalipoproteinemia; and in some, levels of LDL will be normalized completely.

Drug Dosage

In disorders involving moderately high levels of LDL, 20 g of cholestyramine (Questran) or colestipol (Colestid) daily may lower cholesterol levels effectively. The maximum dose of 30 g daily is required in more severe cases. In familial hypercholesterolemia and combined hyperlipidemia, the resins may be used in combination with nicotinic acid.

Side-Effects

Because the resins are confined to the lumen of the intestine, few systemic side-effects are observed. Patients frequently complain of a bloated sensation and constipation, both of which are readily relieved by adding bran to the diet. Malabsorption of fat or fat-soluble vitamins with a daily dose of resin of up to 30 g occurs only in individuals with preexisting bowel disease or with cholestasis. Hypoprothrombinemia has been observed in patients with malabsorption due to these causes. The resins bind thyroxine, digitalis glycosides, and warfarin and impair the absorption of iron and perhaps other drugs. Absorption of all of these substances is assured if they are administered 1 hour before the resin. During long-term treatment with the resins, some patients may complain of dry, flaking skin, perhaps reflecting some effect on cutaneous sterols. This minor symptom responds to local application of lanolin. Because they change the composition of bile micelles, bile acid sequestrants theoretically may increase the risk of cholelithiasis, particularly in obese subjects. In practice, this risk appears to be very small.

NICOTINIC ACID
(Niacin)

Mechanism of Action

Nicotinic acid (but not its amide) is able to effect major reductions in LDL and triglyceride-rich lipoproteins, including chylomicrons. It has been postulated that this is due to decreased flux to liver of free fatty acids, which are normally an important source of VLDL triglyceride fatty acids. Nicotinic acid is a potent inhibitor of the mobilization of free fatty acids from adipose tissue, at least in the short term; its ability to inhibit intracellular lipolysis chronically has not been established. Nicotinic acid inhibits the secretion of VLDL by liver. Incorporation of amino acids into the protein moiety of VLDL is inhibited, and turnover of the lipoprotein is decreased. Nicotinic acid also appears to increase the efficiency of removal of VLDL triglyceride via the LPL pathway. Nicotinic acid has several effects on cholesterol metabolism. It increasessterol excretion acutely and mobilizes cholesterol from tissue pools until a new steady state is established. Although it has no effect on the conversion of cholesterol to bile acids, it decreases cholesterol biosynthesis. That it can cause a continued decrease in cholesterol production even when given with bile acid-binding resins is an important feature of the complementary action of these agents. Because nicotinic acid does not change the turnover of LDL per se, it appears that the reduction of levels of LDL associated with its use must reflect decreased production of LDL resulting from decreased secretion of its precursor, VLDL, from liver. Levels of HDL in plasma, particularly HDL2, are significantly increased by nicotinic acid, reflecting a decrease in the fractional catabolic rate of these lipoproteins.

Drug Dosage

The dose of nicotinic acid required for effective treatment varies with the diagnosis. Complete normalization of LDL levels in heterozygous familial hypercholesterolemia is only achieved when a bile acid-binding resin is combined with 6-7.5 g of nicotinic acid daily (in 3 doses). For other forms of hyperbetalipoproteinemia and for various types of hypertriglyceridemia, a dose of 1.5-3.5 g/d usually has a dramatic effect. Because nicotinic acid causes a flushing sensation, it is usually started at a dosage of 100 mg 3 times daily and increased slowly. Tachyphylaxis to the flushing occurs within a few days at any dose, allowing stepwise increases. Most patients have no flushing or only occasional minimal flushing when stabilized on a given dose. Because the flushing sensation is prostaglandin-mediated, 0.3 g of aspirin given 20-30 minutes before each dose of nicotinic acid will mitigate this symptom in unusually susceptible individuals. It is important to counsel the patient before-hand that the flushing is a harmless cutaneous vasodilatation.

Side-Effects

Moderate abnormalities of liver function often occur if the dosage of nicotinic acid is increased too rapidly. If a daily dose of 2.5 g is not exceeded by the end of the first month, 5 g after the second month, and 7.5 g after the third month, such abnormalities are uncommon. A daily dose of 7.5 g is the maximum under any circumstances. A few patients have mild abnormalities of serum glutamic aminotransferase or alkaline phosphatase activities that do not appear to be clinically significant and revert to normal if the medication is discontinued.

About one-fifth of patients have mild hyperuricemia, which is almost invariably asymptomatic unless the patient has previously had gout. In such cases, a uricosuric agent can be added to the regimen. A few patients will have moderate elevations of blood glucose during treatment. Again, this is reversible except possibly in some patients who have latent diabetes. A more common side-effect is gastric irritation, which responds well to antacids. Rarely, patients develop acanthosis nigricans, which clears if the drug is discontinued. Some patients can have cardiac arrhythmias while taking nicotinic acid. Retinopathy involving macular degeneration has been described rarely.

CLOFIBRATE
(Ethyl Chlorphenoxyisobutyrate)

Mechanism of Action

A number of effects are attributed to this compound. It appears to act chiefly by inreasing the activity of the lipolytic pathway in plasma, specifically increasing levels of LPL activity. It may also increase oxidation of fatty acids by liver. Inhibition of cholesterol biosynthesis in liver during treatment with clofibrate may be secondary to changes in VLDL secretion and catabolism. Decreases in VLDL levels are frequently attended by increased levels of LDL, the so-called "beta shift," such that cholesterol levels in serum may not change despite reductions in triglyceride levels.

Drug Dosage

Clofibrate (Atromid-S) is most useful in treating familial dysbetalipoproteinemia and moderate endogenous hypertriglyceridemia. It is also of value in some patients with mild elevations of LDL. The usual dose is 1 g twice daily, though patients with dysbetalipoproteinemia may respond well to half this dose or less. Clofibrate is usually without benefit in familial hypercholesterolemia and has limited effect when severe chylomicronemia is present. In the latter case, nicotinic acid is usually more effective.

Side-Effects

Side-effects include nausea, diarrhea, skin eruptions, leukopenia, and moderate elevations of serum glutamic aminotransferase activity. Impotence has also been reported. Muscle pain or cramps associated with elevated muscle creatine phosphokinase activity have been noted. Because this syndrome is most likely to occur when serum albumin levels are low, clofibrate should not be given to patients with nephrosis. Clofibrate potentiates the activities of coumarin and indandione anticoagulants, and the hypoglycemic effects of sulfonylureas may be enhanced. It also increases the lithogenicity of bile, particularly in obese patients. Results of a recent multicenter study indicate that clofibrate may have a modest carcinogenic potential for tissues of the gastrointestinal tract.

GEMFIBROZIL

Mechanism of Action

This congener of clofibrate is bound tightly to plasma proteins and is ultimately excreted chiefly by the kidney. It decreases lipolysis in adipose tissue, reduces levels of circulating triglycerides, and causes modest reductions in LDL cholesterol levels. It causes moderate increases in levels of HDL, including the protein moiety. In general, the indications are the same as for clofibrate. It is not yet clear whether the increases in HDL with this drug are greater than those resulting from clofibrate.

Drug Dosage

Gemfibrozil (Lopid) is supplied in 300-mg capsules. The usual dose is 600 mg twice daily. Patients with familial dysbetalipoproteinemia may be managed with smaller doses.

Side-Effects

Skin eruptions and gastrointestinal and muscular symptoms similar to those associated with clofibrate have been described, as well as blood dyscrasias and elevated plasma levels of transaminases and alkaline phosphatase. It is likely that most of the toxic effects associated with clofibrate will be observed as experience with this drug increases. Like clofibrate, it enhances the effects of the coumarin and indandione anticoagulants.

BETA-SITOSTEROL
(Sitosterols)

Betasitosterol is a poorly absorbed plant sterol that appears to impede the absorption of cholesterol. For therapeutic use it is available as sitosterols (Cytellin) in suspension containing 3 g per 15 mL. In doses of 3-6 g/d, sitosterols may effect modest reductions in serum cholesterol in individuals with mild hypercholesterolemia. The drug is well tolerated, and side-effects are rare.

NEOMYCIN

Neomycin is of some usefulness in treating hypercholesterolemia. It apparently acts by inhibiting absorption of cholesterol. The dosage required for this effect is 0.5-2 g/d.

DEXTROTHYROXINE (D-THYROXINE)

Dextrothyroxine is isomeric to the normal thyroid hormone, L-thyroxine. It acts in a fashion similar to L-thyroxine on lipid metabolism, increasing conversion of cholesterol to bile acids. Preparations available in the USA have had hypercalorigenic activity, however, either because of residual L-thyroxine or because of inherent effects of the Disomer on oxidative phosphorylation, and this activity causes increased cardiac work and may induce angina. Dextrothyroxine is no longer indicated, because it has only limited effectiveness in treating hyperlipidemia and because its use may increase the risk of fatal cardiac arrhythmias.

HMG-CoA REDUCTASE INHIBITORS

HMG-CoA reductase mediates a key step in cholesterol biosynthesis. Several potent inhibitors of this enzyme which have been isolated from fungi are now under investigation. Preliminary studies have shown that 2 of these, compactin and mevinolin, cause significant reduction in levels of LDL in plasma.

PROBUCOL

This agent differs structurally from the other hypolipidemic agents, and its mechanism of action remains unclear. It causes only small decreases in LDL levels. Because considerable evidence indicates that it lowers HDL levels, its usefulness is in doubt. The observation that it causes fatal ventricular arrhythmias in animals suggests that it could do so in humans as well under some circumstances.

Probucol (Lorelco) is available as 250-mg tablets.

COMBINED DRUG THERAPY

Combinations of drugs are indicated (1) when LDL and VLDL levels are both elevated; (2) in cases of hypercholesterolemia in which significant increases of VLDL occur during treatment with bile acid-binding resins; and (3) where a complementary effect is required to normalize LDL levels, as in familial hypercholesterolemia.

Clofibrate & Resins

Whereas clofibrate has some effectiveness in decreasing levels of VLDL in plasma in conjunction with bile acid-binding resins, it usually does not complement the effect of the resins on plasma LDL levels in familial hypercholesterolemia. The effectiveness of clofibrate is thus generally limited to disorders in which moderate elevations in levels of plasma VLDL occur. Theoretically, clofibrate should potentiate the lithogenicity of bile, which may already be increased by bile acid-binding resins.

Nicotinic Acid & Resins

Nicotinic acid alone generally has a more potent effect on levels of triglyceride-rich lipoproteins than does clofibrate and also decreases levels of LDL. Nicotinic acid usually normalizes the triglyceride levels in individuals who have increased levels of VLDL while taking resins. Furthermore, the combination of resin and nicotinic acid is the only regimen that will completely normalize levels of LDL in heterozygous familial hypercholesterolemia. This complementarity of action on LDL presumably results from the additive effects of increased catabolism of LDL due to the resin and decreased production of the precursor lipoprotein, VLDL, induced by nicotinic acid. No additional toxicity or side-effects have been described with this regimen beyond those encountered when the agents are used individually. The median daily dose of nicotinic acid required for complete normalization of LDL levels in familial hypercholesterolemia is 6.5 g in conjunction with 24-30 g/d of resin. Levels of HDL cholesterol are significantly elevated on this regimen, reflecting the effect of nicotinic acid. In view of the inverse relationship of HDL levels to the risk of coronary vascular disease, this latter effect may add benefit to the reduction of levels of the atherogenic species, LDL. That this combined regimen is capable of mobilizing cholesterol from tissue sites is supported by the finding that the diameters of tendinous xanthomas, measured by a xeroradiographic technique, are reduced significantly even over a period of only a few months. It appears that this regimen can be used in the long-term treatment of heterozygous familial hypercholesterolemia. Patients who have been on the regimen for 7 years have a sustained effect on lipoprotein levels and have not developed additional side-effects or toxicity.

In treating disorders in which levels of VLDL and LDL are both increased, doses of nicotinic acid used in conjunction with bile acid-binding resins are usually much smaller, 1.5-3 g/d. Despite the fact that nicotinic acid is an anion, it does not bind to the resin in vitro. Its absorption from the intestine is unimpeded by the presence of resin; the 2 medications may therefore be taken together for convenience. Because colestipol has potent acid-neutralizing properties, there is further reason to give the 2 medications together when a patient complains of the gastric irritation that sometimes occurs as an adverse effect of nicotinic acid.

Neomycin & Resins

Another combination reported to have some effectiveness in familial hypercholesterolemia is neomycin plus bile acid resin. Though not normalized, serum cholesterol levels are reduced about 35%. In some cases of hypercholesterolemia, sitosterols may also show some complementary effect with bile acid-binding resins.

SURGICAL TREATMENT OF HYPERLIPIDEMIA

The operation that has received the most study in the treatment of hyperlipidemia is ileal bypass. This procedure is intended to impede the reabsorption of bile acids and not to cause malabsorption of triglycerides. Evidence supporting impeded absorption of cholesterol is weak. The procedure appears merely to mimic the effect of bile acid-binding resins. Serum cholesterol levels are apparently decreased in a variety of disorders, but, as with bile acid sequestrants, levels of triglycerides in serum may be significantly elevated. Thus, the procedure is not indicated in hypertriglyceridemia. Furthermore, in a series of well-documented cases of heterozygous familial hypercholesterolemia, levels of cholesterol in serum were reduced by only 33%. Clearly, ileal bypass is less effective in that disorder than is treatment with bile acid-binding resins and nicotinic acid. A significant number of patients develop diarrhea, which may be difficult to control, and if the procedure is not performed correctly there may be steatorrhea. Absorption of vitamin 312 is impaired in all subjects, necessitating parenteral administration of the vitamin for the remainder of the patient's life. The results obtained from ileal bypass can certainly be duplicated or exceeded in compliant patients with proper diet and drug regimens. This surgical procedure would appear to be indicated only for patients who refuse drug therapy. Ironically, those who develop diarrhea may require treatment with bile acid-binding resins.

In patients with homozygous familial hypercholesterolemia, ileal bypass, even augmented by aggressive multiple drug therapy, has had only a modest effect on the massively expanded pool of circulating LDL. End-to-side portacaval shunts have ameliorated but not normalized LDL levels in some of these patients.

10 Hormones and Cancer

ESTROGENS & CANCER

Breast Cancer

There is no doubt that exogenous estrogens given continuously in large doses to susceptible animal strains such as the Bittner virus-infected C3H mouse can induce breast cancer. Estrogens can also act as cocarcinogens in conjunction with such carcinogens as x-ray or methylcholanthrene. Fortunately, 40 years of human experience show no such relationship in estrogen-treated women. Case reports of sporadic breast cancers in estrogen-treated women are in clear contradiction to numerous large prospective or case control studies in which no increase over the "normal" breast cancer incidence has been found. In several recent series, it has been observed that the mortality rate from breast cancer is actually reduced in estrogen-treated women, probably because they are under medical supervision so that breast cancers, when they do occur, are detected early in a curable stage.

Endometrial Cancer

The relationship of estrogens to endometrial cancer is controversial. At least 11 case control studies report an increased risk ratio in estrogen-treated women, while several others report no such increase or an actual decrease. There are many pitfalls in case control studies of this disease, because of misreading of estrogen-induced hyperplasia as cancer, mismatching of inappropriate untreated controls, protopathic bias (*i.e.*, estrogen treatment erroneously instituted for control of "dysfunctional uterine bleeding" due to preexisting endometrial carcinoma), etc. In addition to an exaggerated reported incidence resulting from the inclusion of endometrial hyperplasia in estrogen-treated women, there is also a false assumption of zero incidence in unexamined "controls" (Horwitz, 1981). It is likely that the risk is exaggerated inasmuch as the reports of greatly increased incidence in estrogen-treated women are not matched by any significant increase in mortality rate. A recent case control study actually claimed that the cancer incidence falls 6 months after estrogen use is reduced! In fact, the 10-year survival rate of women with endometrial "cancers" associated with estrogen treatment not only exceeds that of women with endometrial cancers not associated with exogenous estrogens but even exceeds rates reported in normal life tables. The incidence of endometrial cancer is increasing not only in estrogen-treated women but also in the women of countries such as Norway and Czechoslovakia where estrogen use is either minimal or alleged to be nonexistent.

Clearly, more women are being examined and more endometrial changes read (or misread) as cancer than in the Second and Third National Cancer Surveys, when the maximal rate was 0.8 per 1000 per year at the peak age of 62. Fortunately, 95-98% of reported endometrial "cancers" associated with exogenous estrogen administered are in stage 0 or stage 1. This suggests (1) that they are detected early because of proper supervision of estrogentreated women and prompt endometrial sampling in all episodes of break-through bleeding; (2) that many are probably pseudomalignant hyperplasia rather than true cancers; and (3) that they do not impose an increased risk of death.

A. Deaths Due to Endometrial Cancer : At present, approximately 2800 deaths a year are occurring in the USA from endometrial cancer, and the evidence is that only about 9% of these are associated with exogenous estrogen administration. This number should be contrasted with the 12,000 preventable osteoporotic hip fracture deaths in women each year caused by estrogen deficiency.

B. Benefit of Low-Dose Estrogen and Progesterone : The risk/benefit ratio appears to be strongly in favor of properly supervised low-dose estrogen therapy for menopausal or oophorectomized women in whom such treatment is not contraindicated. For women requiring full replacement doses of estrogen, cycling with a potent progestin such as medroxyprogesterone acetate, 10 mg daily, or norethindrone acetate, 5 mg daily, for 10 days each month minimizes the risk of development of endometrial tissue changes which are read nowadays—rightly or wrongly—as endometrial cancer. In the 1979 study of Gambrell *et al.*, women cycled with such a progestin had one-twelfth the incidence of cancer found in women treated with estrogen alone and one-sixth the incidence of untreated women.

VAGINAL ADENOCARCINOMA

Cancer Risk in "DES Children"

Finally, the vaginal adenosis and rare clear cell carcinoma of the vagina in daughters of women who received very large doses of diethylstilbestrol, hexestrol, or dienestrol in pregnancy require consideration. This carcinoma resulting from a once widely accepted but now discredited form of therapy for threatened abortion is very rare, occurring in approximately 0.014 – 0.14% of young women whose mothers received very large doses of one or more of these 3 synthetic estrogens, but it has not been associated with administration of natural estrogens—probably because they were not given in such large doses. Almost all of these cancers have occurred in females between the ages of 14 and 23. Following the claim in 1947 that diethylstilbestrol was well tolerated in pregnancy and had no adverse effects on the fetus—and the subsequent large but poorly controlled studies claiming efficacy in preventing threatened abortion or in improving the fetal and maternal outlook in diabetes—large-dose synthetic estrogens were widely accepted as good treatment, particularly for women with a history of one or more miscarriages who were very eager to have a child.

In 1977, Herbst et al noted the relationship of the rare clear cell adenocarcinoma of the vagina in young women to the high-dose synthetic estrogen treatment given to their mothers during pregnancy. In the meantime, the efficacy of estrogens in prevention of threatened abortion had been disproved. The lowest dose of estrogens associated with dysplasia or carcinoma of the vagina is 5 mg of diethylstilbestrol daily for most of the mother's pregnancy, particularly for the first trimester.

Examination of Daughters at Risk

Asymptomatic daughters of women who received estrogen treatment during pregnancy should be examined periodically, starting at age 14. Approximately one-third of all exposed young women are found to have adenosis of the vagina, a precancerous lesion. Although adenosis per se is benign and does not require treatment, it does require careful observation. Gynecologic examination at least once a year to age 24 is required. Exposed daughters of any age with symptoms of unusual vaginal discharge, bleeding, etc, require immediate evaluation. Examination should always include visualization of the vagina and cervix, Papanicolaou smear, and biopsy of cervical erosions or other suspicious areas, particularly those that do not stain with iodine. In selected cases, culdoscopy may be necessary. These evaluation procedures are best carried out by an experienced gynecologist.

Identification and follow-up of young women at risk will be necessary for many years, because we still do not know how many have been exposed. A large number of medical centers, recognizing the gravity of this problem, have set up registries to identify those known to be at risk and have established special dysplasia clinics for these patients. Unfortunately, in addition to the rare cases of clear cell adenocarcinoma of the vagina, considerable psychic damage has been done to a large number of women by this prenatal exposure to high doses of synthetic estrogens. It is not surprising, therefore, that feelings of anger are commonly expressed by many of the young women affected by this risk. In addition, the mothers often express guilt and remorse for unwittingly having exposed their daughters to this danger of cancer. It is of the utmost importance, therefore, that these women be provided with both expert gynecologic follow-up and supportive counseling as needed.

ENDOCRINE TREATMENT OF CANCER

Cancers of certain endocrine-sensitive tissues retain the property of those tissues to respond to hormonal influences. This property can be exploited, so that endocrine-responsive disseminated cancers can often be caused to regress by measures that are both more efficient and much easier on the patient than chemotherapy. Until recently, it was difficult or impossible to determine in individual cases whether or not the tumor would respond to endocrine therapy. The demonstration by Jensen and his group a critical level of estrogen receptors must be present in order for breast cancer to respond to oophorectomy or antiestrogen therapy now makes it possible to determine beforehand which disseminated breast cancers are likely to respond to endocrine manipulation and which will require immediate chemotherapy. At the outset it must be emphasized that no method presently available will completely rid the patient of disseminated cancer of the breast, endometrium, prostate, or thyroid. The term "total remission" signifies only that all tumor masses demonstrable by our present somewhat crude methods of tumor detection have disappeared; it does not mean that all cancer cells are destroyed or even rendered nonviable. Despite this shortcoming, endocrine-induced regressions are well worth seeking in endocrine-sensitive tumors, because they are often long-lasting, involve minimal discomfort or toxicity, and can often be utilized serially, so that when a castration-induced regression comes to an end, subsequent regressions can still be obtained by relatively innocuous means.

DISSEMINATED OR RECURRENT BREAST CANCER IN WOMEN

The advantages of hormonal manipulations in endocrine-responsive disseminated breast cancers are as follows: (1) remissions last longer than with chemotherapy; (2) androgenic anabolic steroids stimulate white cell and platelet production, in contrast to the marrow-depressive effects of chemotherapy and radiation; (3) the quality of life for the patient in endocrine-induced remission is often much better than that obtained during cytotoxic chemotherapy; and (4) endocrine manipulations are less costly and lack the toxicity of chemotherapy.

Table 10.1. Effective anticancer endocrine manipulations.

A. Ablative Procedures:

1. Castration.
2. Bilateral adrenalectomy.
3. Hypophysectomy, especially cryohypophysectomy,

B. Endocrine Administration:

1. Antiestrogens: nafoxidine, tamoxifen.
2. Androgenic anabolic steroids: fluoxymesterone, testolactone, dromostanolone.
3. High-dose estrogens: diethylstilbestrol. (Other estrogen preparations in appropriate doses are equally effective pharmacologically but are much more expensive; they are therefore recommended only when the patient cannot tolerate stilbestrol.)

Those endocrine manipulations presently known to be effective in cancer management are summarized in table.

It has now been known for 85 years that certain breast cancers respond to oophorectomy. Paradoxically, it was learned about 35 years ago that administration of large doses of estrogen could also induce regressions, sometimes in tumors that had previously responded to castration. Androgens, modified androgens, and chemical derivatives with little or no hormonal activity were found to be as effective as diethylstilbestrol or testosterone propionate (or more effective); some of these compounds yielded regression rates of 20-25% in well-controlled studies in unselected patients. (These studies antedate selection of candidates for endocrine therapy by estrogen receptor analysis. It is likely that regression rates would double in patients with estrogen-receptor-rich tumors.) Unfortunately, certain very promising compounds have been withdrawn for purely technical reasons (*e.g.*, change in melting point and crystal size in the case of calusterone). Careful review of well-controlled studies with calusterone and testolactone showed (1) that hormonal activity is not necessary for antitumor effect; (2) that regression rates can be increased 2.5-2.8 times by chemical modification of testosterone (in the case of calusterone, by β-hydroxylation at carbon 7); and (3) that previous cytotoxic chemotherapy adversely modifies subsequent response to hormonal treatment.

It is now known that the longer past the menopause, the higher the regression rate, perhaps because estrogen receptor content increases with age. Visceral metastases are less responsive than skeletal deposits, which in turn are less responsive than metastases in skin, lymph nodes, and breast. Thus, in evaluating any study of the effectiveness of endocrine therapy, it is essential that the test group and the group receiving a reference standard (untreated controls are clearly unethical) be similarly randomized for all of these variables. Use of the double-blind technique is also valuable; however, this is often not possible, because endocrine effects (such as masculinization by androgens or vaginal bleeding with estrogens) quickly break the double blindfold. One way to eliminate possible bias on the clinician's part is to have skeletal pulmonary regressions objectively measured by a radiologist who is kept in ignorance of the medication protocol. For lesions not visible by x-ray, review of photographs and caliper measurements of lesions by an external reviewer ensures objectivity.

It is now established that estrogen receptors should be measured in all surgically removed or biopsied breast cancers and in selected metastases. Greene *et al.* (of Jensen's group) have recently shown that monoclonal antibodies to the estrogen receptor in human breast cancer provide a more specific and cheaper radioimmunoassay for these receptors than was hitherto available. However, because the monoclonal antibody test gives higher estrogen receptor values than other tests, the dividing line between estrogen responsiveness and nonresponsiveness must now be redefined in terms of the new technique in each laboratory supplying the analysis.

If fewer than 4 regional lymph nodes contain metastatic deposits—assuming that about 30 nodes have been removed and examined—the prognosis is excellent. If 4 or more nodes are affected, the chances of surgical cure are greatly reduced. In such cases, adjuvant chemotherapy may be considered, though its long-term efficacy is not established. *It is not yet known whether adjuvant chemotherapy interferes with subsequent response to standard endocrine therapy.*

Follow-up of the patient with no evidence of recurrence is not standardized. Most surgeons examine the patient once or twice a year and obtain a chest film annually. For the woman with known recurrence or dissemination of disease, frequency of examination should be increased to every 6-12 weeks, and the examinations must be more detailed. Most patients with disseminated disease are examined every 6 weeks. Women with established prolonged responses to oophorectomy (*eg*, 5-10 years) can safely be seen less often—usually every 6 months unless new symp-toms or lesions appear, in which case immediate examination is indicated.

Evaluation of treatment in disseminated breast cancer requires not only periodic physical examination but also caliper measurements of visible cutaneous metastases, measurable lymph nodes, and palpable viscera, especially liver size. A complete metastatic x-ray bone survey should be done before treatment is started and at each subsequent evaluation. This includes a lateral x-ray film of the skull (always using the same side); lateral views of the cervical, thoracic, and lumbar spine; lateral and posterior-anterior views of the chest, one of which should be overexposed to show the ribs; a single posterior-anterior view of the pelvis, including the upper third of the femurs; and, in patients with widespread osseous metastases, views of the humeri and femurs. In general, long bone films are unrewarding, since metastases distal to the knees and elbows are extremely rare; but in a woman with multiple skeletal lesions, a sudden fracture of the humerus or femur may be the first evidence that these bones are involved unless appropriate examination has been made. Clinically, it is important not to permit a lesion to go on to fracture when early recognition and local x-ray therapy will prevent this catastrophe in almost all cases.

The following laboratory tests should be included at each examination: hematocrit and serum calcium, phosphate, and alkaline phosphatase (*eg*, by SMA-12 panels). For patients receiving chemotherapy, the white count and platelet count should be included also. Scans of bones, liver, and lungs are of little value in monitoring improvement, since lesions may actually become more vascular and hence more pronounced during regression.

When it is determined that the disease is recurrent or metastatic, the most appropriate choice of treatment depends on the extent of the disease, the presence or absence of a critical level of estrogen receptors, and the patient's estrogenic status. If a single osteolytic lesion is detected, it can best be handled by local irradiation; if the disease is more widespread, systemic therapy is usually indicated. Oophorectomy is usually the first choice in *premenopausal* women whose tumors contain sufficient estrogen receptors; this is usually accomplished surgically, although a similar result may be obtained more slowly in selected cases by x-ray radiation of the ovaries. About 60-70% of cancers in women with a high level of estrogen receptors will regress after oophorectomy.

Evaluation of Efficacy of Treatment

It is unfortunate that widely disparate criteria for evaluation of response have been used in various studies, thus making it difficult to choose appropriate therapy on the basis of objective regression rates. In addition, the types of subjects chosen for investigational therapy have varied widely from series to series. It is possible to find reports showing very high regression rates in selected patients who have been evaluated by incomplete or doubtful criteria. Similarly, it is easy to prove that a good hormonal compound is useless by evaluating its efficacy in the following types of patients: (1) those who have already undergone cytotoxic chemotherapy; (2) those with a high incidence of visceral metastases; (3) young women who have failed to respond to castration; and (4) patients with tumors lacking estrogen receptors. In addition, lack of efficacy may be the result of using inappropriate doses. Consequently, the literature is chaotic and confusing, and reported rates of regression are often meaningless.

Evaluation of response requires rigid criteria. A diagnosis of regression is not justified unless more than half of soft tissue lesions have shrunk in size; or unless osteolytic lesions have calcified visceral deposits have shrunk, *eg*, as shown by decrease in liver size on physical examination or direct measurement on x-ray films of the lung parenchyma never if new lesions have appeared. Reduction of pleural effusion does not constitute regression, since effusions can be controlled with diuretics that have no antitumor effects. The patient's status at each evaluation is classified as (1) regression, (2) no change, or (3) progression of disease, *ie*, growth of lesions. A given treatment should not be changed capriciously, as long as the patient is either in regression or shows no change—a cardinal rule when winning is to keep it up!—but progression of disease demands an immediate change of treatment.

Hormonal Manipulations in Treatment of Metastatic Breast Cancer

A. Oophorectomy : For women one or more years postmenopausal, oophorectomy offers virtually no chance of regression, but other endocrine maneuvers are effective for tumors with a high enough level of estrogen receptors. These treatments include androgens, antiestrogens, high doses of estrogens, adrenalectomy, and hypophysectomy. If more than one treatment is used at a time, one cannot know which agent or procedure is responsible for improvement and hence should be continued. This problem arises also when castration is performed prophylactically, because tumor status cannot be evaluated in the absence of measurable lesions. One exception is local x-ray therapy to an expanding osteolytic lesion while waiting for a favorable response to a recently instituted systemic treatment such as oophorectomy. In this case, of course, the response of the irradiated lesion cannot be attributed to the systemic ablative or hormonal therapy.

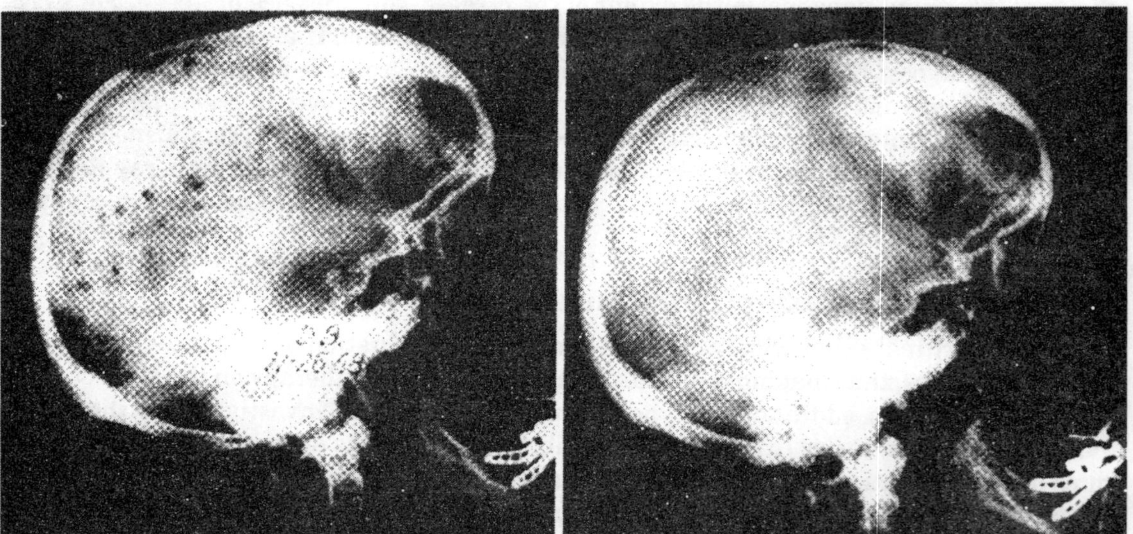

Fig. 10.1. Osteolytic skull metastases in a 63-year-old woman with metastatic breast cancer before (November 1968) and after (June 1969) a 6-month course of calusterone therapy.

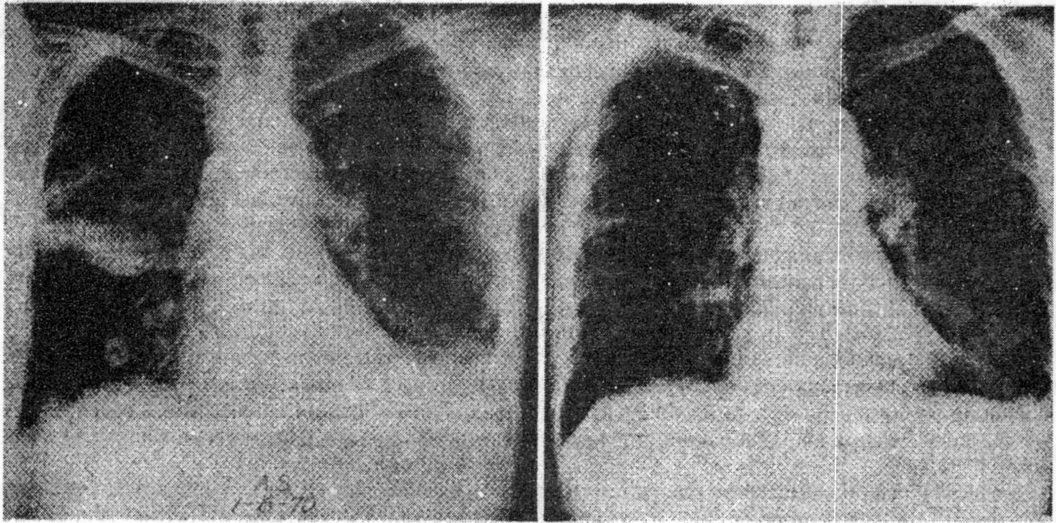

Fig. 10.2. Parenchymal and pleural metastases in a 49-year-old woman with disseminated breast cancer before (January 1970) and after (February 1970) a 6-week course of testolactone, 1000 mg orally daily.

So far, castration produces the longest remissions of all treatments for disseminated breast cancer, averaging 30 months and occasionally exceeding 10 years. Whether similar results can be achieved with the antiestrogens tamoxifen and nafoxidine remains to be seen. When a tumor responds to oophorectomy, subsequent recurrences or metastases will usually respond to further endocrine manipulation: using antiestrogens, androgens, adrenalectomy, or hypophysectomy. Strangely, tumors that have responded to castration may subsequently respond to large doses of estrogen, *eg*, diethylstilbestrol, 15 mg daily; this almost never occurs in the first year after oophorectomy.

B. Antiestrogens : The antiestrogens nafoxidine and tamoxifen (10-20 mg of either drug orally 3 times daily) are of special interest, since they can be used regardless of menopausal status, are well tolerated, and cause little serious toxicity. Complications of nafoxidine therapy include ichthyosis, light sensitivity, and vomiting. Tamoxifen may cause hot flushes, mild thrombocytopenia and leukopenia, vaginal bleeding, and skin rashes. Curiously, nafoxidine has been reported to be effective in some hypophysectomized or adrenalectomized women. Long-acting esters of sex steroids are generally contraindicated, since—in the event of complications such as hypercalcemia—their action cannot be readily terminated.

C. Androgens : Testosterone propionate was the first androgen used in the treatment of advanced breast cancer, but it produced such severe virilization that its use has been abandoned. Reduction of androgenicity without loss of antitumor effect against breast cancer was first achieved by Blackburn et al in 1959. The compound dromostanolone propionate was the first established example of an effective antitumor testosterone derivative with less androgenicity than testosterone propionate. It also had another unique property: Unlike other steroid hormones, it was fully effective when given after a previous course of other *hormonal* therapy, but this effectiveness was considerably diminished when it was given after a course of cytotoxic chemotherapy. It was once mistakenly believed that after a patient had ceased responding to an androgen, she should be given a trial with an estrogen—and, conversely, that after estrogen treatment she should receive androgen treatment. However, it was soon discovered that both testosterone propionate and diethylstilbestrol are very poor agents for secondary hormonal therapy; dromostanolone propionate was the first steroid agent shown to be as effective for secondary as for primary hormonal therapy.

Of the chemically modified testosterone derivatives, testolactone (Teslac), 250 mg orally 4 times daily or 100 mg intramuscularly 3 times weekly, produces no virilization or other toxicity. Also effective are the weak androgens fluoxymesterone, 20 mg orally daily, and dromostanolone propionate (Drolban), 100 mg intramuscularly 3 times weekly. Fluoxymesterone has a mixed androgenic effect. In animals, it is a potent androgen; in hypogonadal or juvenile human subjects, it causes a disproportionate enlargement of the phallus (penis or clitoris) but little other androgenicity. Strangely, it does not increase libido and rarely causes significant acne. It is poor replacement therapy for hypogonadal men, but this lack of visible androgenicity is desirable in the treatment of women with advanced breast cancer. On the other hand, it is a 17α-alkylated steroid and, as such, carries a cholestatic hazard.

Objective regressions produced by primary steroid therapy almost always last more than 6 months and average $11^1/_2$ months in duration. Curiously, when a hormone-induced regression has run its course, withdrawal of therapy may produce another regression.

D. Other Agents : Some combinations of hormonal agents have proved to be of greater value than either agent given alone. For example, prednisone, 30 mg daily, along with triiodothyronine, 50 μg daily, provides potent antitumor activity in some cases of widely disseminated breast cancer, but this dose of prednisone invariably produces iatrogenic Gushing's syndrome. Fortunately, equally good responses can be obtained with the combination of hydrocortisone, 30 mg daily, and triiodothyronine, 50 μg daily, and this dose of the corticosteroid is innocuous.

It should be noted that large doses of corticosteroids—like cytotoxic chemotherapeutic agents—interfere with bone healing. It is rare to see recalcification of osteolytic lesions in corticosteroid-treated patients. It seems likely that corticosteroids and cytotoxic agents both inhibit osteocytic activity. This effect may also explain their ability to reduce the hypercalcemia of malignancy, since it is now believed that the blood calcium level is maintained by osteocytic osteolysis.

E. Ablative Endocrine Therapy : At present, the preferred secondary ablative hormonal treatment is transsphenoidal cryohypophysectomy. This procedure is as effective as—and much easier on the patient than—previous ablative endocrine treatments, including transfrontal hypophysectomy and adrenalectomy, and produces regressions in approximately 30% of unselected patients (presumably more often in receptor-positive tumors). Objective regressions produced by secondary hormonal therapy usually exceed 6 months in duration and commonly average 9 months.

F. Combined Steroid and Cytotoxic Therapy : Little information is available about combinations of steroidal and cytotoxic chemotherapeutic agents. In one study, chlorambucil, 14-20 mg orally daily, was combined with prednisolone, 30 mg orally daily, for 14 days, then tapered to 10 mg twice daily for 7 days and finally to 10 mg once a day as maintenance therapy. It was reported that this treatment produced objective regressions in 34% of patients and that these doses caused no serious disagreeable or toxic effects. The experimental design of this study did not include extramural review or simultaneous control, so it is difficult to know whether these figures are significantly greater than would be obtained with chlorambucil alone in these patients. Goldenberg et al, using the rigorous criteria of the Cooperative Breast Cancer Group, reported in 1973 that 21% of their patients receiving the combination of chlorambucil and prednisone obtained objective remissions. Also using the criteria of the Cooperative Breast Cancer Group, including extramural review, we evaluated the combination of testolactone, 1000 mg orally daily, plus fluorouracil, 500 mg intravenously, once a week, and produced objective regressions in 17 of 40 patients (42%). White blood cell and platelet counts fell slightly in all patients, but sufficiently so for us to withdraw this treatment temporarily in 7 cases and permanently in 2 others. Bone healing did not occur. Cytotoxic agents and corticosteroids both inhibit bone healing, presumably by turning off osteocytes. The same mechanism probably explains their ability to bring down high serum calcium levels. The combination of testolactone and fluorouracil produces a higher regression rate than has previously been obtained with these doses of either agents used alone. Fragmentary data from these and similar studies suggest the possibility of additive effects of the 2 types of treatment, probably because of different modes of action.

Summary

Every woman with breast cancer requires individualized treatment, and particular care is needed in planning serial therapy when the disease has progressed beyond surgical or radiologic control. Sixty to seventy percent of breast cancers contain a high enough level of estrogen receptors to make them responsive to endocrine manipulation. Conversely, almost no estrogen-receptor-poor tumors respond to hormonal treatment, and these are more effectively treated with cytotoxic chemotherapy.

For estrogen-receptor-rich tumors, oophorectomy, antiestrogens, androgens, estrogens, androgens *and* estrogens, and, in appropriate cases, trans-sphenoidal cryohypophysectomy should precede any form of cytotoxic chemotherapy. By serializing treatments, starting with the least drastic and most effective, as suggested in Table 10.2, it is often possible to add years to the lives of women with disseminated breast cancer and—what is more important—to add life to those years.

TREATMENT OF MALE BREAST CANCER

This rare condition occurs more commonly in men with Klinefelter's syndrome, probably as a result of the unremitting estrogenic breast growth stimulus that leads to gynecomastia. Almost all male breast cancers are hormone-dependent and are made worse by androgens. When the cancer is disseminated, at least two-thirds of cases respond to castration, adrenalectomy, hypophysetomy, estrogens, or progestins. These treatments may be serialized, just as in female breast cancer, to produce multiple regressions.

Table. 10.2. Suggested order of treatment for disseminated breast cancer. (ER = estrogen receptor.)

1. X-ray therapy unless cancer is too widely disseminated.
2. Oophorectomy if (*a*) aptient is premmenopausal and (b) tumor is estrogen-rich (ER = postive).
3. First additive hormonal treatment for ER = positive tumors only: antiestrogen.
4. Withdrawal of hormonal treatment-for responders only.
5. Second additive hormonal treatment: testolactone, 1000 mg orally daily, or hydrocortisone, 30 mg orally, plus liothyonine, 100 mg orally daily.
7. Cytotoxic chemotherapy (*a*) initially for ER-poor tumors and (*b*) as last treatment for ER-postive tumors of those not responding to endocrine manipulations.

*In all cases, do not change treatment unless tumor progression is demonstrated.

TREATMENT OF PROSTATIC CANCER

Since the classic work of Huggins in 1941, it has been established that most disseminated prostatic cancers are androgen-dependent, *i.e.*, they respond to castration, estrogens, and antiandrogenic progestins.

Unlike disseminated breast cancer, metastases from prostatic cancer are difficult to measure and evaluate—especially in bone, where they rapidly become osteoblastic. Lacking other measurable end points, serum acid phosphatase levels have proved reliable in evaluating the course of the tumor and its response to treatment. Where measurable lesions are present, they usually change in size only after the serum level of acid phosphatase has fallen in regressions or risen in progression. The development of radioimmunoassay for this enzyme by Foti in 1977 was a considerable advance in what had previously been a difficult technique.

The most important change in treatment of advanced prostatic cancer is the discovery that only 1 mg of diethylstilbestrol daily provides optimal control in most cases. Although this dose does not completely suppress endogenous testosterone secretion, it satisfactorily controls tumor growth in 70—80% of patients. Larger doses, such as the 5 mg of diethylstilbestrol formerly used, actually increased the mortality rate—not from cancer, but from heart disease and stroke. The 1 mg dose does not carry this very serious cardiovascular hazard.

Whether to use castration or estrogens first is undecided. Many authorities who consider castration more effective reserve estrogens for subsequent recurrence. Still later recurrences may respond to the antiandrogens (*e.g.*, flutamide) or progestational agents (*e.g.*, cyproterone, megestrol, medroxyprogesterone).

The course of disseminated prostatic cancer is highly unpredictable. Spontaneous regression or apparent total regressions during treatment may in a few cases last 10 years or more; the average is 2-3 years. The 5-year survival rate of patient with hormonally responsive tumors increases with antiandrogenic treatment from 6% to 40%. Attempts to obliterate all androgens by adrenalectomy or hypophysectomy in patients who had previously responded to orchiectomy have rarely been successful. Medical adrenalectomy using such agents as glutethimide (Elipten) has been more successful, less drastic, and should usually be attempted before going on to cytotoxic chemotherapy. Unfortunately, it is not yet possible to determine by androgen receptor analysis which tumors will respond to antiandrogen treatment. Preliminary data suggest that the dihydrotestosterone content of the primary tumor may predict such a response.

TREATMENT OF ENDOMETRIAL CANCER

Endometrial carcinoma is epidemiologically associated with obesity, diabetes mellitus, infertility, and polycystic ovary syndrome—all conditions of prolonged unopposed estrogen secretion. It usually follows precancerous endometrial hyperplasia, especially of the atypical adenomatous type. Early endometrial carcinoma is difficult—many say impossible—to differentiate from estrogen-induced hyperplasia. As noted earlier in this chapter, progestins are useful in prophylaxis of endometrial carcinoma both in estrogen-induced and spontaneous hyperplasia of the endometrium.

In 1961, Kelley and Baker found that 30% of disseminated endometrial carcinomas also respond to progetins. This work has been widely confirmed, so that progestational therapy is established as the treatment of choice for disseminated endometrial carcinoma. Like other endocrine-responsive tumors, those endometrial carcinomas that respond to progestational therapy usually remain in regression for 1-3 years and, on rare occasion, more that 12 years.

It is not yet possible to identify by receptor analysis which tumors will respond to hormonal manipulation. The agents most commonly used are hydroxyprogesterone caproate (Delalutin), 500 mg intramuscularly twice weekly, and megestrol acetate (Megace), 80 mg orally 4 times daily. These agents are quite well tolerated. Because of the possible association of progestin-containing contraceptive pills with thromboembolic disease, the benefit/risk ratio must be weighed carefully in women with a history of thrombophlebitis.

Index

R

S

□□□